TERRESTRIAL RADIATION EFFECTS IN ULSI DEVICES AND ELECTRONIC SYSTEMS

TERRESTRIAL RADIATION EFFECTS IN ULSI DEVICES AND ELECTRONIC SYSTEMS

Eishi H. Ibe

Chief Researcher, Hitachi Ltd., Japan

IEEE

WILEY

This edition first published 2015

Registered office
John Wiley & Sons Singapore Pte. Ltd., 1 Fusionopolis Walk, #07-01 Solaris South Tower, Singapore 138628.

For details of our global editorial offices, for customer services and for information about how to apply for permission to reuse the copyright material in this book please see our website at www.wiley.com.

Library of Congress Cataloging-in-Publication Data

Ibe, Eishi H.
Terrestrial radiation effects in ULSI devices and electronic systems / Eishi H. Ibe.
pages cm
Includes bibliographical references and index.
ISBN 978-1-118-47929-2 (cloth : alk. paper) 1. Electronic circuits–Effect of radiation on. 2. Integrated circuits–Ultra large scale integration–Reliability. 3. Integrated circuits–Effect of radiation on. I. Title.
TK7870.285.I24 2015
621.3815–dc23

2014022262

Typeset in 10/12pt Times by Laserwords Private Limited, Chennai, India

1 2015

To my daughters, Akane and Hikari

Contents

About the Author

Dr. Eishi Hidefumi IBE received his BS degree in Physics from Kyoto University, Japan in 1975, and his PhD degree in Nuclear Engineering from Osaka University, Japan in 1985.

He has joined the Atomic Energy Research Laboratory, Hitachi Ltd in 1975. He was promoted to chief researcher in the Yokohama Research Laboratory (formerly Production Engineering Research Laboratory), Hitachi Ltd. in 2006.

He has made outstanding accomplishments in nuclear engineering during the first 20 years of his career, in particular radiation effects on water (radiolysis) and component materials, and in single event effects on semiconductor devices during the last 18 years. His expertise covers very wide areas of sciences, such as elementary particle/cosmic ray physics, nuclear/neutron physics, semiconductor physics, mathematics and computing technologies, ion-implantation/mixing and accelerator technologies, electro-chemistry, database handling, RBS (Rutherford Backscattering Spectrometry)/Auger/SEM (Scanning Electron Microscopy)/Laser-beam micro analysis, and so on.

He has carried out pioneering work on simulation techniques of water radiolysis in the coolant of nuclear power plants to reveal that water coolant in the core decomposes into H_2 and H_2O_2. He has also established a theoretical basis for the hydrogen water chemistry techniques used to suppress oxidising H_2O_2, which is now widely applied to Japanese boiling water reactors to mitigate inter-granular stress corrosion cracking of the component materials. He has received awards from the Japanese Atomic Energy Society in 1986 and 1990, and from the American Nuclear Society in 1996.

During the last 18 years, he has dedicated himself to the development of quantification and mitigation techniques for terrestrial neutron-induced soft error in electronic devices and components. He developed the novel soft-error models for CMOS (Complementary Metal Oxide Semiconductor) devices. The models have been utilised to design more reliable semiconductor memory devices and logic gates, bringing in the breakthrough knowledge on the nature of terrestrial neutron soft error. Under his leadership, novel experimental techniques to quantify soft-error susceptibility of the devices and components have been developed and accepted as international standards.

He has contributed to IEEE journals such as *EDS* and *TNS*, conferences such as IRPS, IOLTS, ICICDT, WDSN, NSREC, RADECS, RASEDA, ICITA and SELSE as a program committee member, or a reviewer in the field of neutron-induced faults/errors/failures. He has authored more than 90 international technical papers and presentations including 25 invited contributions in the field of radiation effects. He has reviewed more than 200 technical papers responding to requests from the Chairs of the journals and conferences. This accumulation has given him wide and deep scope in the field of single event effects.

Dr. Ibe was promoted to IEEE Fellow for contributions to analysis of soft errors in memory devices in 2008. Some of his achievements are now accessible worldwide through his recent publications with World Scientific Inc. (2008) and Springer (2010, 2011).

Preface

In everyday life, we do not recognise the presence of terrestrial radiation – secondary particles are produced from cosmic ray and radiation from radioisotopes at ground level. Terrestrial radiation is so weak (low flux) that they do not have any visible or recognisable influence on human tissues, but it does have an impact on LSI (Large Scale Integration), VLSI (Very large scale integration) and ULSI (Ultra large scale integration) devices in electronic systems at ground level.

When I was a fourth grade student of the Kyoto University in 1974, my major subject matter was the measurement of lifetime of terrestrial muon. At that time, no one, including me, knew about or even imagined such impacts from terrestrial neutrons.

Rapid progress in semiconductor industries has forced us to be aware of the impacts of terrestrial radiation on semiconductor devices. First, alpha-ray soft error from contaminated radioisotopes on/in the DRAM (Direct Random Access Memory) and SRAM (Static Random Access Memory) devices. As the readers will see in this book, terrestrial neutron-induced soft error has been unacknowledged up until the late 1990s for many reasons. As device scaling has nosedived into below 100 nm, the impacts of terrestrial radiation has spread very widely and deeply. Not only terrestrial neutrons but also other terrestrial radiative particles such as protons and muons are recently among the focus of scientific investigations. Beyond memories, sequential and combinational logic devices and circuits are also being scrutinised. Concerns over failures have broadened from servers/routers to the automobile industry.

It is commonly recognised now that failures in electronic systems due to faults or errors introduced in devices/circuits by terrestrial radiation can only be mitigated by the combination or cooperation of mitigation techniques in two or more stack layers such as substrate, cell, circuit, CPU (Central Processing Unit), middleware, OS (Operating System) and application. This is a very challenging task that requires a wide variety of scientific fields like astronomy, cosmic ray physics, nuclear physics, accelerator physics, semiconductor physics, circuit theory, computer theory, numerical simulation, EDA (Electric Design Automation) tools, coding theory, reliability physics, database handling, and so on.

Meanwhile, this task is fascinating. During my research in this field, I have learned a number of exciting facts about the Earth.

We cannot live without air that is only a 50 km thick layer above the Earth – 1/250 of the diameter of the Earth. An astronaut has a limit to how long he can stay in the inner/outer space due to the limit of radiation exposure by cosmic rays. We, humankind, cannot live on a planet without air and have been protected from harsh cosmic radiation in outer space by only this very thin layer of air in the Earth.

Beautiful aurora australis and borealis are the outcome of interactions between cosmic rays and the atmosphere.

Carbon-14 that is used for radiocarbon dating is produced by nuclear reaction of nitrogen-14 and cosmic ray proton in the atmosphere. Even clouds in the sky have recently been revealed to be mostly triggered by cosmic rays according to CERN's team report.

The author hopes that this book will trigger the readers' interest in the impact of cosmic rays on the Earth and our everyday lives.

16 April 2014
Eishi H. Ibe
Enjoying scuba diving in Saipan, USA

Acknowledgements

I gratefully acknowledge Professors Emeritus T. Nakamura, M. Baba and Professor Y. Sakemi for helpful discussions and support for the database on nuclear reactions and high-energy neutron experiments at CYRIC, Tohoku University. We also acknowledge Dr Alexander Prokofiev for cordial support in high-energy neutron experiments at TSL, Uppsala University. Communicative discussions with Drs. C. Slayman, S.-J. Wen of Cisco Systems Inc., N. Seifert of Intel, R. Baumann of TI, M. Nicolaidis of TIMA Laboratory, D. Alexandrescu and A. Evans of iRoc, T. Uemura of Fujitsu Laboratory and H. Kobayashi of SONY are deeply acknowledged. I am also grateful to Professors K. Kobayashi of Kyoto Institute of Technology, H. Onodera of Kyoto University, Drs. M. Yoshimura and Y. Matsunaga of Kyushu University for giving valuable information on SEU tolerant flip-flops and EDA tools. Invaluable discussions and information are given by Drs. Kuboyama, and D. Kobayashi of JAXA, Professor Y. Takahashi of Nippon University, and Ms. A. Makihara of HIREC.

Acronyms

ACE Architectural Correct Execution
ALLS Aligned Laboratory System
ALPEN ALpha Particle source/drain PENtration
ALS Absolute Laboratory System
ALU Arithmetic-Logic Unit
AMUSE Autonomous MUltilevel emulation system for Soft Error evaluation
ANITA Atmospheric-like Neutrons from thIck TArget
AOI Area Of Interest
ASIC Application Specific Integrated Circuit
ASIL Automotive Safety Integrity Level
ASTEP Altitude Single event effects Test European Platform
AVF Architectural Vulnerability Factor
AVP Architectural Verification Program
BAN Body Area Network
BCDMR Bistable Cross-coupled Dual Modular Redundancy
BICS Built-In Current Sensor
BISER Built-In Soft Error Resilience
BIPS Built-in Pulse Sensor
BIST Built-In Self Test
BL Bit Line
BNL Brookhaven National Laboratory
BOX Buried Oxide
BPSG Boron Phosphor Silicate Glass
BUT Board Under Test
CAM Content Addressable Memory
CAN Controller Area Network
CCD Charge Coupled Device
CHB CHecker Board
CHBc CHecker Board complement
CL Confidence Level

CLR	Cross-Layer Reliability
CM	Center of Mass
CMOS	Complementary Metal Oxide Semiconductor
CMP	Chemical Mechanical Polishing
CNL	UC Davis Crocker Nuclear Laboratory
CNRF	Cold Neutron Research Facility
CORIMS	COsmic Radiation IMpact Simulator
CPU	Central Processing Unit
CRAM	Configuration Random Access Memory
CRC	Cyclic Redundancy Code
CYCLON	Cyclotron of Louvain la Neuve
CYRIC	CYclotron and RadioIsotope Center
DCC	Duplication + Comparison + Checkpointing
DF	Derating Factor
DICE	Dual Interlocked storage CEll
DLL	Delay Locked Loop
DMR_1	Dual Modular Redundancy
DMR_2	Dynamic Memory Reconfiguration
DOA	Design On Average
DOAV	Design On Average and Variation
DOUB	Design On Upper Bound
DPM	Defects Per Million
DRAM	Dynamic Random Access Memory
DSP	Digital Signal Processor
DUE	Detected Unrecoverable Error
DUT	Device Under Test
ECC	Error Correction Code/Error Checking and Correction
EDA	Electric Design Automation
EDAC	Error Detection And Correction
EMI	Electro-Magnetic Interference
EX	Execution
FBE	Floating Body Effect
FDSOI	Fully Depleted SOI
FF	Flip-Flop
FFDA	Field Failure Data Analysis
FIT	Failure In Time
FPGA	Field Programmable Gate Array
FRAM	Ferroelectric Random Access Memory
GDS	Graphic Data System
GEM	Generalized Evaporation Model
GPS	Global Positioning System
GPU	Graphic Processing Unit

GPGPU	General Purpose GPU
GPU	Graphic Processing Unit
GTO	Gate Turn-Off Thyristor
HA	High Altitude
HHC	Hierarchical Hardware Checkpointing
HHFL	Heavy Halt Failure
ICICDT	International Conference on IC Design and Technology
ICITA	International Conference on Information Technology and Applications
ID	Instruction Decode
IF	Instruction Fetch
IGBT	Insulated Gate Bipolar Transistor
IOLTS	International On-Line Testing Symposium
INC	Intra-Nuclear Cascade
IRPS	International Reliability Physics Symposium
IUCF	Indiana University Cyclotron Facility
JAXA	Japan Aerospace Exploration Agency
JESD	JEDEC StanDard
J-PARC	Japan Proton Accelerator Research Complex
LABIR	inter LAyer Built-In Reliability
LAMPF	Los Alamos Meson Physics Facility
LANSCE	Los Alamos NationalScience Center
LBNL	Lawrence Berkeley National Laboratory
LEAP	Layout design through Error Aware Placement
LENS	Low-Energy Neutron Source
LET	Linear Energy Transfer
LFSR	Linear Feedback Shift Register
LHFL	Light Halt Failure
LIN	Local Interconnect Network
LINAC	LINear particle ACcelerator
LNL	Laboratori Nazionali di Legnaro
LSI	Large Scale Integration
LTFL	Latency Failure
LUT	Lookup Table
MA	Memory Access
MBU	Multi-Bit Upset
MCBI	Multi-Coupled Bipolar Interaction
MCU[1]	Multi-Cell Upset
MCU[2]	Micro Control Unit
MF	Masking Factor
MFTF	Mean Fluence To Failure
MNFL	Marginal Failure

MOSFET	Metal Oxide Semiconductor Field Effect Transistor
MPR	Memory Page Retire
MTTF	Mean Time To Failure
MTTR	Mean Time To Repair
NBTI	Negative Bias Temperature Instability
NCAP	European New Car Assessment Programme
NIST	National Institute of Standards and Technology
NMIJ	National Metrology Institute Japan
NoC	Network on Chip
NSAA	Nonstop Advanced Architecture
NSREC	Nuclear and Space Radiation Effects Conference
NYC	New York City
OS	Operating System
PC[1]	Program Counter
PC[2]	Power Cycle
PC[3]	Personal Computer
PCB	Printed Circuit Board
PCSE	Power Cycle Soft-Error
PDSOI	Partially Depleted SOI
PHITS	Particle and Heavy Ion Transport Code System
PIPB	Propagation Induced Pulse Broadening
PLL	Phase Locked Loop
PVF	Program Vulnerability Factor
QMN	Quasi-Monoenergetic Neutron
RADECS	RAdiation Effects on Components and Systems
RAM	Radom Access Memory
RAP	Resilience Articulation Point
RAS	Reliability, Availability and Serviceability
RASEDA	RAdiation effects on SEmiconductor Devices for space Application
RCNP	Research Center for Nuclear Physics
RHBD	Radiation Hardened-By-Design
RIIF	Reliability Information Interchange Format
RILC	Radiation Induced Leakage Current
RMA	Return Material Authorisation
ROM	Read Only Memory
RTL	Register Transfer Level
RTOS	Real Time Operating System
SAW	Surface Acoustic Wave
SBRM	Symptom Based Redundant Multithreading
SBST	Software-Based Self-Test
SBU	Single Bit Upset

SDC	Silent Data Corruption
SEALER	Single Event Adverse and Local Effects Reliever
SEB	Single Event Burnout
SECDED	Single Error Correction and Double Error Detection
SEE	Single Event Effect
SEFI	Single Event Functional Interrupt
SEFR	Single Event Fault Rate
SEGR	Single Event Gate Rupture
SEILA	Soft Error Immune LAtch
SEL	Single Event Latchup
SELSE	Silicon Errors in Logic – System Effects
SEM	Soft Error Mitigation
SER	Soft-Error Rate
SES	Single Event Snapback
SESB	Single Event SnapBack
SET	Single Event Transient
SEU	Single Event Upset
SEUT	Single Event Upset Tolerant
SHE	Software Hardening Environment
SIL	Safety Integrity Level
SILC	Stress Induced Leak Current
SIMS	Secondary Ion Mass Spectrometry
SITR	Self-Imposed Temporal Redundancy
SLC	Single Level Cell
SLFL	Silent Failure
SOI	Silicon On Insulator
SPFD	Sets of Pairs of Functions to be Distinguished
SPICE	SimulationProgram with Integrated Circuit Emphasis
SRAM	Static Random Access Memory
SRIM	Stopping and Range of Ions in Matter
STEM	Soft and Timing Error Mitigation
STI	Shallow Trench Isolation
TAMU	Texas A&M University
TID	Total Ionizing Dose effect
TCAD	Technology Computer-Aided Design
TID	Total Ionisation Dose
TISS	Trusted Interface Subsystem
TMR	Triple Module Redundancy
TRIUMF	Tri-University Meson Facility
TSL	The Svedberg Laboratory
TTA	Time Triggered Architecture
TTNoC	Time-Triggered Network-on-Chip

TVF	Timing Vulnerability Factor
UG	Under Ground
ULSI	Ultra Large Scale Integration
VLA	Very Low Alpha
VLSI	Very Large Scale Integration
WB	Write Back
WL	Word Line

1

Introduction

1.1 Basic Knowledge on Terrestrial Secondary Particles

Cosmic rays, which have extremely high energies, come from the galactic core and the sun to the atmosphere of the Earth. Primary cosmic rays in outer space consist mainly of protons (about 90%). Since cosmic rays are charged particles they twine around lines of geomagnetic or heliomagnetic forces as illustrated in Figure 1.1. Some of them are trapped by geomagnetic force to form the Van Allen radiation belts. Cosmic rays with energies less than the *geomagnetic rigidity cutoff* are deflected before entering the geomagnetic field. On the other hand, some are attracted into geomagnetic poles along with lines of geomagnetic force sometimes accompanied by the aurora borealis or australis. Cosmic rays are deflected more strongly near the equator since the lines of geomagnetic force are parallel to the surface of the Earth. Therefore, the strength of cosmic rays that reach the atmosphere differs depending on the geomagnetic latitude of the Earth.

When the energetic protons enter the atmosphere (troposphere and stratosphere) of the Earth, some protons undergo *nuclear spallation reaction* with nuclei (mainly nitrogen and oxygen nuclei) in the atmosphere to produce a number of light particles including neutrinos, photons, electrons, muons, pions, protons and neutrons as illustrated in Figure 1.2. Since secondary neutrons have longer ranges in the atmosphere compared to protons, they release cascades of spallation reactions in the atmosphere to make *air showers* that reach the surface of the Earth. Figure 1.3 shows an estimated differential neutron spectrum at NYC (New York City) sea level based on measurements in different locations in the USA [1]. The neutron energy at the ground ranges over 1 GeV and its flux beyond 1 MeV is around 20 n/cm^2/h in average. As the air can shield neutrons, strength (flux and energy) of neutrons depends upon altitude and to a slight extent atmospheric pressure [2]. Compared to the neutron flux at ground level, the neutron flux at avionics altitude is much higher by a factor of 100.

Furthermore, as cosmic rays are also deflected by the heliomagnetic field or the sun's activity which has about an 11-year cycle, the strength of the neutron flux at ground level also has about an 11-year cycle as shown in Figure 1.4 [3]. At the *solar maximum*, the neutron flux at ground level is almost at its weakest, while it is at its

Terrestrial Radiation Effects in ULSI Devices and Electronic Systems, First Edition. Eishi H. Ibe.

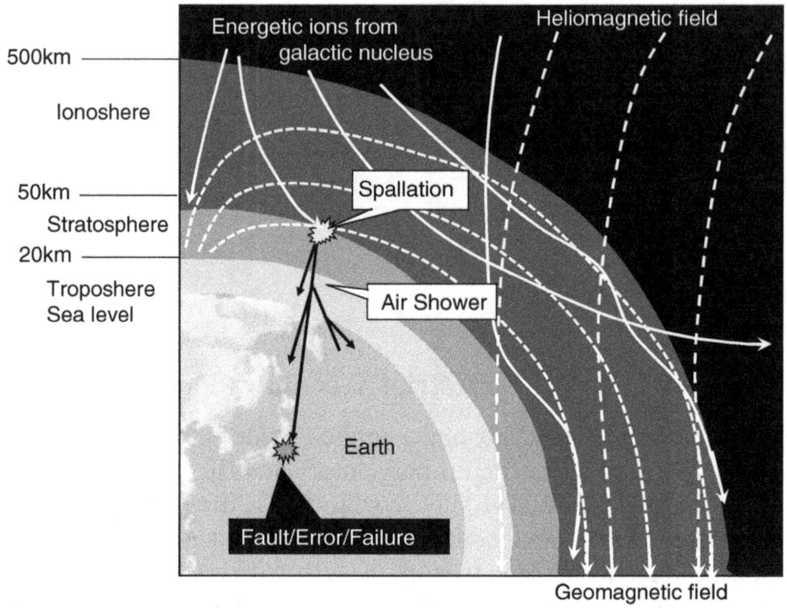

Figure 1.1 Overall scheme of terrestrial radiation-induced single event effects

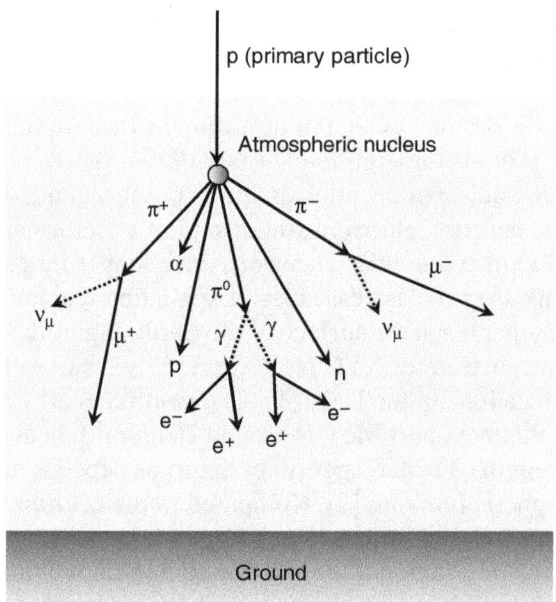

Figure 1.2 Initial stage of secondary particle production

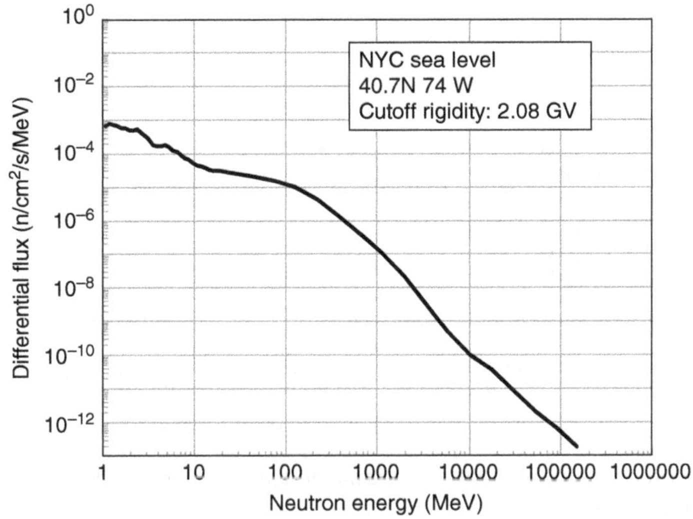

Figure 1.3 Differential high-energy neutron spectrum at NYC sea level based on JESD89A

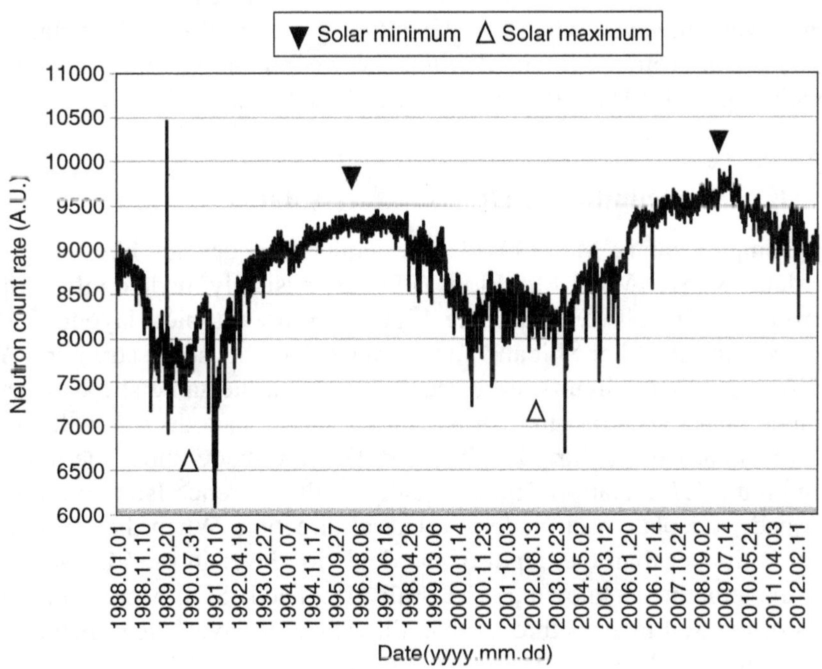

Figure 1.4 Long-term cyclic variation in neutron flux measured at Moscow Neutron Monitor Centre (http://cr0.izmiran.rssi.ru/mosc/main.htm)

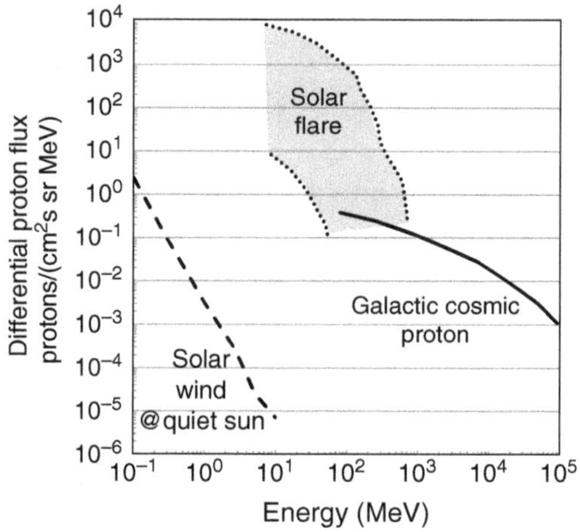

Figure 1.5 Differential proton spectra originated from solar-minimum sun, big flares on the sun and the galactic core

strongest at the *solar minimum*. Under normal activity, the sun emits a large quantity of protons, but their energy level is relatively low as shown in Figure 1.5 for the solar maximum period [4] since protons from the sun do not cause air showers directly at ground level. However, when big flares take place on the sun's surface, a much larger quantity of protons is emitted with energy comparative to the galactic protons as shown in Figure 1.5 [5] and this can cause air showers.

1.2 CMOS Semiconductor Devices and Systems

CMOS (Complementary Metal Oxide Semiconductor) devices like Static Random Access Memory (SRAM) or Flip Flops (FFs) are basically made on the stripe structure of p and n-dual wells. For example, Figure 1.6 shows typical layouts of diffusion layers (nodes) in SRAM one bit and an OR gate cell on the stripe structure. All nodes in memories and logic circuits are basically made on the same stripe structure in a chip. Unlike dual well structure, triple well structure has a deep n-well. As for Silicon On Insulator (SOI), Buried OXides (BOXs) are made under the dual wells as shown in Figure 1.7. Isolation oxides, usually Shallow Trench Isolation (STI) oxides are also made to isolate each node in a lateral direction. When the thickness of the SOI layer is thinner than the depth of the depletion layer in the SD (Source-Drain) channel, the structure is known as FD (Fully-Depleted) SOI. Meanwhile, when the thickness of the SOI layer is thicker than the depletion layer, the structure is known as PD (Partially-Depleted) SOI. Since the upper surface of BOX is completely covered by the depletion layer, parasitic capacitance can be largely reduced compared to Bulk/PDSOI, resulting in steep sub-threshold characteristics, reduction in latency and power consumption.

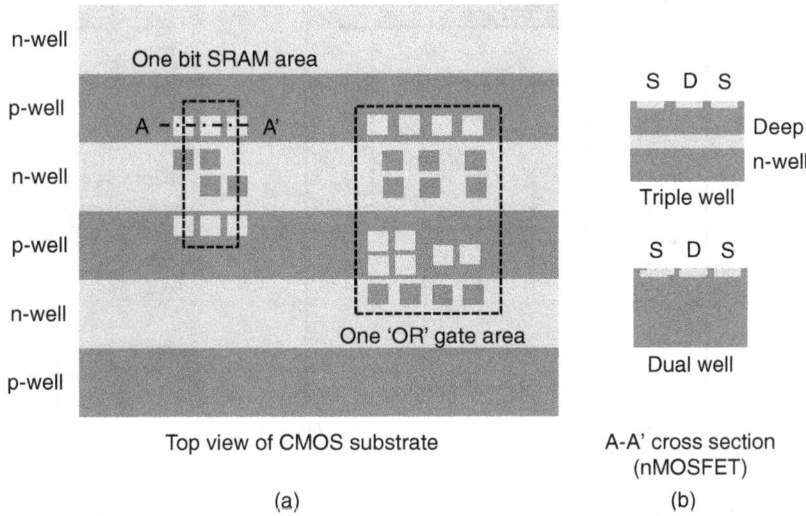

Figure 1.6 Basic layouts of CMOSFET devices on the stripe structure of p- and n-wells and cross sections of triple and dual well. (a) Top view of CMOS substrate and (b) A-A′ cross section (nMOSFET)

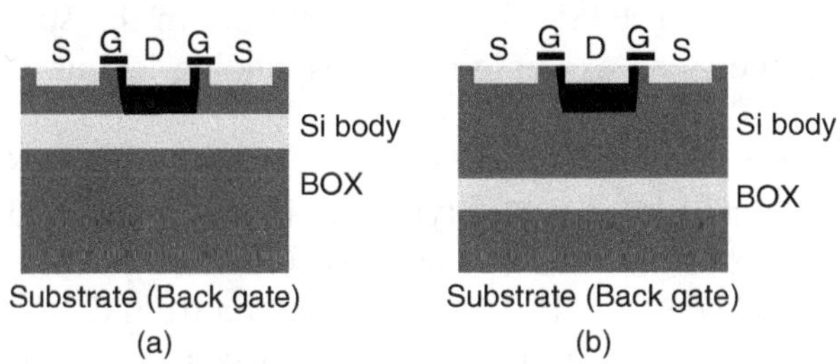

Figure 1.7 Structure of SOI device (image). (a) FDSOI and (b) PDSOI

Sugii et al. develop Silicon on Thin BOX (SOTB) structure by which back-gate bias can be applied in a silicon layer below the bottom of the thin (about 10 nm thickness) BOX in order to control V_{th} [6]. This structure is also known as *double-gate* structure that originates from FinFET [7].

Circuits are made by electrically connecting some nodes by wires above the well structure. Figure 1.8 shows a typical circuit for a SRAM. Figure 1.8a illustrates a simplified expression of a SRAM, in which two inverters are connected as a ring. The circuit is stable when the data in both nodes are reversed, and thus can store data. More specifically, as shown in Figure 1.8b, two nodes Q and Q(bar) store data '1' (high/V_{cc}) or '0' (low/V_{ss}). When the transfer transistors Tr_5 and Tr_6 is 'ON', data on the nodes are written or read. When the state of n_1/n_2 is high, the pMOSFET (p-channel Metal

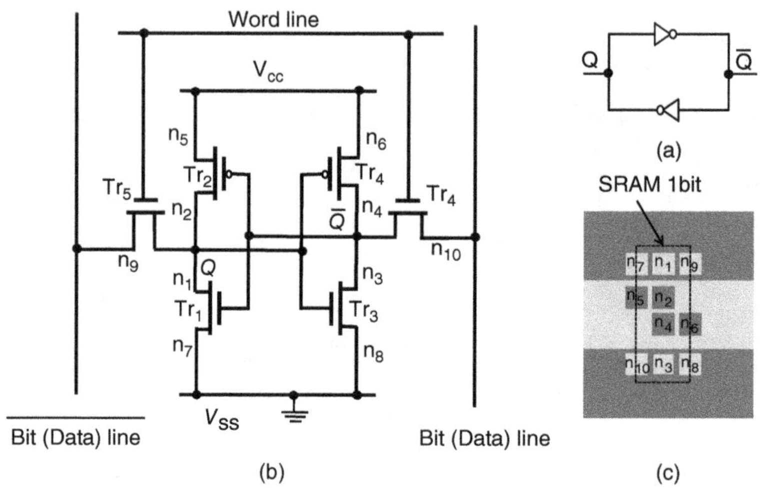

Figure 1.8 SRAM function and layout. (a) Equivalent circuit, (b) circuit and (c) node layout on p- and n-wells

Oxide Semiconductor Field Effect Transistor) transistor Tr_4 is 'OFF' and nMOSFET (n-channel MOSFET) Tr_3 is 'ON', making the state of n_3/n_4 low. This makes pMOSFET transistor 'ON' and nMOSFET 'OFF', resulting in the states of n_1-n_4 being stable. Figure 1.7c shows an example of corresponding layout of source and drain nodes n_1-n_{10}.

Logic circuits consist of a number of combinational logic gates (AND, OR, NAND, NOR, ...), and sequential logic gates, typically FFs as illustrated in Figure 1.9. In synchronous logic circuits, the total operation is controlled by clock signals produced typically in Phase Locked Loop (PLL). An instruction from CPUs is executed in one interval of clocks by controlling inputs of gates between two subsequent FFs. Execution results are captured to each FF when the nsubsequent clock is sent to the clock input of each FF.

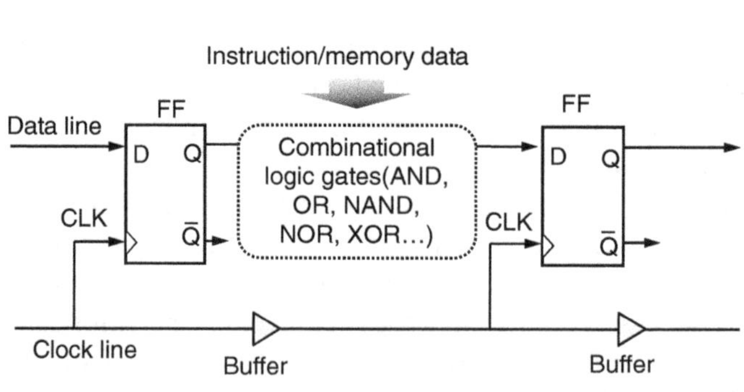

Figure 1.9 Example of logic circuit

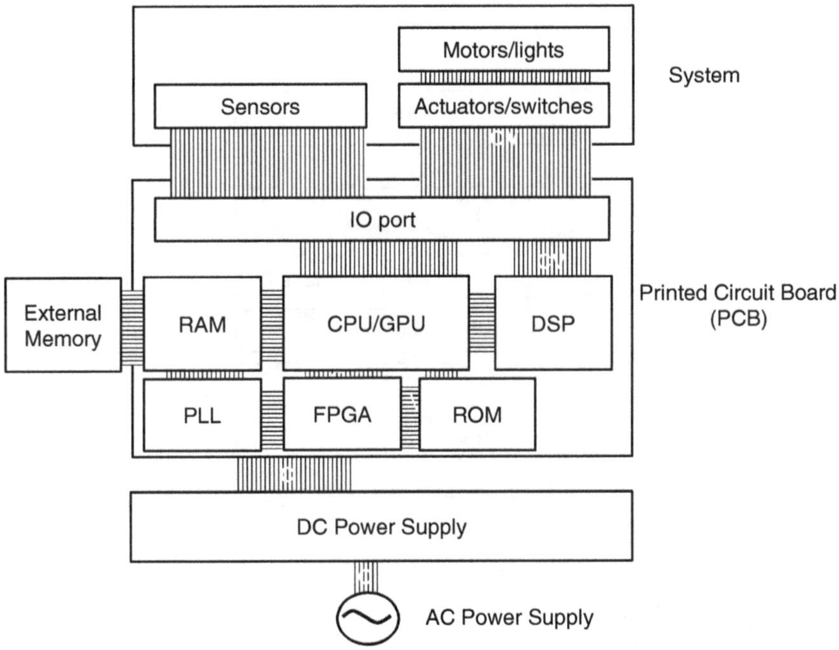

Figure 1.10 Example of electronic system implementation

Figure 1.10 illustrates the typical architecture of an electronic system that consists of power supply, printed circuit boards (PCBs) with various chips including CPU/GPU (Graphic Processing Unit), RAM, ROM (Read-Only Memory), FPGA (Field Programmable Gate Array), PLL, DSP (Digital Signal Processor), PLL and IO (Input/Output) port, and a system such as a server, router, automobile, train, aircraft, and so on. Such a system may have sensors to monitor the system condition, actuators and motors to control the operation of the system.

As a whole, any electronic system consists of a number of stack layers as illustrated in Figure 1.11 from the lowest layer, well/substrate to the highest layer, application or final hardware products.

1.3 Two Major Fault Modes: Charge Collection and Bipolar Action

When a charged particle passes through semiconductor devices, electron-hole pairs are produced along with the particle path (we call this 'track'). When the particle passes through the depletion layer or pn-junction under an off-state n-diffusion layer with potential V_{cc}, electrons in the depletion layer are collected to the n-diffusion layer to cause a *single event fault* and holes are repulsed out of the depletion layer due to the potential in the n-depletion layer (we call this 'node' or 'storage node') and flow out through a ground contact. This movement of electrons and holes causes elongation

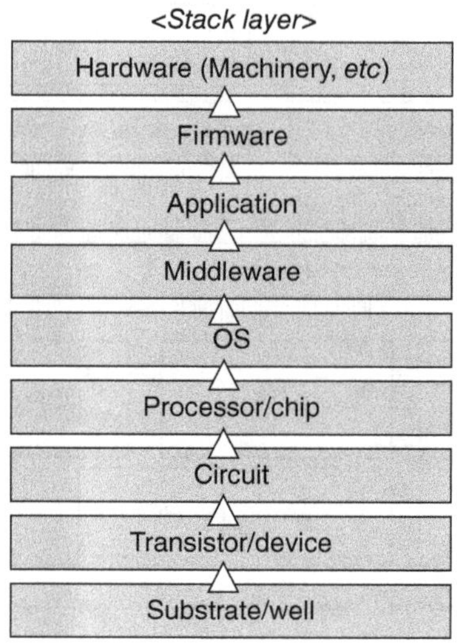

Figure 1.11 Example of stack layers in an electronic system

of the potential field along with the track to collect additional electrons produced initially outside the depletion layer. This phenomenon is known as 'funnelling' [8]. As for triple-well structure as shown in Figure 1.12, funnelling also takes place in the pn-junction adjacent to the deep n-well surface. When the charged particle passes through both pn-junctions on the diffusion layer and the deep n-well, electrons are collected on both sides. In this case, about a half of electrons produced along with the particle path may be collected into the storage node. This is the conventional soft-error model, or *charge collection model*, under which soft-error takes place

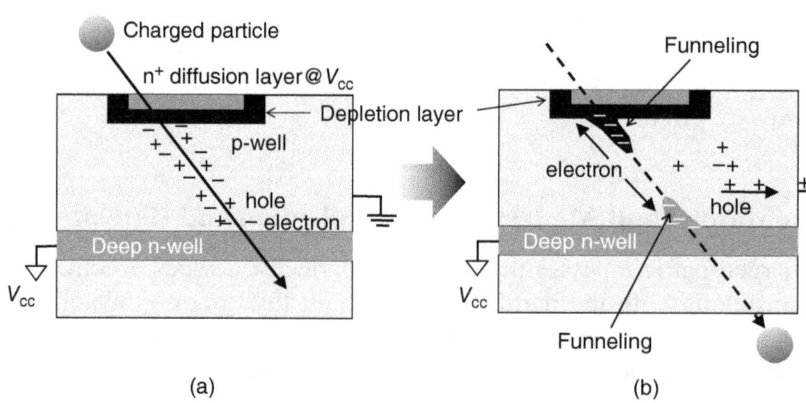

Figure 1.12 Charge collection model in a semiconductor structure by funnelling. (a) Charged particle hit and charge deposition and (b) charge collection to an n+ diffusion layer

only when the charged particle passes through the off-state n-diffusion layer and the amount of charge exceeds the critical charge Q_{crit} necessary to flip the data in the diffusion layer.

As device scaling nosedives to 130 nm process, Ibe et al. have pointed out a novel soft-error mechanism, Multi-coupled Bipolar Interaction (MCBI) [9], under which the condition that a charged particle passes through the off-state diffusion layer is not necessary for the occurrence of soft-error as illustrated in Figure 1.13. When a charged particle passes through pn-junction(s) in a p-well wall not of the storage node, electrons produced in the well flow out of the well through the same funnelling mechanism and holes are left in the well to make its level high, resulting in the parasitic transistor in the well being 'On'. As a substantial number of nodes are made in a single common well, faults due to MCBI can take place in multiple nodes to cause a SET (Single Event Transient) or MNT (Multi-Node-Transient).

There are a number of sources of faults other than SET or MNT in electronic devices as summarised in Table 1.1. Namely, electronic systems may fail due to cross-talk

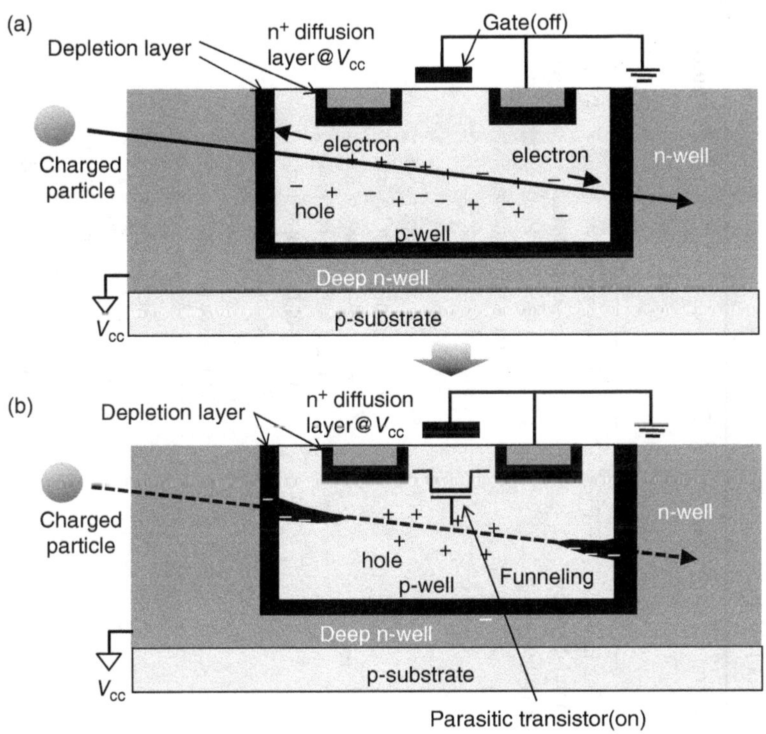

Figure 1.13 Noble bi-polar action model in a triple well n-MOSFET structure (MCBI: Multi Coupled Bipolar Interaction model). (a) Penetration of a charged particle through p-n junctions on the p-well. (b) Elevation of potential in the p-well by residual holes to turn on the parasitic transistor

Table 1.1 Modes of faults

Class	Definition	Name	Characteristics	Source	Affected area	In-situ detection method	In-situ recover/mitigation method
Transient/noise	Transient in electric potential and/or current in a chip	SET[a]	Single transient due to charge collected to the diffusion layer in the chip. Pulse width is below a few nano seconds, and can last longer than two clock pulses	Well/substrate	Random but limited to single well	Time and/or space redundancy such as DMR[b], TMR[c]	Time and/or space redundancy
		MNT[d]	Simultaneous SETs in more than two diffusion layers. Mainly, MNTs take place in a single well due to charge sharing or bipolar action. Space redundancy techniques such as DICE[e]. TMR may not work against MNTs	Well/substrate	Random but limited to single well	Monitoring the well potential and/or current	None
		RILC[f]	When a charged particle passes through tunnel oxide of a floating gate memory (flash memory), a leakage path is formed in the tunnel oxide and the potential in the floating gate may shifts resulting in change of V_{th} to cause soft-error	Tunnel oxide of a floating gate memory	Random but limited to tunnel oxide	ECC/parity	ECC

		Cross-talk	Noise propagation between close wires via parasitic capacitance	Wire	Random but limited to wire(s)	Time and/or space redundancy	Time and/or space redundancy
		Disturbance in power supply		Power supply line	Unlimited area	Monitor in power supply	None
		EMI[g]	Electromagnetic noise including burst noise	Anywhere	Unlimited area	Electro-magnetic probe	None
Trap in oxide	Holes trapped in oxide. They may cause leakage current and may disappear in time	TID[h]	Parasitic level by the trap may cause V_{th} shift, or potential due to holes in oxide may cause leakage path formation in adjacent semiconductor portion	Oxide	Random in a small area	Vth measurement	Annealing may work
Defect	Lattice defects	–	Damage or interstitials in crystal, which may deteriorate device functionality. They may cause stack at '0/1' error and can be permanent error	Anywhere/ tunnel oxide	Random in a small area	None	None

[a]Single event transient.
[b]Double modular redundancy.
[c]Triple modular redundancy.
[d]Multi-node transient.
[e]Dual interlocked storage cell
[f]Radiation induced leakage current.
[g]Electro-magnetic interference.
[h]Total ionization dose effects.

between two or more wires with high parasitic capacitance [10], noise produced in the power supply [11], Electro-Magnetic Interference (EMI) [11], hole traps in oxides which may cause the V_{th} shift [12] of the transistor or high potential due to accumulated holes in oxide which may cause leakage path formation in the adjacent Si channel (Total Ionisation Dose effects: TID. For more details, see Chapter 3), and a lattice defect may deteriorate the device's functionality [13]. RILC (Radiation Induced Leakage Current) in the tunnel oxide of the floating gate causes a V_{th} shift of flash memory, resulting in soft-errors [14, 15]. The most important characteristics of single event failure is that the source is basically limited to one single well or substrate, whereas other fault sources are not limited to anything other than the well/substrate. The faults that do not originate from terrestrial radiation are not within the scope of this book, we refer to any fault caused by terrestrial radiation as 'fault' in the remaining text of this book.

1.4 Four Hierarchies in Faulty Conditions in Electronic Systems: Fault – Error – Hazard – Failure

Starting from faults in the well or substrate, failures in electronic systems take place in a kind of hierarchy of faulty conditions as illustrated in Figure 1.14. As illustrated in Figure 1.6, faults are only produced in the wells and faults do not always cause errors

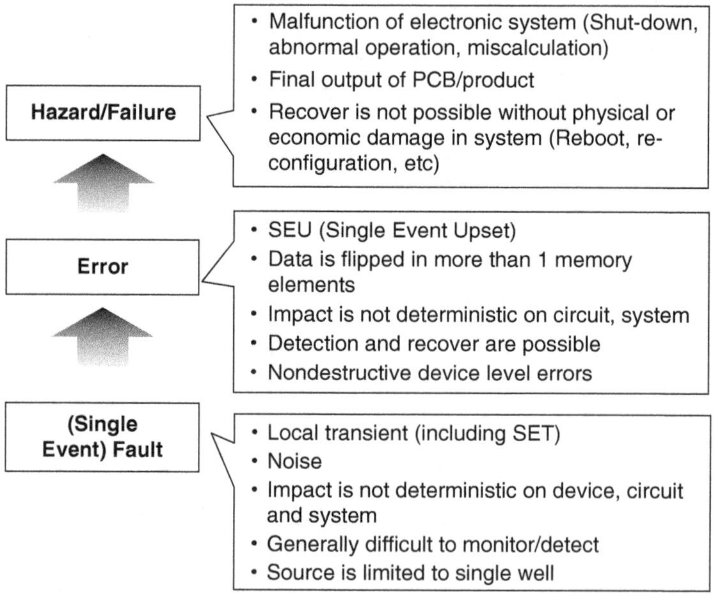

Figure 1.14 Hierarchy of faulty conditions: fault-error-failure

(data flips in memory elements). They are generally difficult to detect and may actually disappear or may not be strong enough to cause errors. Only when fault is captured and causes data flips in memory devices such as SRAMs, DRAMs, flash memories and FFs, is it regarded as an *error* at the device or circuit level. When a single particle penetrates into a device(s), it can cause multiple faults or multiple errors. In definition, physical consequence due to one single particle including a neutron is called a '*Single Event Effect (SEE)*'. When an SEE causes an error(s), we refer to this phenomenon as '*SEU (Single Event Upset)*'. Again, SEU can consist of multiple errors. *The important thing is that Soft Error Rate (SER) is defined by the number of SEUs, not by the number of errors.*

One more important concept used in this book is the *SEU cross section* σ_{seu} that is defined by,

$$\sigma_{\mathrm{seu}} = \frac{N_{\mathrm{seu}}}{\Phi_{\mathrm{p}}}, \tag{1.1}$$

where,

N_{seu}: the number of SEUs (not errors!)/count;
Φ_{p}: fluence of particles (neutrons)/(n/cm^2).

Fluence means the total number of particles passed through a unit area.

σ_{seu} can be measured using the accelerator experiment (see Chapter 4) and one can calculate SER by using the σ_{seu} as follows:

$$SER = \sigma_{\mathrm{seu}} \times \phi_{\mathrm{p}} \times 1 \times 10^9/\mathrm{FIT}, \tag{1.2}$$

where,

ϕ_{p}: flux of the particle (count h^{-1} cm^{-2});
FIT: Failure In Time, SER in 10^9 hour.

Flux means the number of particles which pass through a unit area per unit time.

An error does not always cause failure, depending mainly on the location and the functionality of the system. Only when an error(s) propagates to the final output and causes malfunction of the system canwe call this consequence a '*failure*'. We may call incorrect output of the controller or PCB, which does not affect normal operation of the system, a '*hazard*'. An error does not always cause a system hazard or failure, because it may disappear or may be *masked* during propagation in the chip or board by some masking effects. Some mitigation techniques like parity, Error Checking and Correction (ECC) and interleaving techniques may be applied to reduce SER. Failure is not compensated for by the system without physical or economic damage. Failures include shutdown and abnormal operation of the system. Incorrect calculation by using super computers can also be categorised into failure.

1.5 Historical Background of Soft-Error Research

Scaling down of semiconductor devices to sub-100 nm technology encounters a wide variety of technical challenges including: V_{th} variation [6], Negative Bias Temperature Instability (NBTI) [16], short-channel effect [17], gate leakage [18], and so on. Terrestrial neutron-induced SEU has become one of the key issues that can present a major setback in scaling. Before going into detail about the current situation concerning soft-error problems, let us look at the historical background on which the soft-error problem has escalated.

Table 1.2 summarises the history of soft-error research with SRAM design rules and densities. Ever since α-ray soft error was first discovered in DRAMs, we have experienced a number of distinctive *paradigm shifts*. Five or six paradigm shifts in soft-error research from 1979 to 2013 are highlighted in Table 1.2. It is well known that α-ray soft error in DRAM was discovered by May and Woods in 1979 [19]. In the same year, the possibility of soft-error due to a terrestrial neutron is pointed out by Ziegler and Lanford [20]. As the impact of α-ray was much larger than neutron induced soft-error, intensive efforts are focused on α-ray soft error. Until the early 1990s, alpha-ray soft error in DRAMs was overcome by several countermeasures such as triple-well structure, usage of low-alpha materials and shielding by package materials [21–23]. Scaling of semiconductor devices potentially has beneficial effects toward reduction of α-ray soft error. Namely, reduction in the probability of hitting the storage node by alpha particles due to shrinkage of the area of the storage node and reduction of the charge collected to the nodes due to reduction in the volume for charge collection [24, 25]. Thus, soft error has not mattered at the ground level for the time being. As the device design rule is shrunk down further to around 130 nm, soft error in SRAMs by a terrestrial neutron has become significant. It is surprising that trends of literature data of SER in DRAM and SRAM by nucleon (neutron and proton) turn out to be reversed and the soft-error rate of SRAM has become much higher than DRAMs when memory density exceeds 4 Mbit/device or the SRAM design rule has become smaller than 250 nm as indicated in Figure 1.15. Due to the beneficial effects of scaling as mentioned previously, soft error in DRAMs that have capacitors has naturally decreased. Meanwhile, the parasitic capacitor of SRAMs inevitably has decreased resulting in the reduction of the critical charge Q_{crit} as the tradeoff of the beneficial effects of scaling.

As this malicious trend in SRAM soft error by terrestrial neutrons is becoming obvious, major network vendors since 2000 have begun to request that SRAM venders carry out neutron irradiation tests to clarify the susceptibility for soft error [27]. The reversal phenomena and network vendor action are believed to trigger the paradigm shift in soft-error research from DRAM to SRAM. Filing neutron irradiation data reports to users have been almost mandatory for memory vendors, and this has triggered discussions on neutron standard testing methods worldwide. As a consequence, JESD89 [28] for neutron, proton and α-ray SER testing methods was issued as the de

Table 1.2 Five paradigm shifts in the history of soft-error research

No.	Years	Main features in paradigm shift	Events	SRAM design rule (nm)	SRAM density (Mb)
1	1979 to early 1990s	✓ Discover a-ray soft error and development of mitigation techniques for DRAMs ✓ α-ray soft error in DRAMs is overcome	✓ First finding by May (1979)	>250	<64 K
2	Late 1990s to 2000	✓ Impact of terrestrial neutron induced soft-error is widely recognized in SRAMs [9]	✓ An SRAM vender was blamed for soft-error of SRAMs	130	128 K–4 M
3	2000–2005	✓ MCU (multi-cell upset) in SRAMs become major concerns [26] ✓ Bipolar mode soft-error was discovered (*idem.*) ✓ Concerns spread over logic devices	✓ JESD89 (2001) ✓ Polishing method: BPSG → CMP	90	8 M
4	2006–2009	✓ Memory SER was basically overcome [9] ✓ Mitigation of MNT in flip flop pursued	✓ JESD89A (2006) ✓ AEC Q100 G (2008)	65	16 M
5	2010–2013	✓ Concerns spread over redundant large and real-time systems ✓ Power/cost effective mitigation techniques are becoming the top priority in terrestrial industries ✓ Concerns spread over terrestrial particles other than neutrons	✓ ISO26262 (2011) ✓ IEC62396 ✓ Cloud/exa-scale computing/big data	<40	>32 M

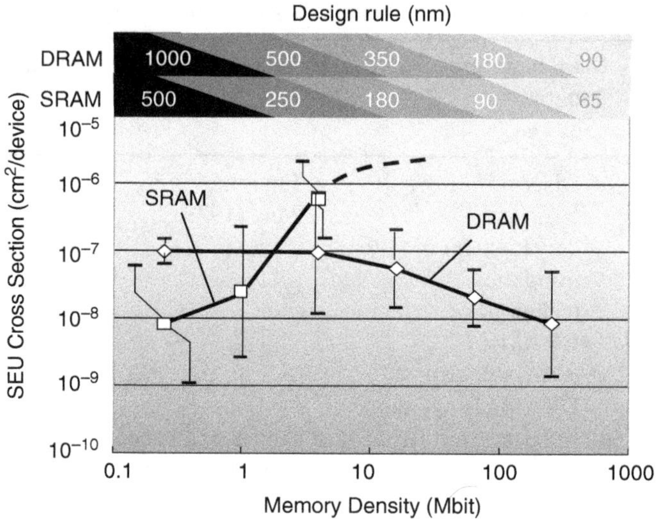

Figure 1.15 Reverse trends in SEU cross section in SRAMs and DRAMs as scaling proceeds

facto soft-error testing standard in 2001. In the third paradigm shift in 2000–2005, concerns over neutron soft error further spread in two directions from around 2004. One direction was about the concerns over multi cell upset (MCU) that emerged from the 130 nm process due to bipolar [26] or charge sharing effects [29]. When MCU takes place over more than two bits in the same word of SRAM, it cannot be recovered using EDAC (Error Detection and Correction) or ECC or it results in system crash. EDAC or ECC can detect two-bit errors and correct one-bit error, but cannot correct two-bit errors in the same word. It was found that almost all MCUs take place in one MOSFET well and align along with only two adjacent bit lines. Based on this finding, MCUs in memory devices were basically overcome by applying both ECC and interleaving [9]. Another direction is over concerns on SET (noise) in sequential and combinational logic devices. SER in flip-flops is predicted to be close to that in SRAM beyond the 90 nm design rule [30], but there have been no effective detection and correction methods in logic devices except for redundancy techniques. Moreover, obvious threats due to MNT are found in most space redundancy techniques, like TMR (Triple Module Redundancy) [31] and DICE (Dual Interlocked storage Cell) [32]. In TMR, instructions are executed in three identical modules and two out of three voting is carried out to obtain reliable output for the next set of modules. DICE has two input nodes, and only when the values of two nodes are identical is the input transferred to output. MNT appears in logic devices down to the same mechanism as MCU in memory devices. If this happens in two modules of TMR [33] or two input nodes of DICE [34], the recovery mechanism fails and may result in SDC

(Silent Data Corruption) that can cause un-recognisable system failure [35]. Several ideas to harden DICE-like flip-flops are proposed by Uemura et al. [36] and Lee et al. [37] separately in 2010.

In addition to the two distinct directions, Baumann *et al.* pointed out soft error caused by thermal neutron caption reaction by ^{10}B that has a natural abundance of 19.9% in Boron contained in the BPSG (Boron Phosphor Silicate Glass) used for flattening wafer surface [38]. When thermal neutrons, which typically have an energy level of 25 meV equilibrium with atmospheric molecular dynamics, are captured in ^{10}B nuclei, He ion (1.47 MeV) and ^{7}Li ion (0.84 MeV) are released and cause soft error by direct ionisation. This type of soft error appeared to be overcome by changing the polishing process from BPSG to CMP (Chemical Mechanical Polishing) for the time being.

In the fourth paradigm shift in 2006–2009, discussions over the revision of JESD89 started in 2003 because the original JESD89 has a number of limitations: the testing facility assigned as the standard facility is the only spallation neutron source in Los Alamos National Laboratory, for example. The new version JESD89A [39] was issued in 2006 with more reliable testing and analysis methods including the *quasi-monoenergetic neutron* test method. A differential spectrum of terrestrial neutron was revised and a certain number of neutron irradiation facilities were added as the standard facilities. An IEC60749-38 standard that is consistent with JESD89A was issued in 2008 as the de jure standard [40].

In the fifth paradigm shift during 2010–2013, the impact by soft error spread over large electronic systems. For big data-centres [41] or exa-scale supercomputers [42], power reduction is one of most important key designs. The space redundancy techniques that have a large area and power overheads will therefore not be applied to such big systems, in principle. Real-time (safety critical) systems like avionics or micro-control units in automobiles are also attracting serious concern and in-depth studies are being widely pursued [43, 44]. AEC Q100 G [45] and ISO26262 [46] for automobiles were also issued in 2008 and 2011, respectively.

With common recognition that soft-error induced system failure cannot be suppressed to a satisfactory level by applying mitigation techniques only to single stack layer (device, circuit, chip, board, firmware, OS, middleware, and so on), communication and combined mitigation techniques among stack layers are encouraged as an unavoidable option [47, 48]. It has already been generally accepted from around 2000 that mitigation techniques applied to a single stack layer cannot be an effective nor promising solution against system failures, and collaboration among stack layers must be realised [49, 50]. In reality, such collaboration has turned out to be very difficult since the basic engineering skills in each stack layer are essentially and significantly different. Most engineers/researchers cannot expand their specialties beyond their stack layers. Novel strategies to overcome this situation are needed in order to allow further exploration. A *built-in* communication scheme among the stack layers is proposed by Ibe *et al.* in their LABIR (inter Layer Built-In Reliability) concept

[51, 52]. Evans *et al.* are proposing the RIIF (Reliability Information Interchange Format) as the common format or protocol to be used in system design among stack layers [53].

In the possible sixth paradigm shift, other terrestrial particles, such as muons, low energy neutrons and protons, are being pointed out as possible SER threats at ground level. Due to mainly scaling semiconductor devices under 100 nm, a number of new radiations/particles are being pointed out as the source of soft error. Alpha particles with VLA (Very Low Alpha)-level packaging can again cause soft error in SRAMs [54]. Sub-100 nm SRAMs are sensitive to soft error due to terrestrial muons [55]. Low (thermal) energy neutrons cause soft error due to neutron capture reaction of B-10 in devices without BPSG processes [56, 57]. Concerns about electrons (beta rays) and gamma rays have been boosted after the severe nuclear accident in the Fukushima nuclear plant site [58, 59].

1.6 General Scope of This Book

In Chapter 2, sources of terrestrial radiation and their properties are introduced. In Chapter 3, mechanisms of radiation effects are described in depth. In Chapter 4, fundamentals of electronic devices and systems such as memories, logic gates, FPGAs, OS and processors necessary to understand the following chapters are introduced. Chapter 5 summarises a wide range of experimental facilities for irradiation tests. In Chapter 6, simulation techniques for soft-error and typical simulation results are introduced, particularly for neutron irradiation effects on several error/fault modes. Chapter 7 describes detection and classification techniques of faults, errors and failures. In Chapter 8, mitigation techniques against failures in electronic systems and challenges are summaried. Chapter 9 summaries this book. In the Appendices, some additional information together with some coding techniques in Visual Basic and some sample codes are provided. Those who are not interested in simulation may skip Chapter 6 and the Appendices.

References

[1] JEDEC Standard JESD89A. (2006) *Measurement and Reporting of Alpha Particle and Terrestrial Cosmic Ray-Induced Soft Errors in Semiconductor Devices*, JEDEC.
[2] Nakamura, T., Baba, M., Ibe, E. *et al.* (2008) *Terrestrial Neutron-Induced Soft-Errors in Advanced Memory Devices*, World Scientific, Hackensack, NJ.
[3] http://cr0.izmiran.rssi.ru/mosc/main.htm (accessed August 17, 2014)
[4] Solar Energetic Particles and Cosmic Rays http://www.solar-system-school.de/lectures /marsch/7.pdf (accessed 14 February 2013)
[5] http://www.kusastro.kyoto-u.ac.jp/~kamaya/ISP/chap7.ppt (accessed 14 February 2013)

[6] Sugii, N., Tsuchiya, R., Ishigaki, T. *et al.* (2008) Comprehensive study on Vth variability in silicon on thin BOX (SOTB) CMOS with small random-dopant fluctuation: finding a way to further reduce variation. IEEE International Devices Meeting, San Francisco, CA, 15–17 December, pp. 249–253.
[7] Hisamoto, D., Lee, W.-C., Kedziersk, J. *et al.* (2000) FinFET-a self-aligned double-gate MOSFET scalable to 20 nm. *IEEE Transactions on Electron Devices*, **47** (12), 2320–2325.
[8] Hu, C. (1982) Alpha-particle-induced field and enhanced collection of carriers. *IEEE Electron Device Letters*, **EDL-3** (2), 31–34.
[9] Ibe, E., Chung, S., Wen, S. *et al.* (2006) Spreading diversity in multi-cell neutron-induced upsets with device scaling. The 2006 IEEE Custom Integrated Circuits Conference, San Jose, CA, 10–13 September, 2006, pp. 437–444.
[10] Walker, C.S. (1990) *Capacitance, Inductance and Crosstalk Analysis (Artech House Antennas and Propagation Library)*, Altech House Publisher.
[11] Kanekawa, N., Ibe, E., Suga, T. and Uematsu, Y. (2010) *Dependability in Electronic Systems-Mitigation of Hardware Failures, Soft Errors, and Electro-Magnetic Disturbances*, Springer, New York.
[12] Crain, S.H., Mazur, J.E., Katz, R.B. *et al.* (2001) Analog and digital single-event effects experiments in space. *IEEE Transactions on Nuclear Science*, **48** (6), 1841–1848.
[13] Lacoe, R.C. (2004) Fabricating radiation-hardened digital components at commercial CMOS foundries using hardness-by-design techniques. The 6th International Workshop on Radiation Effects on Semiconductor Devices for Space Application, Tsukuba, 6–8 October 2004, pp. 227–234.
[14] Cellere, G., Pellati, P., Chimenton, A. *et al.* (2001) Radiation effects on floating-gate memory cells. *IEEE Transactions on Nuclear Science*, **48** (6), 2222.
[15] Butt, N.Z. and Alam, M. (2008) Modeling single event upsets in floating gate memory devices. IEEE International Reliability Physics Symposium, Anaheim, CA, No. 5D.1, pp. 547–555.
[16] Wen, S., Wong, R. and Silburt, A. (2008) IC component SEU impact analysis. IEEE Workshop on Silicon Errors in Logic – System Effects, University of Texas at Austin, 26 March, p. 27.
[17] Villanueva, D., Pouydebasque, A., Robilliart, E. *et al.* (2003) Impact of the lateral source/drain abruptness on MOSFET characteristics and transport properties. IEEE International Electron Devices Meeting, Washington, DC, 7–10 December 2003 (9.4).
[18] Clark, L.T., Moh, K.C., Holbert, K.E. *et al.* (2007) Optimizing radiation hard by design SRAM cells. *IEEE Transactions on Nuclear Science*, **54** (6), 2028–2036.
[19] May, T.C. and Woods, M.H. (1979) Alpha-particle-induced soft errors in dynamic memories. *IEEE Transactions on Electron Devices*, **ED-26** (1), 2–9.
[20] Ziegler, J.F. and Lanford, W.A. (1979) Effect of cosmic rays on computer memories. *Science*, **206**, 776–788.
[21] Takeuchi, K., Shimohigashi, K., Takeda, E. and Yamasaki, E. (1987) Experimental characterization of α-induced charge collection mechanism for megabit DRAM cells. IEEE International Solid-State Circuits Conference, 10 February 1987, pp. 99–100.

[22] Sai-Halasz, G.A., Wordeman, M.R. and Dennard, R.H. (1982) Alpha-particle-induced soft error rate in VLSI circuits. *IEEE Transactions on Electron Devices*, **ED-29** (4), 725–731.

[23] Thompson, C.E. and Meese, J.M. (1981) Reduction of α-particle sensitivity in dynamic semiconductor memeories (16k d-RAMs) by neutron irradiation. *IEEE Transactions on Nuclear Science*, **28** (6), 3987–3993.

[24] Ibe, E. (2001) Current and future trend on cosmic-ray-neutron induced single event upset at the ground down to 0.1-micron-device. The Svedberg Laboratory Workshop on Applied Physics, Uppsala, Sweden, May, 3 (1).

[25] Ibe, E. Yahagi, Y. Kataoka, F. *et al.* (2002) A self-consistent integrated system for terrestrial-neutron induced single event upset of semiconductor devices at the ground. 2002 International Conference on Information Technology and Application, Bathurst, Australia, 25–28 November 2002, pp. 273–221.

[26] Ibe, E., Kameyama, H., Yahagi, Y. *et al.* (2004) Distinctive asymmetry in neutron-induced multiple error patterns of 0.13umocess SRAM. The 6th International Workshop on Radiation Effects on Semiconductor Devices for Space Application, Tsukuba, Japan, 6–8 October 2004, pp. 19–23.

[27] Cataldo, A. (2001) SRAM Soft Errors Cause Hard Network Problems, http://eetimes .com/electronics-news/4042377/SRAM-soft-errors-cause-hard-network-problems- (accessed 18 February 2013).

[28] JEDEC Standard JESD89. (2001) *Measurement and Reporting of Alpha Particle and Terrestrial Cosmic Ray Induced Soft Errors in Semiconductor Devices*. JEDEC, pp. 1–63.

[29] Seifert, N., Gill, B., Zhang, M. *et al.* (2007) On the scalability of redundancy based SER mitigation schemes. International Conference on IC Design and Technology, Austin, Texas, 18–20 May (G2), pp. 197–205.

[30] Shivakumar, P., Kistler, M.S., Keckler, A. *et al.* (2002) Modeling the effect of technology trends on the soft error rate of combinational logic. International Conference on Dependable Systems and Networks, pp. 389–398.

[31] Pilotto, C., Azambuja, J.R. and Kastensmidt, F.L. (2008) Synchronizing triple modular redundant designs in dynamic partial reconfiguration applications. Proceedings of the 21st Annual Symposium on Integrated Circuits and System Design, September 2008, pp. 199–204.

[32] Calin, T., Nicolaidis, M. and Velazco, R. (1996) Upset hardened memory design for submicron CMOS technology. *IEEE Transactions on Nuclear Science*, **43** (6), 2874–2878.

[33] Quinn, H., Morgan, K., Graham, P. *et al.* (2007) Domain crossing events: limitations on single device triple-modular redundancy circuits in Xilinx FPGAs. International Nuclear and Space Radiation Effects Conference, Honolulu, Hawaii, July 23–27 (C-5).

[34] Seifert, N. and Zia, V. (2007) Assessing the impact of scaling on the efficacy of spatial redundancy based mitigation schemes for terrestrial applications. IEEE Workshop on Silicon Errors in Logic – System Effects 3, Austin, TZ, April 3, 4.

[35] Quinn, H., Tripp, J., Fairbanks, T. and Manuzzato, A. (2011) Improving microprocessor reliability through software mitigation. IEEE Workshop on Silicon Errors in Logic – System Effects, Champaign, Illinoi, 29–30 March, pp. 16–21.

[36] Uemura, T., Tosaka, Y., Matsuyama, H.K. and Shono, K. (2010) SEILA: soft error immune latch for mitigating multi-node-SEU and local-clock-SET. IEEE International Reliability Physics Symposium 2010, Anaheim, California, 2–6 May, pp. 218–223.

[37] Lee, H.-H. Lilja, K. and Mitra, S. (2010) Design of a sequential logic cell using LEAP: layout design through error aware placement. IEEE Workshop on System Effects of Logic Soft Errors, Stanford University, 23 March.

[38] Baumann, R.C. and Smith, E.B. (2000) Neutron-induced boron fission as a major source of soft errors in deep submicron SRAM devices. IEEE International Reliability Physics Symposium Proceedings, San Jose, CA, 10–13 April, pp. 152–157.

[39] JEDEC Standard JESD89A. (2006) *Measurement and Reporting of Alpha Particle and Terrestrial Cosmic Ray Induced Soft Errors in Semiconductor Devices*. JEDEC, pp. 1–93.

[40] IEC IEC60749-38. (2008) *Part 38: Soft Error Test Method for Semiconductor Devices With Memory. Semiconductor Devices. Mechanical and Climatic Test Methods*, pp. 1–9.

[41] Pecchia, A., Cotroneo, D., Kalbarczyk, Z. and Iyer, R.K. (2011) Improving log-based field failure data analysis of multi-node computing systems. International Conference on Dependable Systems and Networks, Hong Kong, China, 28–30 June, pp. 97–108.

[42] Bronevetsky, G. and deSupinski, B. (2007) Soft error vulnerability of iterative linear algebra methods. IEEE Workshop on Silicon Errors in Logic – System Effects 3, Austin, TX, April 3, 4.

[43] Abella, J.F., Cazorlal, J., Gizopoulos, D. *et al.* (2011) Towards improved survivability in safety-critical systems. 17th IEEE International On-Line Testing Symposium, Athens, Greece, 13–15 July (S3), pp. 242–247.

[44] Baumeister, D. and Anderson, S.G.H. (2012) Evaluation of chip-level irradiation effects in a 32-bit safety microcontroller for automotive braking applications. IEEE Workshop on Silicon Errors in Logic – System Effects, Champaign-Urbana, Illinois, 27–28 March (2.2).

[45] Automotive Electronics Council (2007) Failure Mechanism Based Stress Test Quantification for Integrated Circuits, AEC-Q100 Rev.G, pp. 1–35.

[46] ISO ISO26262. (2011) *Road Vehicles-Functional Safety*, International Organization for Standardization.

[47] Quinn, H. (2011) Study on cross layer reliability. IEEE Workshop on Silicon Errors in Logic – System Effects, Champaign, Illinoi, 29–30 March.

[48] Carter, N. (2010) Cross-layer reliability. IEEE Workshop on System Effects of Logic Soft Errors, Stanford University, 23 March.

[49] Slayman, C. (2003) Eliminating the threat of soft errors – a system vendor perspective. IRPS SER Panel Discussion, Eliminating the Threat of Soft Error, Dallas, TX, 2 April 2003, (6).

[50] Ibe, E., Kameyama, H., Yahagi, Y. and Yamaguchi, H. (2005) Single event effects as a reliability issue of IT infrastructure. 3rd International Conference on Information Technology and Applications, Sydney, Australia, I, 3–7 July 2005, pp. 555–560.

[51] Ibe, E., Shimbo, K., Toba, T. *et al.* (2011) LABIR: inter-LAyer built-in reliability for electronic components and systems. Silicon Errors in Logic – System Effects, Champaign, IL, 27 March.

[52] Ibe, E., Shimbo, K., Taniguchi, H. *et al.* (2011) Quantification and mitigation strategies of neutron induced soft-errors in CMOS devices and components-the past and future. IEEE International Reliability Physics Symposium, Monterey, CA, 12–14 April (3C2).
[53] Evans, A., Nicolaidis, M., Wen, S.-J. *et al.* (2012) RIIF – reliability information interchange format. IEEE International On-Line Testing Symposium, Sitges, Spain, 27–29 June 2012 (6.2).
[54] Kobayashi, H., Kawamoto, N., Kase, J. and Shiraishi, K. (2009) Alpha particle and neutron-induced soft error rates and scaling trends in SRAM. IEEE International Reliability Physics Symposium 2009, Montreal, Quebec, 28–30 April (2H4), pp. 206–211.
[55] Sierawski, B.D., Mendenhall, M.H., Reed, R.A. *et al.* (2010) Muon-induced single event upsets in deep-submicron technology. *IEEE Transactions on Nuclear Science*, **57** (6), 3273–3278.
[56] Wen, S., Wong, R., Romain, M. and Tam, N. (2010) Thermal neutron soft error rate for SRAMs in the90nm-45nm technology range. 2010 IEEE International Reliability Physics Symposium, Anaheim, CA, 2–6 May (SE5.1), pp. 1036–1039.
[57] Wen, S., Pai, S.Y., Wong, R. *et al.* (2010) B10 findings and correlation to thermal neutron soft error rate sensitivity for SRAMs in the Sub-micron technology. IEEE International Integrated Reliability Workshop, Stanford Sierra, CA, 17–21 October, pp. 31–33.
[58] Baumann, R.C. (2011) Determining the impact of alpha-particle-emitting contamination from the Fukushima Daiichi disaster on Japanese manufacturing sites. 12th European Conference on Radiation and its Effects on Component and Systems, Sevilla, Spain, 19–23 September, pp. 784–787.
[59] Ibe, E., Toba, T., Shimbo, K. and Taniguchi, H. (2012) Fault-based reliable design-on-upper-bound of electronic systems for terrestrial radiation including muons, electrons, protons and low energy neutrons. IEEE International On-Line Testing Symposium, Sitges, Spain, 27–29 June, 2012 (3.2).

2

Terrestrial Radiation Fields

2.1 General Sources of Radiation

Figure 2.1 depicts two simplified mechanisms by which photons and/or secondary particles are produced.

When a high-energy particle penetrates into a nucleus, various nuclear reactions take place with to the end result of decomposing the nucleus into fragments as shown in Figure 2.1a. The nuclear reactions are likely to be: neutron capture [1], fission [2], fusion [3] and spallation reaction [4]. As a result of the reactions, energetic photons, electrons, neutrons, protons, muons, pions and heavier particles are released with a residual nucleus. Even high-energy photons and electrons can produce neutrons, protons, deuteriums, tritiums and alpha particles through photonuclear reaction [5]. When these particles are charged particles, they are known as secondary ions. Photons (gamma rays), electrons (beta-rays) and helium ions (alpha-rays) are emitted through radioactive decay of unstable radioisotopes as shown in Figure 2.1b.

2.2 Backgrounds for Selection of Terrestrial High-Energy Particles

As shown in Chapter 1, most terrestrial high-energy particles are studied in this book because those particles listed in Table 2.1 have a charge or at least the probability of nuclear reaction and therefore are likely to cause faulty conditions in electronic devices, components and systems. Symbols, mass, charge, spin, mean lifetime and mean decay mode for each particle are listed in the table. The 'bar' in the parenthesis means the corresponding anti-particle: ν_e stands for anti-neutrino emitted with beta-ray; ν_μ stands for neutrino associated with muon; π^0 meson decays immediately to produce two high-energy photons with a half-life of 8.4×10^{-17} seconds. These photons are a source of electron-pairs which produce cascade electron showers or sources of muons by interaction with atmospheric nuclei. Among the particles, positrons disappear immediately by recombination with electrons. Pions (π^+, π^0, π^-) has such short life spans that they never reach the ground or cause faulty conditions. Neutrinos do reach the ground, but they have very weak or no interactions with matter. Positrons, pions and neutrinos, therefore, are not evaluated in any further detail.

Terrestrial Radiation Effects in ULSI Devices and Electronic Systems, First Edition. Eishi H. Ibe.

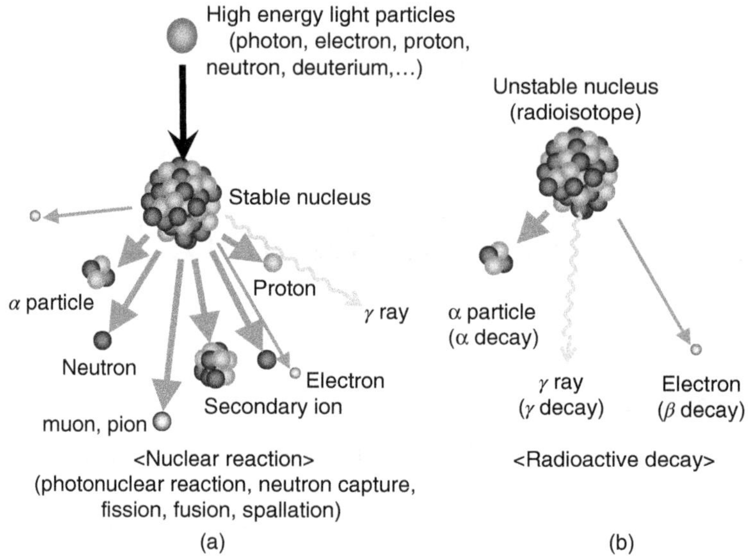

Figure 2.1 Sources of terrestrial radiation: (a) nuclear reactions (photonuclear reaction, neutron capture, fission, fusion and spallation) and (b) radioactive decay

Table 2.1 Properties of terrestrial high-energy particles

Particle	Symbol	Mass (MeV)	Charge	Spin	Mean life time (s)	Mean decay mode[a]
Photon	γ	0	0	1	–	–
Electron	$e^-(\beta)$	0.51	1	1/2	–	–
Positron	e^+	0.51	−1	1/2	–	–
Muon	μ^-	105.66	−1	1/2	2.2×10^{-6}	$e^- + \nu_\mu + \overline{\nu}_\varepsilon$
	μ^+	105.66	1	1/2	2.2×10^{-6}	$e^+ + \overline{\nu}_\mu + \nu_e$
Pion	π^+	139.57	1	0	2.6×10^{-8}	$\mu^+ + \nu_\mu$
	π^-	139.57	−1	0	2.6×10^{-8}	$\mu^- + \nu_\mu$
	π^0	134.98	0	0	8.4×10^{-17}	2γ
Proton	p	938.27	1	1/2	–	–
Neutron	n	939.57	0	1/2	886.7	$p + e^- + \overline{\nu}_e$
Alpha	α	3733	2	0	–	–

[a] ν: neutrino and Top bar means anti-particle.

Nuclear spallation reaction takes place only with high-energy neutrons and protons of a terrestrial energy level, but low-energy neutrons may cause a nuclear capture reaction with Boron-10 contained in electronic devices and will definitely cause faulty conditions. Low-energy protons have direct ionisation effects which cause the faulty conditions as well. Muons have relatively short life spans, but reach the ground

Table 2.2 Databases for terrestrial spectra of secondary cosmic

Particle	Database for spectrum	Memo
Photon	Not applied	–
Electron	EXPACS	–
Muon	EXPACS	–
Pion	[a]	Not evaluated because of poor data and extremely fast decay.
Proton	EXPACS	Cause nuclear spallation reaction and direct ionisation
Thermal, epi-thermal neutron	[b]	<1 MeV, cause neutron capture reaction with Boron-10
High energy neutron	JESD89A/EXPACS	>1 MeV, cause nuclear spallation reaction. No direct ionisation
Helium	EXPACS	Not evaluated due to low flux in the field

[a]Ziegler, J.F. and Puchner, H., eds. (2004).
[b]Nakamura, T., *et al.* (2008).

as air showers. In this chapter, the strength (flux and energy) of each particle at ground level is summarised. To analyse the radiation effects of the terrestrial high-energy particles, their energy spectra are primarily important. Table 2.2 summarises the database or evaluation methods used for the spectra of each particle. Photon and helium are actually γ- and α-rays from contaminated radioactive isotopes in the field so their spectrum cannot be defined. Helium produced in an air shower is found in the packaging material of electronic devices, components and systems so that they do not have to be evaluated. The terrestrial and avionics altitude spectra of electrons, muons, protons and helium are calculated by using the database EXPACS [6]. Pion's spectrum data cannot be obtained in EXPACS and there is very limited literature on it [7]. Thermal and epi-thermal (higher energy than thermal neutrons) neutron spectrum are obtained from Ref. [8]. High-energy neutrons (>1 MeV) cause nuclear spallation reactions with materials in the device to produce charged particles in the device, which can cause faulty conditions. The spectrum of terrestrial high-energy neutrons is obtained in JESD89A [9] or EXPACS.

2.3 Spectra at the Avionics Altitude

Figure 2.2 shows calculated differential (differentiate by energy) flux at 10 km altitude above NYC sea level for neutrons, protons, muons and electrons by using EXPACS. In this energy range (1 – 10 000 MeV), neutrons have the highest differential flux. In general, electrons have about 1 order of magnitude less than the neutron's differential flux. Muons' fluxes are much less than neutron's flux being below 100 MeV, but are

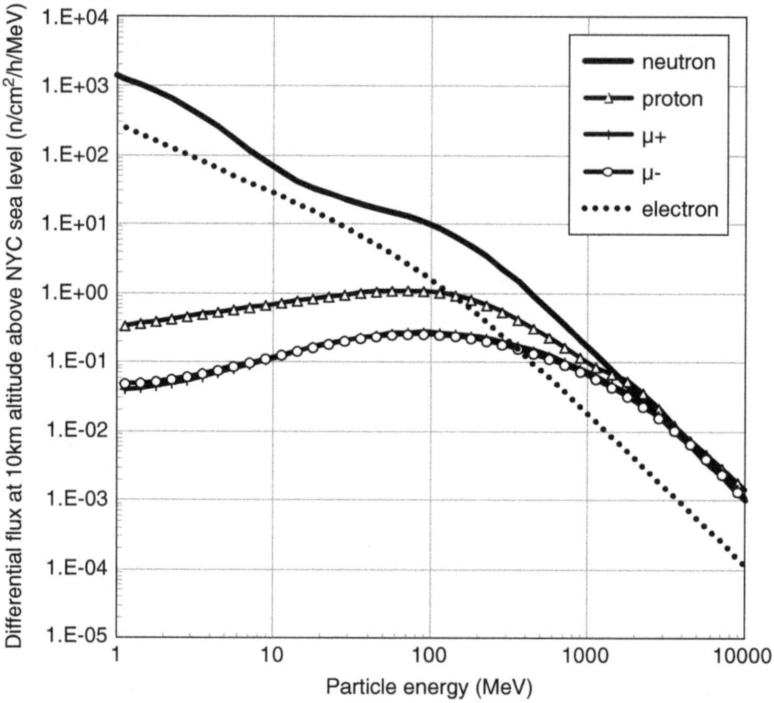

Figure 2.2 Differential flux of secondary cosmic rays at avionics altitude

close to or even slightly exceed neutron's flux above 2000 MeV. Figure 2.3 shows differential flux at NYC sea level calculated by using EXPACS and literature data for pions [7]. Using Figure 2.2 as a comparison, it is clearly seen that the flux levels are much lower than those at avionics altitude. Neutron and proton fluxes decrease much faster than electron and muons. Muons' flux becomes much higher than neutron's flux when its energy exceeds about 300 MeV. Pion and helium fluxes are much lower than other particles' fluxes.

Figure 2.4 summarises measured neutron spectra in an energy range of thermal to 100 GeV together with a JESD89A fitting curve [8, 9] at the ground and avionics level. Neutron flux at avionics altitude is higher than that at the ground level by a factor of 10–1000. The factor becomes larger at higher energy levels.

When looking at the differential flux curve in log-log scale as in Figure 2.4, one may feel that the contribution of the low-energy neutron to total flux is much higher than that of the high-energy neutron. This is somewhat tricky. In order to make this point clear, a cumulative flux plot as in Figure 2.5 is recommended. Cumulative flux over particle energy E_p means there is a total flux of particles that have energy of more than E_p. For example, a neutron curve at 10 MeV gives 13 n/cm^2/h. This means that neutrons that have energy more than 10 MeV have a flux of 13 n/cm^2/h in total.

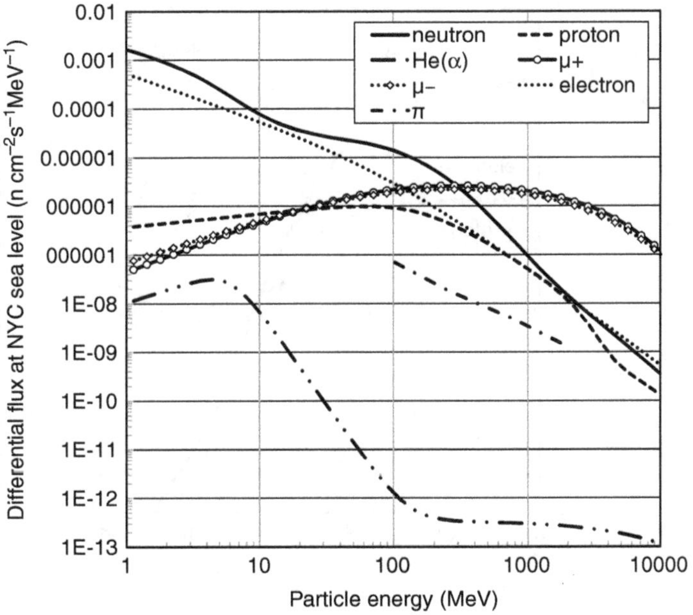

Figure 2.3 Differential flux of terrestrial radiation at NYC sea level

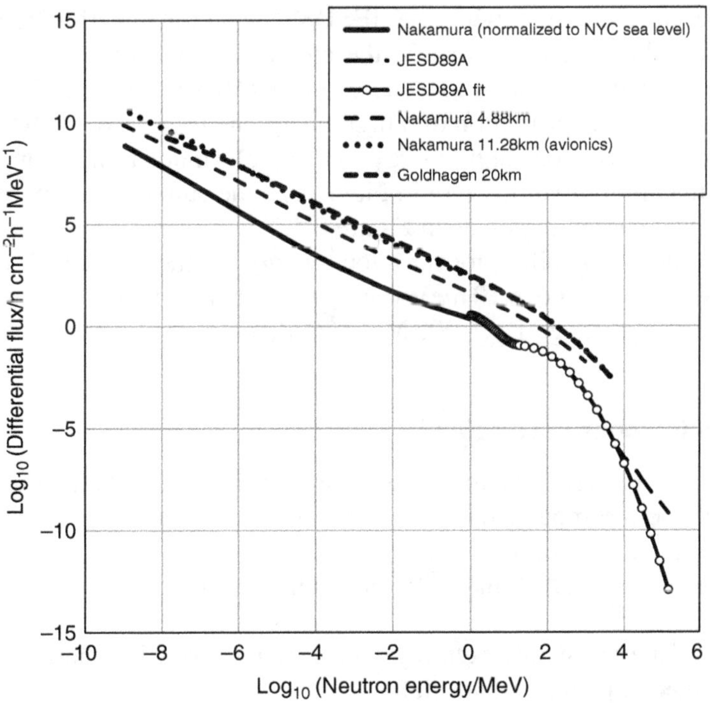

Figure 2.4 Measured differential flux of high-energy neutrons at NYC sea level and avionics altitude

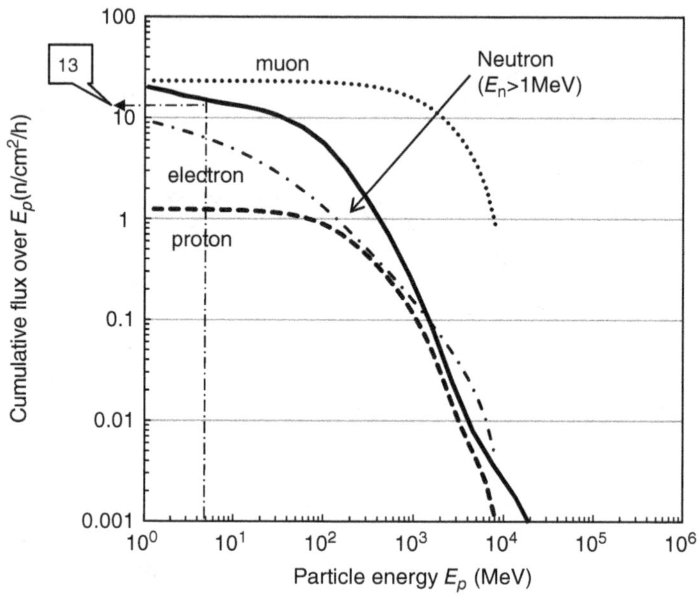

Figure 2.5 Cumulative flux of terrestrial radiation at NYC sea level

This number is consistent with the JESD89A recommendation for the neutron flux beyond 10 MeV. If E_p is 1 MeV, the cumulative flux over 1 MeV gives about 20 n/cm^2/h. The contribution of low energy (1 – 10 MeV) neutron is 7 n/cm^2/h, which is comparable to the contribution of a high-energy neutron (>10 MeV).

Since the contribution of a high-energy (>100 MeV) muon is high compared to other particles in the differential flux plot in Figure 2.3, the cumulative flux of muons stays high from a low-energy region as in Figure 2.5.

Figure 2.6 shows a similar plot for a low-energy (10 meV to 1 MeV) neutron. It is seen that the total flux above 10 meV is about 30 n/cm^2/h, which is a comparable flux to that in the energy range 1 – 10 000 MeV, 20 n/cm^2/h.

2.4 Radioisotopes in the Field

Table 2.3 summarises radioisotopes observable at ground level along with their properties. These radioisotopes emit high-energy photons (gamma-rays), electrons (beta rays) and helium nuclei (alpha rays).

Radioisotopes such as ^{40}K and ^{222}Rn are well-known natural sources of radiation at ground level.

^{40}K and ^{222}Rn emit mainly beta rays and gamma rays with energies up to 2 MeV. ^{40}K is contained in plants and vegetables.

Light radioisotopes, in general, emit beta ray with gamma ray(s).

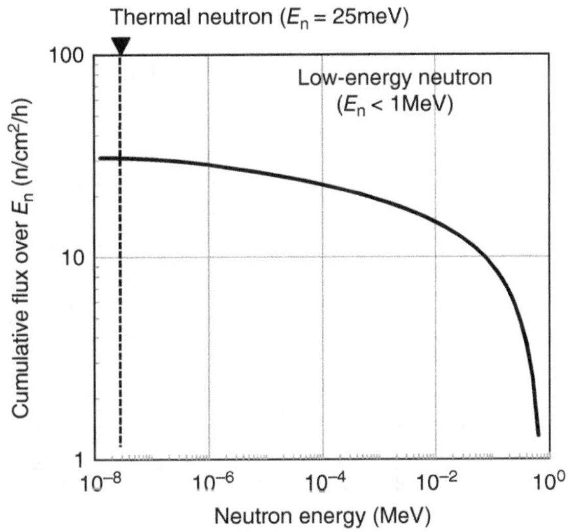

Figure 2.6 Cumulative flux of low-energy (<1 MeV) neutron flux at NYC sea level

^{14}C is produced by the nuclear spallation reaction of the secondary neutrons in an air shower with a nitrogen nucleus,

$$n + {}^{14}N \rightarrow {}^{14}C + p. \tag{2.1}$$

^{14}C decays with a half-life of 5715 years to become ^{14}N again,

$$ {}^{14}C \rightarrow {}^{14}N + e^- + \nu_{e(bar)} \tag{2.2}$$

Radiocarbon dating is made by measuring the ratio of $^{12}C/^{14}C$.

^{16}N is produced in water coolant in a fission nuclear reactor by the nuclear reaction of neutron and ^{16}O in water molecules,

$$n + {}^{16}O \rightarrow {}^{16}N + p. \tag{2.3}$$

^{16}N is transported to the turbine of a boiling water reactor to cause radiation in the turbine buildings [10], and decays with a half-life of 7.13 s to become ^{16}O again in the following reaction, emitting beta and gamma rays,

$$ {}^{16}N \rightarrow {}^{16}O + e^-. \tag{2.4}$$

In severe nuclear power plant accidents (Three Mile Island in 1979, Chernobyl in 1986 and Fukushima in 2011), a number of radioisotopes are released from the reactor pressure vessels. ^{137}Cs, ^{134}Cs and ^{131}I were major radioisotopes released from the 3.11 nuclear severe accident site in Fukushima in 2011. ^{131}I has a relatively short half-life (8.04 days) and decays rapidly so that almost all radioisotopes remaining

Table 2.3 Radioisotopes in the terrestrial fields

Element	Radio-isotope	Natural abundance (%)	Decay mode	Half life	Gamma-ray energy (MeV/%)	Beta ray energy (MeV/%)	Alpha-particle energy (MeV/%)	Remark
C	^{14}C		β^-	5715 yr		0.156/100		Utilised for radiocarbon dating
N	^{16}N		β^-	7.13 s	6.13/68.8, 7.11/4.7	10.42		Source of sky-shine radiation from turbine buildings of boiling water reactors
K	^{40}K	0.0117	β^-	1.26×10^9 yr		1.31/89		Source of environmental radiation exposure
			β^+, EC		1.46/10.5	1.5/10.5		
Co	^{60}Co		β^-	5.27 yr	1.173/100			Artificial gamma-ray source
					1.335/100	0.315/99.7		
I	^{131}I		β^-	8.04 d	0.08, 0.28, 0.36, 0.64	0.606		Released from nuclear accidents
Cs	^{134}Cs		β^-/EC	2.0652 yr	1.22(EC)	2.059		EC: electron capture
	^{137}Cs		β^-	30.2 yr	0.66	0.514/95		Released from nuclear accidents
Pb	^{210}Pb		α	22.6 yr			3.72	Daughter isotope from ^{238}U
			β^-			0.017/81, 0.061/19		
Po	^{210}Po		α	138.4 d			4.516/0.001, 5.304/100	Daughter isotope from ^{238}U
Rn	^{222}Rn		α	3.824 d			5.59	Daughter isotope from ^{238}U. Main source of environmental radiation exposure
	^{220}Rn		α	55.6 s			6.404	Daughter isotope from ^{232}Th
Ra	^{226}Ra		α	1599 yr			4.87	Daughter isotope from ^{238}U
Th	^{232}Th	100	α	1.40×10^{10} yr	0.059		4.08	Contained in monazite
U	^{234}U	0.0055	α	2.455×10^5 y	0.05		4.856	
	^{235}U	0.72	α	7.04×10^8 yr			4.679	Nuclear fission material by thermal neutron
	^{238}U	99.3	α	4.47×10^9 yr			4.04	Source of ^{239}Pu
Pu	^{239}Pu		α	2.4×10^4 yr			5.244	Produced under nuclear fission reactions. Nuclear fission material by high-energy neutron
Am	^{241}Am		α	432.7 yr			5.637	Artificial alpha-ray source
Cf	^{252}Cf		Fission	2.65 yr				0.1–20 MeV neutron source

in the contaminated area in Japan were ^{134}Cs and ^{137}Cs in December 2011. In the case of Fukushima, adverse impacts on electronic systems from this radiation due to contamination of the released radioisotopes were of initial concern, but it turned out to be a negligible impact following careful quantitative evaluation [11–13].

Nuclear fuel material such as ^{253}U and ^{258}U and their daughter isotopes can also be radiation sources. Heavy radioisotopes have, in general, very long life spans and emit alpha rays with energies around 5 MeV. Many are daughter isotopes (radioisotopes produced as a resultant of alpha- and/or beta-decay) starting from ^{232}Th (Th chain), ^{241}Pu (Np chain), ^{238}U (U chain) and ^{235}U (Ac chain). ^{232}Th is generally contained in monazite.

^{253}U(0.72%), ^{258}U(99.3%) and ^{232}Th(100%) are naturally available in abundance so they can be contained in materials such as plastic moulding for semiconductor devices. Since ^{210}Po is a daughter isotope of ^{238}U, a very small amount of ^{210}Po is stably contained in the solder bumps of semiconductor devices. These radioisotopes emit alpha rays and cause α-ray soft error.

High-energy particles can also be generated by using high-energy particle accelerators and nuclear reactors. Artificial radioisotopes such as ^{60}Co, ^{252}Cf, ^{241}Am are produced by using these accelerators or nuclear reactors such as CANDU reactors [14] by neutron activation of isotopes ^{59}Co, ^{238}U and ^{239}Pu, respectively [15–17].

The artificial radioisotopes are mainly used for industrial purposes. ^{60}Co is used in gamma irradiation facilities to inhibit sprouting potatoes [18]. ^{252}Cf emits 0.1–20 MeV neutron [19] and is used as a neutron source. ^{241}Am is an alpha ray source and is used for alpha ray soft-error experiment or used in smoke detectors in order to charge-up aerosols.

2.5 Summary of Chapter 2

Properties of secondary cosmic particles produced by the nuclear spallation reaction of a cosmic proton with atmospheric nuclei in the inner space are summarised. Spectra of terrestrial radiation such as protons, electrons, muons, pions, neutrons and alpha are evaluated. Radioisotopes observable at ground level are surveyed as radiation sources of alpha (helium nucleus), beta (electron) and gamma (photon) rays.

References

[1] Martin, R.C., Knauer, J.B. and Balo, P.A. (1999) Production, distribution, and applications of californium-252 neutron sources. To be presented at the IRRMA '99 4th Topical Meeting on Industrial Radiation and Radioisotope Measurement Applications Raleigh, NC, 3–7 October 1999, http://www.ornl.gov/~webworks/cpr/pres/102606.pdf (accessed 9 June 2014).

[2] Windows To Universe http://www.windows2universe.org/sun/Solar_interior/Nuclear_Reactions/Fusion/Fusion_in_stars/ncapture.html (accessed 9 June 2014).

[3] Atomicarchive http://www.atomicarchive.com/Fission/Fission1.shtml (accessed 9 June 2014).
[4] ELSA (2014) Particle Accelerators Around the World, http://www-elsa.physik.uni-bonn.de/accelerator_list.html.
[5] Reda, M.A. and Harmon, J.F. (2004) A Photoneutron Source for Bulk Material Studies, International Centre for Diffraction Data 2004, Advances in X-ray Analysis, 47, 212–217.
[6] EXPACS http://phits.jaea.go.jp/expacs/jpn.html (accessed 9 June 2014).
[7] Ziegler, J.F. and Puchner, H. (eds) (2004) *SER-History, Trends, and Challenges*, SER-History, Trends, and Challenges, pp. 4–8.
[8] Nakamura, T., Baba, M., Ibe, E. *et al.* (2008) *Terrestrial Neutron-Induced Soft-Errors in Advanced Memory Devices*, World Scientific, Hackensack, NJ.
[9] JEDEC JESD89A. (2006) *Measurement and Reporting of Alpha Particle and Terrestrial Cosmic Ray Induced Soft Errors in Semiconductor Devices*, JEDEC, pp. 1–94.
[10] Ibe, E., Karasawa, H., Nagase, M. *et al.* (1989) Chemistry of radioactive nitrogen in BWR primary system. *Journal of Nuclear Science Technology*, **26** (9), 844–851.
[11] Technical Committee on Semiconductor Reliability, Semiconductor Jisso & Product Technology Committee, Japan Electronics and Information Technology Industries Association (2011) EITA View Concerning Effects of Radioactive Materials Released from Fukushima Nuclear Power Plant on Semiconductor LSI Products, http://semicon.jeita.or.jp/hp/srg/docs/JEITA-SERPG-View_en.pdf (accessed 9 June 2014).
[12] Baumann, R.C. (2011) Determining the impact of alpha-particle-emitting contamination from the Fukushima Daiichi disaster on Japanese manufacturing sites. 12th European Conference on Radiation and its Effects on Component and Systems, Sevilla, Spain, 19–23 September, pp. 784–787.
[13] Ibe, E., Toba, T., Shimbo, K. and Taniguchi, H. (2012) Fault-based reliable design-on-upper-bound of electronic systems for terrestrial radiation including muons, electrons, protons and low energy neutrons. IEEE International On-Line Testing Symposium, Sitges, Spain, 27–29 June 2012 (3.2).
[14] Candu http://www.candu.org/candu_reactors.html (accessed 9 June 2014).
[15] Environmental Science Division http://www.evs.anl.gov/pub/doc/Cobalt.pdf (accessed 15 February 2013)
[16] U.S. Department of Energy http://bmd.ans.org/pdf/isotopesplenary/AtcherANS.pdf (accessed 15 February 2013)
[17] Argonne National Laboratory http://www.evs.anl.gov/pub/doc/Americium.pdf (accessed 27 February 2013)
[18] Growing Acceptance of Food Irradiation http://www.iaea.org/Publications/Magazines/Bulletin/Bull102/10205701314.pdf (accessed 27 February 2013)
[19] Mannhart, W. (2008) Status of the evaluation of the neutron spectrum of ^{252}Cf(sf). AEA Consultants' Meeting, 13–15 October 2008, http://www-nds.iaea.org/standards-cm-oct-2008/6.PDF (accessed 27 February 2013).

3

Fundamentals of Radiation Effects

3.1 General Description of Radiation Effects

Figure 3.1a illustrates a simplified mechanism of one of the effects of radiation on matter. When a charged particle (secondary ion) passes close to an atom, some electrons are extracted by the Coulomb force of the secondary particle to produce electron-hole pairs. Such electron-hole pair production results in the charge-up of inorganic solid matters, ionization for gases, radiation decomposition/synthesis/polymerization for organic matter [1, 2] and radiation decomposition for liquid (water radiolysis in water [3]). The effects for organic matters and liquids are the main causes of radiation exposure of human tissues.

Another radiation effect starts with the displacement of lattice atoms through the direct collision of particles including photons (very rare) [4] and neutrons [5] as illustrated in Figure 3.1b. Displacement of lattice atoms causes further defect/vacancy, dislocation loops and interstitials which results in changes in the properties of matters [6].

Since the present book focuses on the effects of radiation on ULSIs (Ultra Large Scale Intergration devices) and electronic systems, only the radiation effects on solids are discussed. Neither radiation effects of organics and liquids, where chemical reactions have an important role to play, nor lattice displacement, that may cause cumulative effects, are taken into consideration.

When a charged particle penetrates into solid matter, electron-hole pairs are produced along with the ion track due to Coulomb interaction as illustrated in Figure 3.2a. As for metals, the mobility of both electron and hole is very high and there is no band gap, so they recombine immediately without any observable radiation effect. Displacement mechanisms may cause some cumulative damage in solid material, such as radiation hardening [7], and irradiation embrittlement [8] in structural materials in nuclear power plants.

Even in dielectric materials like SiO_2, an electron-hole pair can be produced with an energy of 17 eV for one pair, but the pair recombines and disappears when the electric field is weak as illustrated in Figure 3.2b(i). When the electric field is strong, electrons easily migrate along with the electric field, resulting in accumulation of holes in the dielectrics as shown in Figure 3.2b(ii). Such holes can make high electric fields or

Terrestrial Radiation Effects in ULSI Devices and Electronic Systems, First Edition. Eishi H. Ibe.

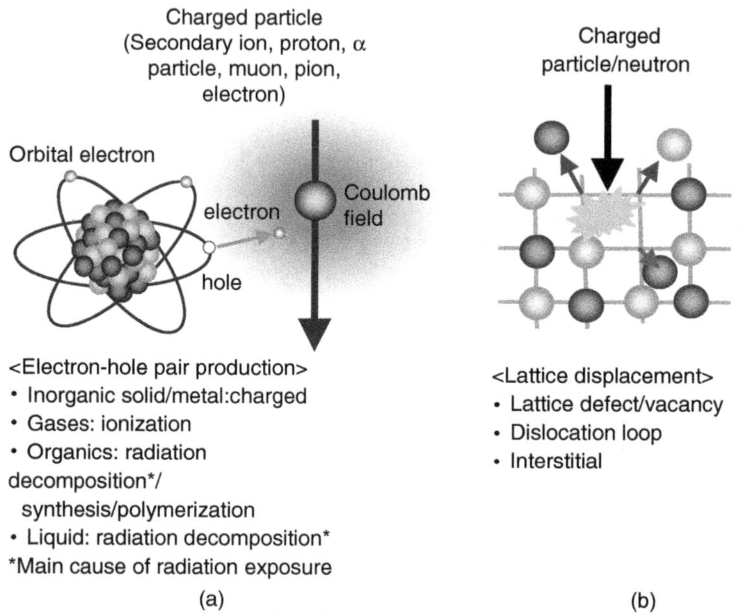

Figure 3.1 Fundamental mechanisms of radiation effects: (a) Coulomb interaction and (b) lattice displacement

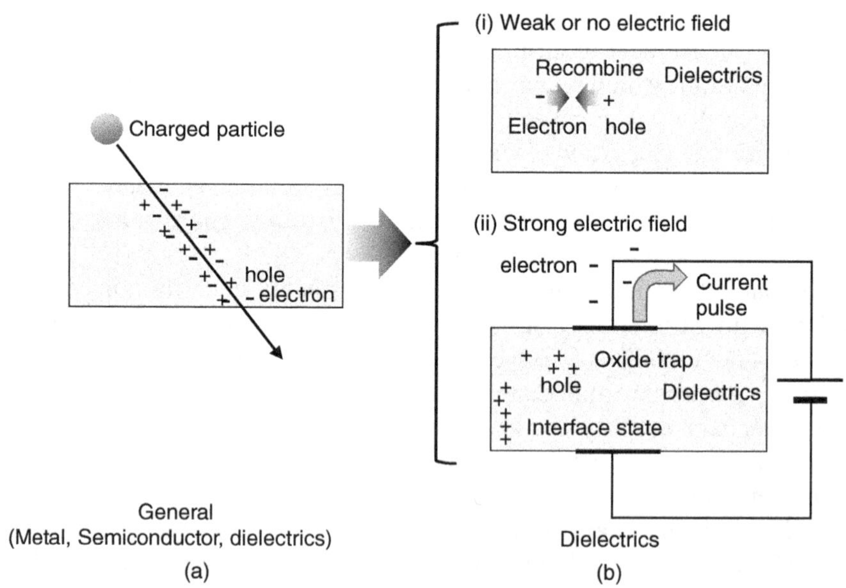

Figure 3.2 Physical phenomena in general solid and dielectrics induced by a high-energy charged particle. (a) General description and (b) dielectrics

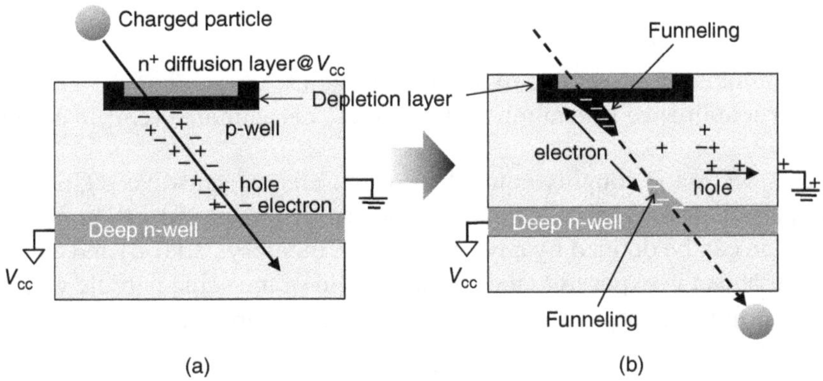

Figure 3.3 Physical phenomena in basic nMOSFET structure with deep n-well induced by a high-energy charged particle. (a) Charged particle penetration through a n+ diffusion layer (storage node) and charge deposition. (b) Charge collection to the diffusion layer

parasitic energy levels inside or on the surface of the oxide resulting in TID (Total Ionisation Dose) effects [9–13].

TID matters only in space applications with the exception of robots or sensors which might be used in the decommissioning of a nuclear reactor, so that TID is not taken into consideration in the present book.

In semiconductor material, electron-hole pairs are produced along with the ion track in the same manner as dielectrics as illustrated in Figure 3.3a. Figure 3.3 depicts a deep n-well layer that is applied to triple well structure in CMOSFET (Complementary Metal Oxide Semiconductor Field Effect Transistor). When the ion penetrates into the depletion layer in the off-state pn-junction of a diffusion layer or storage node, electrons and holes flow in the opposite direction to the electric field in the depletion layer. Electrons flow into the diffusion layer and holes flow out to the ground, resulting in elongation of the electric field, which causes collection of more charge than that initially contained in the original depletion layer as illustrated in Figure 3.3b. This mechanism is called the funnelling effect [14]. The same mechanism is used if the ion penetrates into the deep n-well layer as shown in Figure 3.3b. Electrons produced outside the initial depletion layer can be collected in the diffusion layer to produce a transient pulse and/or lower the potential of the diffusion layer. This is a basic Single Event Effect (SEE) mechanism in semiconductor materials.

A certain amount of charge is collected within the diffusion layer and observable effects like soft error in memory elements can take place. This amount of charge necessary to flip the stored data is called critical charge Q_{crit}.

3.2 Definition of Cross Section

Single event cross section is one of the most fundamental metrics in particle physics, which is defined by,

$$\sigma = \frac{N_{\text{event}}}{\Phi} \tag{3.1}$$

where,

N_{event}: the number of events due to incident particles;
Φ: total particle fluence (the number of particles) per unit area (cm^2 in common).

The cross section is roughly equivalent to the area of sensitive region to reaction with the particle unless there is a strong and far-field (distant) reaction mechanism. Cross section can be defined by any object like a memory cell (bit) and a component.

When the object is exposed to the radiation field of the same particle with the same energy spectrum, the rate of the event R (/s) can be estimated by,

$$R = \phi_p \times \sigma \tag{3.2}$$

where,

ϕ_{p}: flux (the number of particles passing through a unit area per unit time) of particles (cm^{-2} s^{-1}).

In the case of soft error, R in Eq.(3.2) gives SEU (Single Event Upset = Soft Error) cross section σ_{SEU}, which is defined in terms of the number of single events. It must be noted that σ_{SEU} is not defined in terms of the number of errors. Multiple errors (referred to as Multi Cell Upset later) may take place in a single event.

3.3 Radiation Effects by Photons (Gamma-ray and X-ray)

As photons (gamma ray) are neutral, they have very limited effects on the matter.

It causes radiation effects only when it collides with an orbital electron or when it passes near the nucleus. Photons can displace an orbital electron giving all its energy to the electron through the photon-electron effect (Figure 3.4a) [15]. The kinetic energy E_{p} of the electron is $E_{\text{p}} = h\nu - \phi$, where ν is the quantum frequency of the photon, h is Planck's constant (= 6.63×10^{-34} J s), and ϕ is the work function of the target material. This effect mainly takes place for electrons in the inner shell of an atom. Electrons in the outer shell may be displaced by the Compton scattering of a photon (Figure 3.4b) [16]. Photons are scattered by the electron, giving some of their kinetic energy to the electron. Photon-electron effects and Compton scattering make electron-hole pairs in semiconductor or dielectrics, which may cause TID. When a photon with kinetic energy of more than 1.02 MeV passes very close to the nucleus, the photon disappears and causes electron pair (electron and positron) production (Figure 3.4c) [17]. Positrons may disappear immediately by recombination reaction with massive surrounding electrons and an electron is eventually retained.

The radiation effects of photons, therefore, can be interpreted as the radiation effect by electrons with an energy rating of 0.51 MeV (electron pair production) or lower than gamma ray energy (normally a few megaelectron volts).

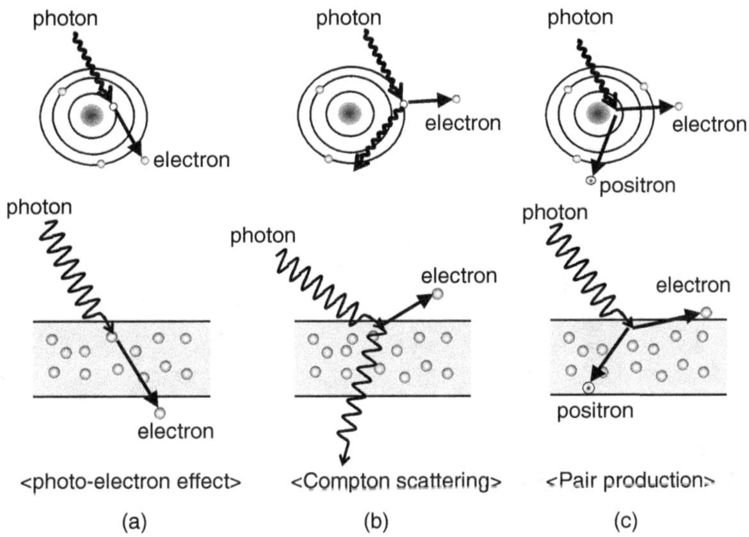

Figure 3.4 Radiation effects induced by photons in materials. (a) Photo-electron effect, (b) Compton scattering and (c) electron-positron pair production

Photon loses its energy through the three processes, and the energy can be represented as,

$$E(x) = E_0 \exp(-\mu x), \tag{3.3}$$

where,

μ: linear absorption coefficient,
x: path length from the point where the photon energy is E_0.

The linear absorption coefficient is a function of energy as shown in Figure 3.5 for Si [18].

The photoelectron effect is dominant for low energy (<100 keV), Compton scattering is dominant in the middle range (0.1 – 10 MeV) and for higher photon energy (>10 MeV), electron pair production is dominant.

In addition to the scattering mechanisms explained above, high-energy photons can produce neutrons, protons, deuteriums, tritiums and alpha particles through photonuclear reaction by bombarding photons with a heavy target like Pb [19]. This reaction is utilised to produce high-energy neutrons by using a high-energy electron LINAC (Linear particle Accelerator), but is negligible as a source of terrestrial radiation.

3.4 Radiation Effects by Electrons (Beta-ray)

When electrons penetrate into solids, electron-hole pairs can be produced. Linear Energy Transfer (LET) by electrons in Si can be calculated by using ESTAR [20].

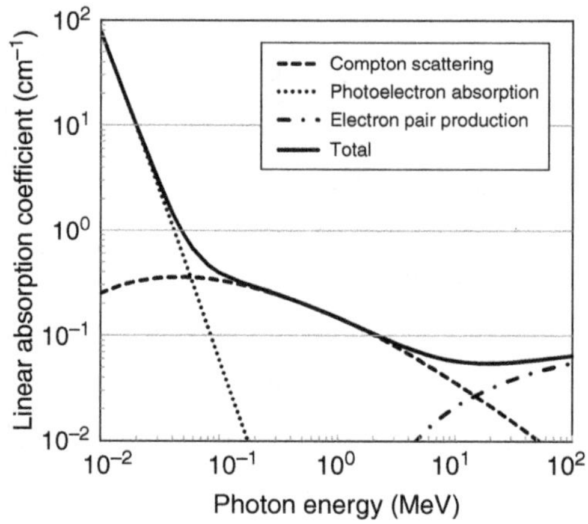

Figure 3.5 Dominant energy loss mechanisms of photons in Si as a function of energy

Since an energy rating of 3.6 eV is required for production of one pair of electron and hole in Si, LET (in a unit of $\text{MeV mg}^{-1}\text{ cm}^{2}$) ΔE_{dep} can be replaced by charge deposition density D_{dep} using a density of Silicon ρ_{Si} (2.33 g/cm^{3}).

$$D_{\text{dep}} = 1.91 \times 10^{-9} \Delta E_{\text{dep}} (\text{C/cm}). \tag{3.4}$$

Figure 3.6 shows the calculation results of ρ_{dep} in Si as a function of electron energy.

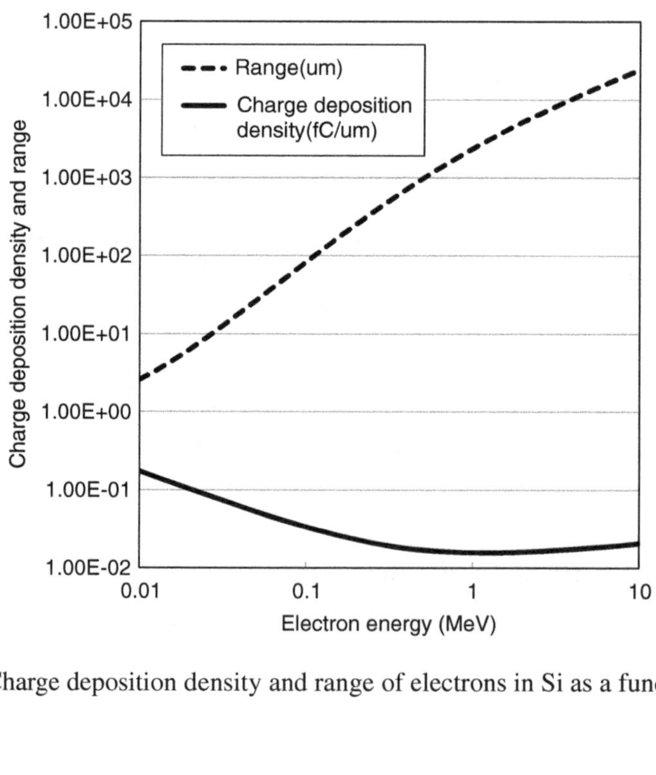

Figure 3.6 Charge deposition density and range of electrons in Si as a function of energy

In principle, the range of charged particles can be calculated numerically as flows:Starting with initial energy E_0, a particle loses ΔE while it passes Δx in Si, decreasing its energy to E_1. The following equation holds,

$$\Delta x = \frac{\Delta E}{\rho_{\mathrm{Si}}} \left(\frac{dE}{dx}\right)^{-1}_{@x=0} \quad (\mathrm{cm}). \tag{3.5}$$

Range R can be the total sum of Δx until the particle energy becomes zero.

The calculated range of electrons in Si is also shown in Figure 3.6.

In the normal energy range of beta rays, the range can be as long as a few centimetres in Si, but the ρ_{dep} is very low compared to other charged particles introduced later.

3.5 Radiation Effects by Muons

The LET can be calculated using the Bethe-Bloch equation [21],

$$-\frac{dE}{dx} = 2\pi N_{\mathrm{a}} r_{\mathrm{e}}^2 m_{\mathrm{e}} c^2 \rho_{\mathrm{Tar}} \frac{Z_{\mathrm{T}}}{A_{\mathrm{T}}} \frac{Z_{\mathrm{q}}^2}{\beta_{\mathrm{q}}{}^2} \left[\ln\left(\frac{2m_{\mathrm{e}}\gamma^2 v_{\mathrm{q}}{}^2 W_{\max}}{I_{\mathrm{av}}{}^2}\right) - 2\beta_{\mathrm{q}}{}^2 - \delta_{\mathrm{c}} - \frac{2C}{Z_{\mathrm{T}}} \right] \tag{3.6}$$

where,

$$W_{\max} = \frac{2m_{\mathrm{q}} c^2 \eta^2}{1 + 2s\sqrt{1+\eta^2} + s^2} \tag{3.7}$$

Average excitation potential I_{av} is given by,

$$\frac{I_{\mathrm{av}}}{Z_{\mathrm{T}}} = 12 + \frac{7}{Z_{\mathrm{T}}} \quad (\mathrm{eV}) Z_{\mathrm{T}} < 13 \tag{3.8}$$

or,

$$\frac{I_{\mathrm{av}}}{Z_{\mathrm{T}}} = 9.76 + 58.8 Z_{\mathrm{T}}{}^{-1.19} \quad (\mathrm{eV}) \quad Z_{\mathrm{T}} \geq 13. \tag{3.9}$$

With $X = \log_{10}(\beta\gamma)$, the correction term δ_{c} and C are given by,

$$\delta_{\mathrm{c}} = \begin{cases} 0 & X < X_0 \\ 4.6052X + C_0 + a(X_1 - X)^m & X_0 < X < X_1 \\ 4.6052X + C_0 & X > X_1 \end{cases} \tag{3.10}$$

C_0, a, m, X_0, X_1 are constants given for each element, and, $C_0 = -4.44$, $a = 0.1492$, $m = 3.25$, $X_0 = 0.2014$ and $X_1 = 2.87$ for Si. C is given by,

$$\begin{aligned} C = {} & (0.422377\eta^{-2} + 0.0304043\eta^{-4} - 0.00038106\eta^{-6}) \times 10^{-6} I_{\mathrm{av}}{}^2 \\ & + (3.850190\eta^{-2} - 0.1667989\eta^{-4} + 0.00157955\eta^{-6}) \times 10^{-9} I_{\mathrm{av}}{}^3. \end{aligned} \tag{3.11}$$

With energy q_{p} (3.6 eV for Si) necessary to produce an electron-hole pair, $(dE/dx)/q_{\mathrm{p}}$ gives the number density of electron-hole pairs produced along with the electron path.

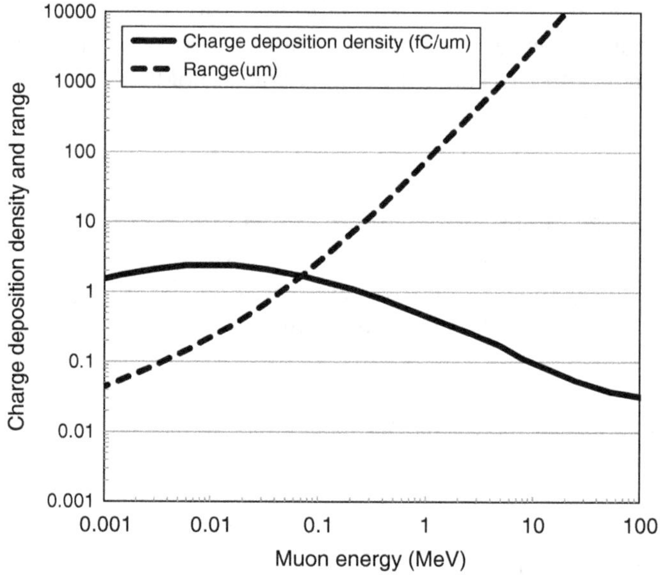

Figure 3.7 Charge deposition density and range of muons in Si as a function of energy

The range R_q of electrons with initial energy E_{q0} can be calculated using the following equation,

$$R_q = \int_0^{K_{q0}} \left(\frac{dK}{dx}\right)^{-1} dK \tag{3.12}$$

where,

$$\beta = v/c;$$

v: velocity of electron;
m_e: mass of electron;
E: energy of electron;
Z: atomic number of target material;
n: atomic number density of target material;
I: mean excitation potential of target (176.3 eV for Si).

Calculated results of ρ_{dep} and range in Si are shown in Figure 3.7.

3.6 Radiation Effects by Protons

As a proton is a charged particle, direct ionisation by proton must be considered [22, 23]. As a proton is a nucleon at the same time, spallation nuclear reaction can take place simultaneously. Both effects have to be considered as radiation effects of proton.

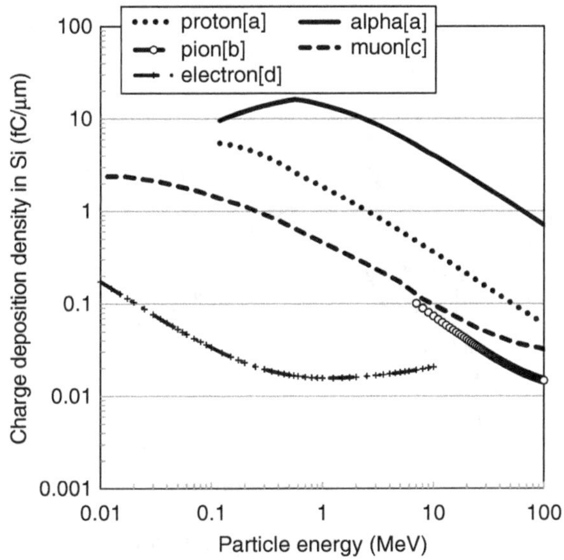

Figure 3.8 Charge deposition density of various particles in Si as a function of particle energy [25]

As for direct ionisation effects, LET and range in Silicon can be calculated by using SRIM (Stopping and Range of Ions in Matter) [24] and the results are summarised in Figure 3.8 together with other terrestrial particles. Since proton produced by nuclear spallation reaction with cosmic ray in the atmosphere has a kinetic energy as high as a few hundred megaelectron volts, its range in Silicon can be as long as a few metres as explained in more detail later.

A certain number of secondary ions can be produced through nuclear spallation reaction by energetic proton with a Silicon nucleus. The total reaction cross section of nucleon (neutron and proton) is shown in Figure 3.9 as a function of particle energy. It can be seen that a proton with an energy rating of 2.5 MeV may cause both direct ionisation effects and nuclear spallation reactions and a proton with an energy rating of less than 2.5 MeV can only cause direct ionisation effects. When a proton penetrates into Si substrate, it causes a direct ionisation effect initially and may cause a spallation reaction later. The actual probability of direct ionisation effects can be estimated by Monte Carlo Simulation with a device structure model, whereas the probability of nuclear reaction $P_{\mathrm{nuc,p}}$ can be estimated by,

$$P_{\mathrm{nuc,p}} = \int_0^\infty \sigma_{\mathrm{tot}}(E) N_{\mathrm{Si}} \frac{d\varphi}{dE} dE \tag{3.13}$$

where,

σ_{tot}(E): total cross section of nuclear reaction with proton (cm^2) [26];
N_{Si}: number density of Si (cm^{-3});
φ: proton flux (h^{-1} cm^{-2}).

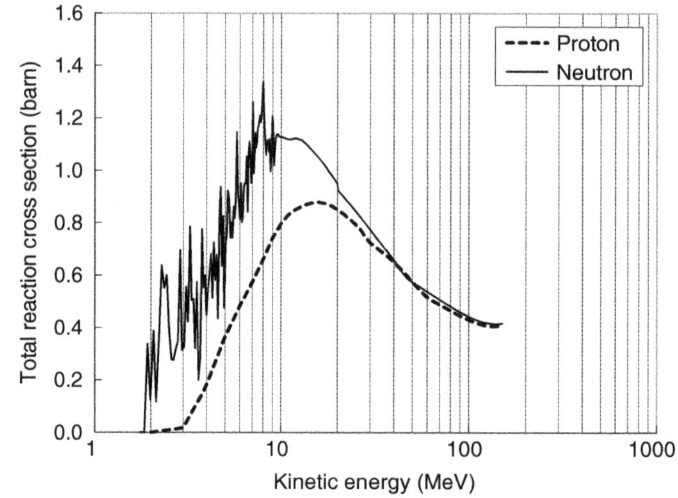

Figure 3.9 Total nuclear reaction cross section of high-energy proton and neutron in Si

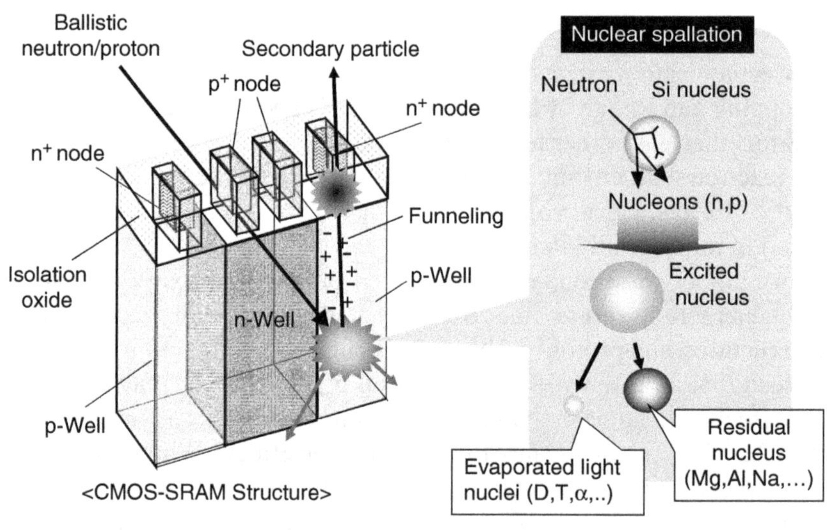

Figure 3.10 Microscopic fault mechanism due to spallation reaction of high energy neutron and proton in a SRAM cell

Figure 3.10 illustrates nuclear spallation reaction between a proton and a Si nucleus, which may take place in an SRAM (Static Random Access Memory) bit. When a proton penetrates into a Si nucleus, many-body scattering takes place among nucleons (neutrons and protons in the Si nucleus). This process can be regarded as a sequence of binary collisions between two nucleons. Nucleons that have enough energy compared to the potential energy of the nucleus can be emitted from the nucleus. This model is known as INC (Intra Nuclear Cascade) model [27]. After total kinetic energy becomes lower than the potential barrier, the evaporation process [28] of light particles (p, n, D, T and alpha) takes place. This process can also be regarded as a sequence of evaporation of a light particle. The probability of such evaporation can be calculated from an inverse binary collision cross section for each combination of an evaporated particle and an excited residual nucleus [29] as explained in Chapter 6 in more detail.

In each binary collision, the relativistic centre of a mass system must be used to calculate the direction, moment and energy of each particle. The particles produced by this reaction have their own kinetic energies and may cause charge deposition in the SRAM bit.

3.7 Radiation Effects by Alpha-Particles

LET of alpha particles are also shown in Figure 3.8. Very high-energy alpha particles can cause nuclear spallation reactions as well as heavier ions, but it is sometimes left out of the energy range of terrestrial radiation. Only direct ionisation is considered for an alpha particle. Again, an alpha ray that comes from outside a semiconductor device can be easily shielded, so only inner alpha contaminated sources like ^{238}U, ^{232}Th and ^{210}Po are of concern.

3.8 Radiation Effects by Low-Energy Neutrons

A neutron is a neutral particle so that it has no direct ionisation effect at all. In some manufacturing processes of semiconductor devices such as ion implantation, polishing and etching, processing gas that contains Boron is used. ^{10}B is contained in about 10 per cent of natural Boron and has some likelihood of reacting with low-energy (roughly <1 MeV) neutrons in the atmosphere. Thermal neutron has an energy rating of 25 meV in equilibrium with room temperature, and produces an alpha particle (1.47 MeV) and a Li ion (0.84 MeV) after a neutron capturing reaction with ^{10}B takes place.

Actually, ^{10}B does not only react with the thermal neutron, but also with neutrons with higher energies. The reaction cross section is shown in Figure 3.11, which covers the energy range up to 1 MeV. If the kinetic energy of a neutron is negligibly low, compared to *mass defect* after the reaction, ^{4}He and ^{7}Li have a kinetic energy rating of 1.47 and 0.84 MeV, respectively. For higher energy neutrons, the kinetic energy of these ions varies due to the kinetic energy of an incident neutron as shown in Figure 3.12 (see Chapter 6 for nuclear reaction model).

Baumann and Smith [30] first point out the impact on thermal neutron induced soft-error. Initially, ^{10}B is contained in Boron Phosphor Silicate Glass (BPSG) that is used for polishing in semiconductor manufacturing processes. The problem is

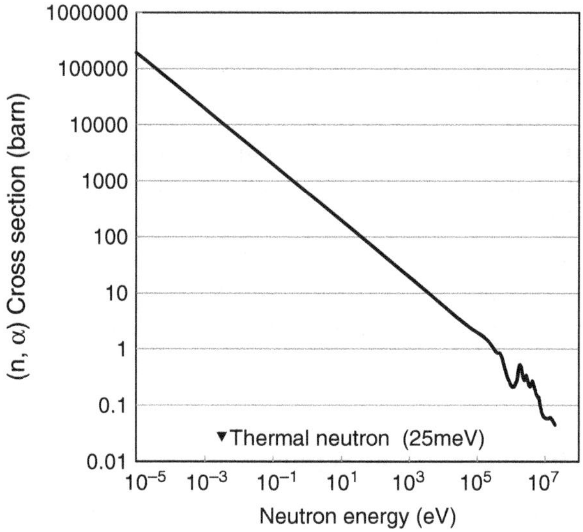

Figure 3.11 (n, α) reaction cross section of low-energy neutrons with B-10

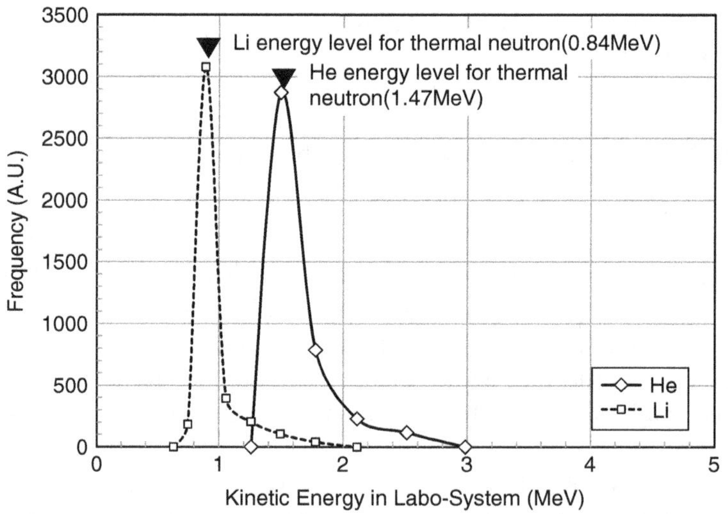

Figure 3.12 Calculated energy spectra of Li and He produced by neutron capture reaction $^{10}B(n,\alpha)^{7}Li$ reaction with low-energy terrestrial neutron spectrum in Figure 3.11

believed to be overcome by replacing the BPSG process by the Chemical Mechanical Polishing (CMP) process. However, the problem is being pointed out as revived in SRAMs without BPSG processes as scaling proceeds [31, 32]. ^{10}B is reported to exist in metal layers by SIMS (Secondary Ion Mass Spectrometry) measurement and the origin of ^{10}B is supposed to be B_2H_6 etching gas for metal wire/via process. ^{10}B

may also be originated from the BCl_3 etching gas used recently for ultra-fine metal process [33].

3.9 Radiation Effects by High-Energy Neutrons

High-energy (roughly >1 MeV) neutrons have almost the same probability of nuclear reaction as proton, as shown in Figure 3.9. The total nuclear cross section of neutron with Si is slightly higher than that of proton below 50 MeV. This is due to the Coulomb barrier of Si nucleus for protons. High-energy neutrons only cause a spallation reaction, and a cross section of the capture reaction of ^{10}B is negligibly small. The spallation reaction can be modelled in the same manner as proton.

3.10 Radiation Effects by Heavy Ions

LETs and ranges for each heavier ion that can be produced through the nuclear spallation reaction are summarised in Figures 3.13 and 3.14, respectively. Due to the law of conservation of momentum, most kinetic energy is given to light particles such as protons and alpha particles as can be seen in Figure 3.13. Protons may have kinetic energies as high as 1 GeV under the terrestrial field in Tokyo. Since kinetic energies

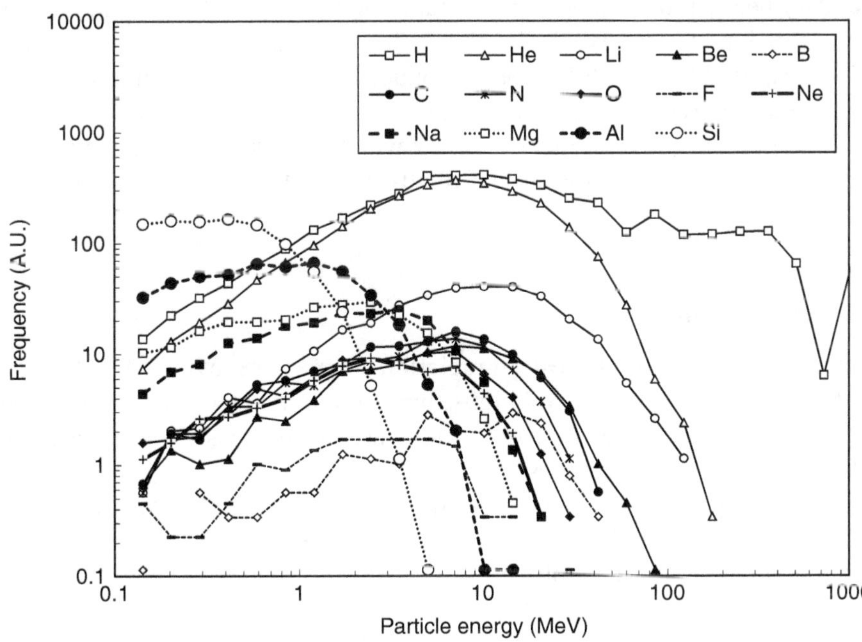

Figure 3.13 Calculated energy spectra of elements produced by nuclear spallation reaction of high energy neutrons in Si at Tokyo sea level

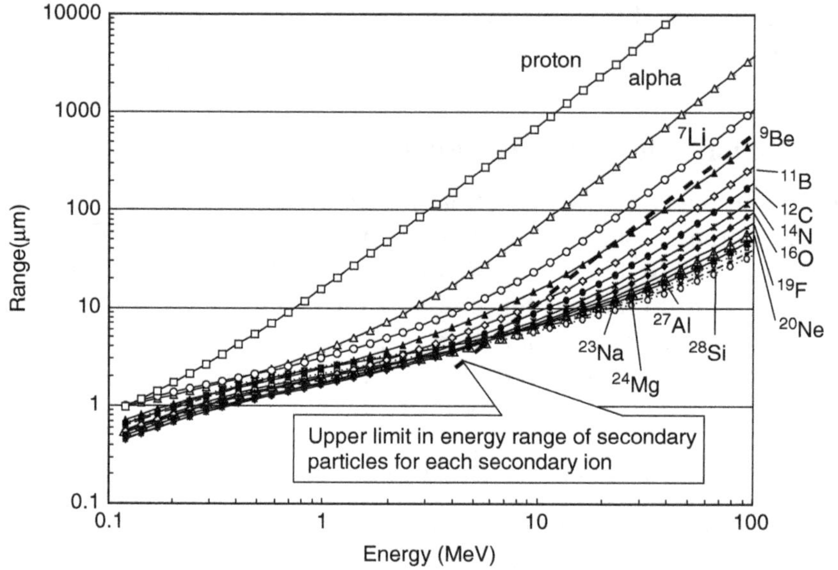

Figure 3.14 Ranges of typical isotopes produced by nuclear spallation reaction of high energy neutron in Si

of secondary heavy ions produced by a spallation reaction in Si are relatively low, an upper limit of kinetic energy can be shown for a given range in Si as illustrated in Figure 3.14 (dotted line). While proton produced from nuclear spallation reaction of terrestrial neutrons with Si may have a range of much more than a few centimetres in Si, ^{24}Mg have only 5 MeV maximum, by which the maximum range in Si is only about 5 μm. Only direct ionization in a relatively short range may be considered for heavier ions.

3.11 Summary of Chapter 3

Fundamentals of effects by terrestrial radiation sources are reviewed and explained.

Physics and radiation effects of high-energy photons, electrons, muons, protons, alpha rays and neutrons are discussed. Pions are not discussed in depth because of lack of terrestrial spectrum data as explained in Chapter 2.

TID effect is explained to some extent in conjunction with the radiation effects on dielectrics. As the present book focuses on single events at ground level, TID is not referred to in the rest of this book.

References

[1] Atomicarchive. Nuclear Fusion, http://www.atomicarchive.com/Fusion/Fusion1.shtml (accessed 9 June 2014).

[2] Kharissova, O.V., Kharisov, B.I. and Mendez, U.O. (2012) *Radiation-Assisted Synthesis of Composites, Materials, Compounds, and Nanostructures*, John Wiley & Sons, Inc., Hoboken, NJ.
[3] Ibe, E. and Uchida, S. (1985) Numerical techniques for quantitative evaluation of chemical reaction systems with volatile species and their applications to water radiolysis in BWRs. *Journal of Nuclear Materials*, **130**, 45–50.
[4] Messenger, G.C. (1992) A summary review of displacement damage from high energy radiation in silicon semiconductors and semiconductor devices. *IEEE Transactions on Nuclear Science*, **39** (3), 468–473.
[5] Oen, O.S. and Holmes, D.K. (1959) Cross sections for atomic displacements in solids by gamma rays. *Journal of Applied Physics*, **30** (8), 1289–1295.
[6] Vizkelethy, G., Bielejec, E.S., Doyle, B.L. *et al.* (2006) Simulation of Neutron Displacement Damage in Bipolar Junction Transistors Using High-Energy Heavy Ion Beams. Sandia Report, SAND2006-7746.
[7] Stiegler, J.O. and Mansur, L.K. (1979) Radiation effects in structural materials. *Annual Review of Materials Science*, **9**, 405–454.
[8] Steele, L.E. ASTM STP1011. (1989) *Radiation Embrittlement of Nuclear Reactor Pressure Vessel Steels, American Society for Testing and Masterials..*
[9] Wang, J.-A. (2010) Lessons Learned from Developing Reactor Pressure Vessel Steel Embrittlement Database. ORNL/TM-2010-20.
[10] Winokur, P.S. (2000) Why semiconductors must be hardened for space deployment. *COTS Journal*, 45–47.
[11] Crain, S.H., Mazur, E., Katz, R.B. *et al.* (2001) Analog and digital single-event effects experiments in space. IEEE Transactions on Nuclear Science, Honolulu, Hawaii, 23–27 July, Vol. 48 (6), pp. 1841–1848.
[12] Schwank, J.R., Shaneyfelt, M.R., Felix *et al.* (2005) Effects of total dose irradiation on single-event upset hardness. 2005 Radiation and Its Effects on Components and Systems, 19–23 September, Palais des Congres, Cap d'Agde, France (B-1).
[13] Schrimpf, R., Warren, K.M., Weller, R.A. *et al.* (2008) Reliability and radiation effects on IC technologies. IEEE International Reliability Physics Symposium, Anaheim, CA, 15–19 April (2C.1), pp. 97–106.
[14] Hu, C. (1982) Alpha-particle-induced field and enhanced collection of carriers. *IEEE Electron Device Letters*, **EDL-3** (2), 31–34.
[15] Hufner, S. (1994) *Photoelectron Spectroscopy: Principles and Applications*, Springer.
[16] NDT http://www.ndt-ed.org/EducationResources/CommunityCollege/Radiography /Physics/comptonscattering.htm (accessed 9 June 2014).
[17] Leptons http://hyperphysics.phy-astr.gsu.edu/hbase/particles/lepton.html (accessed 9 June 2014).
[18] http://physics.nist.gov/PhysRefData/XrayMassCoef/ElemTab/z14.html (accessed 9 June 2014).
[19] Reda, M.A. and Harmon, J.F. A Photoneutron Source for Bulk Material Studies, http://www.icdd.com/resources/axa/vol47/V47_30.pdf (accessed 17 April 2014).
[20] NIST http://physics.nist.gov/PhysRefData/Star/Text/ESTAR.html (accessed 17 April 2014).
[21] Nagamine, K. (2003) *Chapter 3 Muon Inside Condensed Matter*, Introductory Muon Science, Cambridge, pp. 40–50.

[22] Heidel, D.F., Marshall, P.W., Pellish, A., Rodbell, K.P., LaBe, K.A., Schwank, R., Rauch, S.E., Hakey, M.C., Berg, M.D., Castaneda, C.M., Dodd, P.E., Friendlich, M.R., Phan, A.D., Seidleck, C.M.. Shaneyfelt, M.R. and Xapsos, M.A. (2009) Single-event upsets and multiple-bit upsets on a 45 nm SOI SRAM. *IEEE Transactions on Nuclear Science*, **56** (6), 3499–3504.

[23] Sierawski, B.D., Pellish, A., Reed, R.A. *et al.* (2009) Impact of low-energy proton induced upsets on test methods and rate predictions. *IEEE Transactions on Nuclear Science*, **56** (6), 3085–3092.

[24] SRIM www.srim.org (accessed 9 June 2014).

[25] Nagamine, K. (2003) *Introductory Muon Science*, High Energy Accelerator Research Organization, Tsukuba, p. 43.

[26] NEA http://www.oecd-nea.org/janis/ (accessed 9 June 2014).

[27] Tang, H.H.K., Srinivasan, G.R.a. and Azziz, N. (1990) Cascade statistical model for nucleon-induced reactions on light nuclei in the energy range 50-MeV-1GeV. *Physical Review C*, **42** (4), 1598–1622.

[28] Cole, A.J. (2000) *Statistical Models for Nuclear Decay: From Evaporation to Vaporization*, Institute of Physics Publishing, Bristol, Philadelphia, PA.

[29] Furihata, S. and Nakashima, H. Analysis of Activation Yields by INC/GEM, http://wwwndc.jaea.go.jp/nds/proceedings/2000/p22.pdf (accessed 9 June 2014).

[30] Baumann, R.C. and Smith, E.B. (2000) Neutron-induced boron fission as a major source of soft errors in deep submicron SRAM devices. 2000 IEEE International Reliability Physics Symposium Proceedings, San Jose, CA, 10–13 April, pp. 152–157.

[31] Wen, S., Pai, S.Y., Wong, R. *et al.* (2010) B10 findings and correlation to thermal neutron soft error rate sensitivity for SRAMs in the sub-micron technology. IEEE International Integrated Reliability Workshop, Stanford Sierra, CA, 17–21 October, pp. 31–33.

[32] Wen, S., Wong, R., Romain, M. and Tam, N. (2010) Thermal neutron soft error rate for SRAMs in the90nm-45nm technology range. 2010 IEEE International Reliability Physics Symposium, Anaheim, CA, 2–6 May (SE5.1), pp. 1036–1039.

[33] Kang, T.Y., Lim, W.L. and Hong, S.J. Process Optimization of BCl_3/Cl_2 Plasma Etching of Aluminum with Design of Experiment (DOE), http://www.ineer.org/Events/ICEEiCEER2009/full_papers/full_paper_189.pdf (accessed 27 February 2013).

4

Fundamentals of Electronic Devices and Systems

4.1 Fundamentals of Electronic Components

In this section, the fundamentals of electronic devices necessary to understand Chapters 6–8 are introduced. Those who are familiar with this subject can skip this section.

4.1.1 DRAM (Dynamic Random Access Memory)

A dynamic random access memory (DRAM) is made on an nMOSFET (n-channel Metal Oxide Semiconductor Field Effect Transistor) structure that consists of n^+ diffusion layers (the source and drain) and p^+ substrate as illustrated in Figure 4.1 [1]. The drain is connected to a capacitor to retain a certain amount of charge. A cylindrical stack-type capacitor is illustrated in Figure 4.1a. Since the critical charge Q_{crit} of a DRAM is roughly equal to $CV_{cc}/2$, DRAMs are not sensitive to SEE (Single Event Effect) with high capacitance C as previously shown in Figure 1.15. Read and write functions can be simply made by opening the gate through a WL (Word Line) as illustrated in Figure 4.2b.

4.1.2 CMOS Inverter

A CMOS (Complementary Metal Oxide Semiconductor) inverter inverts the input signal from 0 to 1 or vice versa and is the most basic element in electronics [2]. It consists of a set of pMOSFET (p-channel Metal Oxide Semiconductor Field Effect Transistor) and nMOSFET as illustrated in Figure 4.2a. As shown in Figure 4.2b, the source of pMOSFET is connected to V_{cc} and the drain of nMOSFET is connected to the ground. When the input of the inverter is '0', the gate of pMOSFET is open and the gate of nMOSFET is closed, resulting in the output '1'. In a similar manner, the output is '0' when the input is '1'. The symbol of an inverter is shown in Figure 4.2c.

Terrestrial Radiation Effects in ULSI Devices and Electronic Systems, First Edition. Eishi H. Ibe.

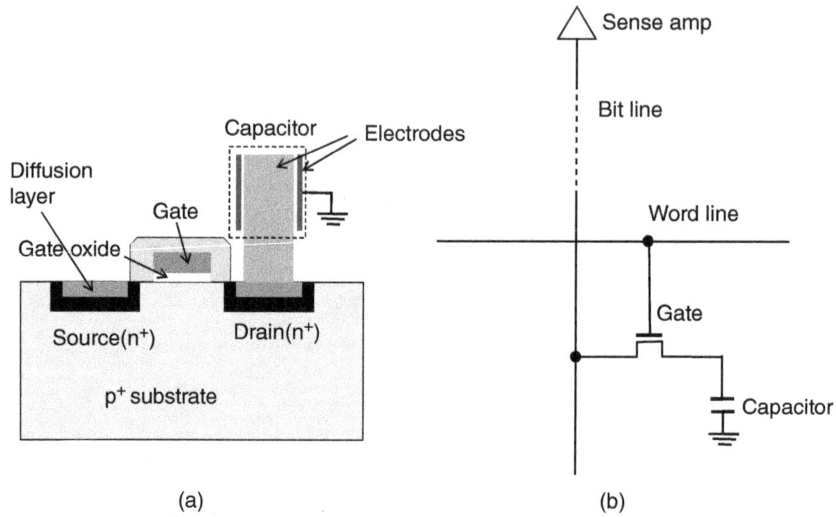

Figure 4.1 (a) Structure of DRAM and (b) its equivalent circuit

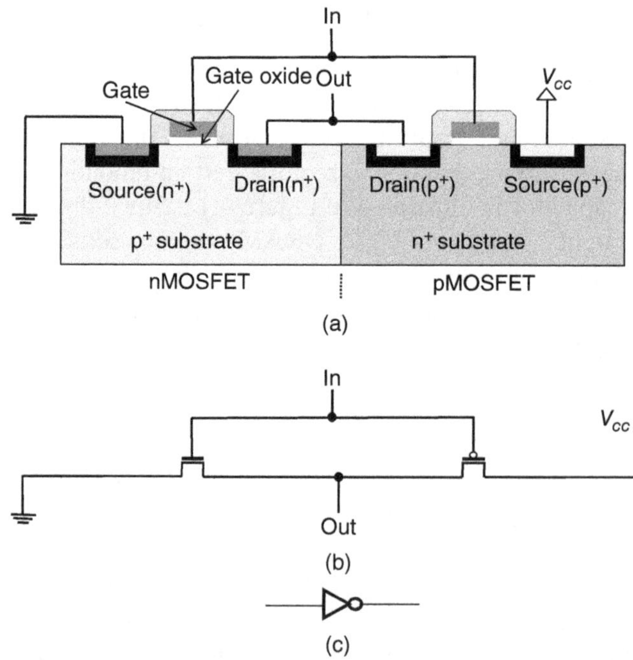

Figure 4.2 (a) Structure of an inverter, (b) circuit and (c) its symbol

4.1.3 SRAM (Static Random Access Memory)

Although the basic structure and circuitry has already been shown in Figure 1.6, the relation between the typical layout and circuitry of static random access memory (SRAM) is described in Figure 4.3a,b [3]. As in Figure 4.3b, the transistor Tr_1 (nMOSFET) and the transistor Tr_2 (pMOSFET) is composed of an inverter and their layout is shown in Figure 4.3a. The $gate_1$ is common for Tr_1 and Tr_2 as shown in Figure 4.3a and connected to $\overline{Q}$. The $gate_2$ is open when WL potential is '1' to read or write data from or to the node Q. As shown in Figure 4.3c as an equivalent circuit of a SRAM, when the 'high' node of SRAM is hit by a charged particle, electrons flow into the node, lowering the potential of the node. As can be seen in Figure 4.3b, the 'High' node is open to V_{cc} which drives to the node 'High'.

If the speed of lowering the potential is high enough compared to the speed of recovery, the condition of the SRAM is flipped. The flipping mechanism of SRAM is thus dynamic and Q_{crit} cannot be determined from the static data. Dynamic circuit simulation is necessary to obtain Q_{crit} from a SRAM.

4.1.4 Floating Gate Memory (Flash Memory)

Figures 4.4a,b illustrate the structure of floating gate memory that consists of diffusion layers (the source and drain), a floating gate and a control gate with a thin gate oxide

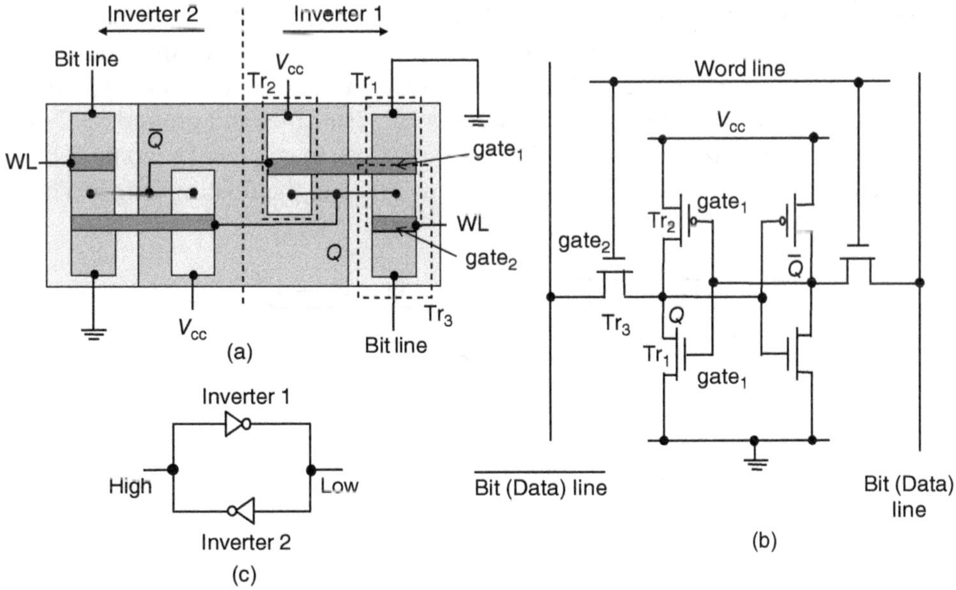

Figure 4.3 Relation between (a) SRAM layout and (b) its circuit. (c) Equivalent circuit of a SRAM

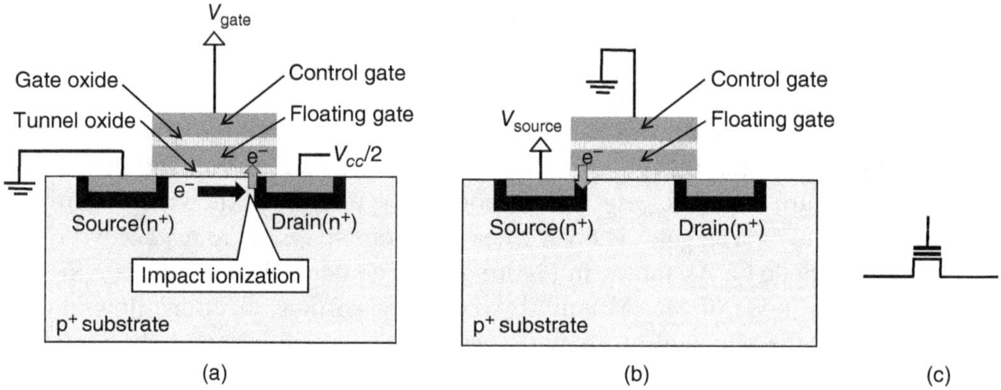

Figure 4.4 Structure and functionality of floating gate memory. (a) 'Write' mode, (b) 'Erase' mode and (c) its symbol

and a tunnel oxide which is a few nanometers thick [4]. Data '0' can be written by injecting hot electrons produced by impact ionization to the floating gate through the tunnel oxide. Impact ionization is made by keeping the potential of the control gate V_{gate} high (e.g. 12 V).

Data '1' state is made by extracting the electrons ('erase') in the floating gate by keeping the source potential V_{source} high as shown in Figure 4.4b.

The symbol of a floating gate memory is shown in Figure 4.4c.

There are two types of floating gate memories known as NOR and NAND types depending on the method for reading data. Figure 4.5a,b show the circuitries of NOR type and NAND type floating gate memories, respectively. As for NOR type floating gate memory, only one device is set between a BL (Bit Line) and ground. Only when there is no electron (state '1') in the floating gate, does the channel become 'ON' and the current flows between the BL and ground and data in the floating gate can be read

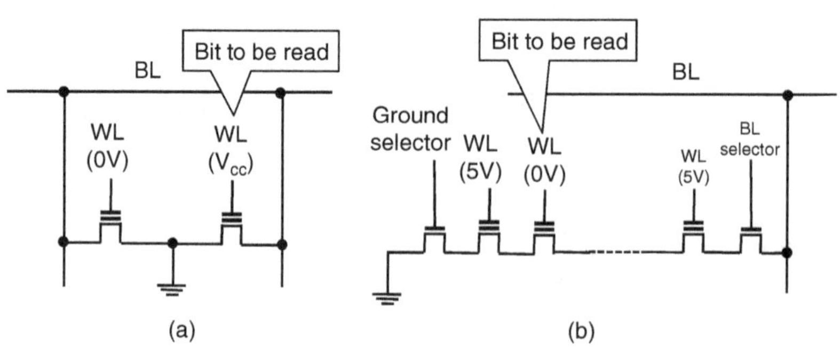

Figure 4.5 'Read' mode of floating gate memory. (a) NOR type and (b) NAND type

by sensing the current. When the floating gate is filled with electrons to the necessary amount, the potential of the control gate is cancelled by the negative potential of the floating gate so that the source-drain channel cannot be opened. A Single Level Cell (SLC) floating gate memory has only two potential levels ('0' and '1') depending on the number of electrons in the floating gate, while Multi Level Cell (MLC) floating gate memory has more than two states, typically four charge states that give two digits, '00', '01', '10' and '11'.

A series of floating gates are connected in a sequence with a BL selector (transistor) and a ground selector at both ends in NAND type floating gate memory as illustrated in Figure 4.5b. When data in one of the floating gates is to be read, the potential of the control gate is kept at '0' and the potential of the control gates of all other floating gates are kept high enough (5 V, for example) to make the corresponding channel 'ON' regardless of the potential of those floating gates. Only when the state of the SLC floating gate to be read is '1', does current flow through the BL, otherwise the current does not flow and the data to be read is '0'.

4.1.5 Sequential Logic Devices

Sequential logic circuits can hold data and are known as memory elements as well [5]. Figure 4.6a,b illustrate simplified circuitries of typical sequential logic circuits, a D-latch and a FF, respectively. D-latch keeps the input in a SRAM when the gate

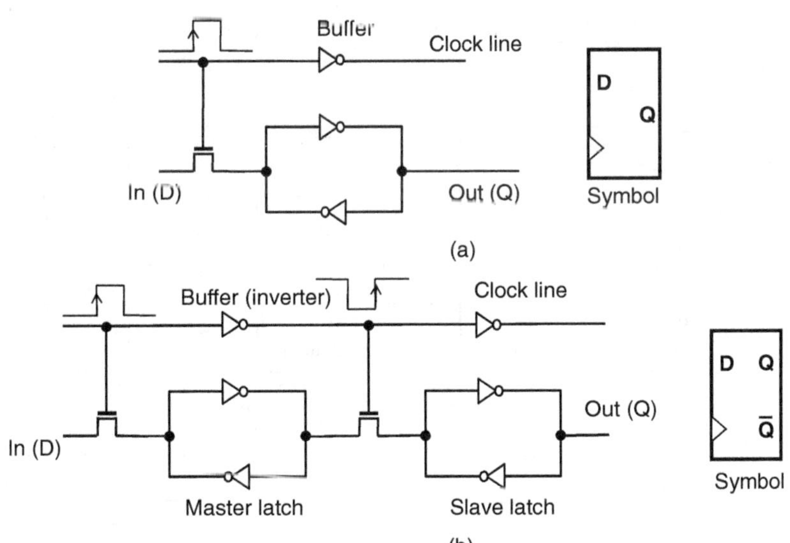

Figure 4.6 Equivalent circuits of sequential logic devices. (a) D-latch and (b) flip-flop

is open. The output of the D-latch simply propagates downstream. FF consists of a master latch and a slave latch. Input data is kept in the master latch by sensing an up-edge of a clock but the data does not propagate to the slave latch. After the inverted clock signal reaches the gate of the slave latch, the gate is open by the second edge of the same clock pulse. Thus, the data appears at the output of a FF after one clock pulse goes through two gates in a FF.

4.1.6 Combinational Logic Devices

Combinational logic circuits contain a number of logic circuits that do not retain data by themselves. Instead, they can be sources of SETs (Single Event Transients) which are transient noises produced by a charged particle.

Figure 4.7 summarizes typical combinational devices such as (i) AND, (ii) NAND, (iii) OR, (iv) NOR, (v) XOR (Exclusive OR) and (vi) MUX (Multiplexer) with their symbols, expressions of their functions and typical output (O) with A = 1 and B = 0.

In general, a combinational logic stage consists of a number of combinational circuits and their final output is kept in a FF and transferred to the next stage according to the clock signal as illustrated at the bottom of Figure 4.7.

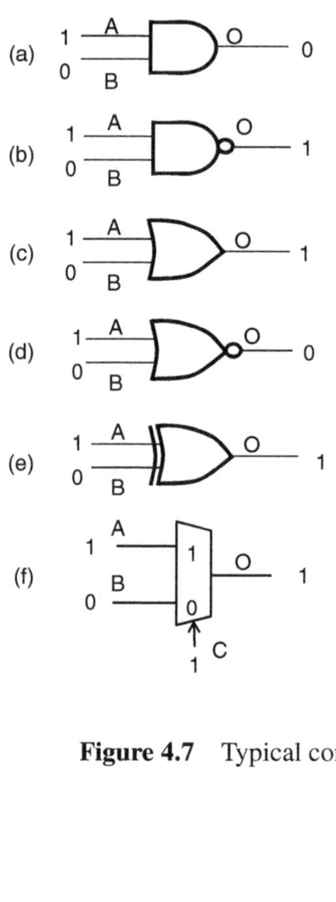

Symbol	Name	Output	Function
a	AND	$A\cdot B$	O=1 only when A=1 and B=1
b	NAND	$\overline{A\cdot B}$	O=0 only when A=1 and B=1
c	OR	$A+B$	O=0 only when A=0 and B=0
d	NOR	$\overline{A+B}$	O=1 only when A=0 and B=0
e	XOR	$A\cdot\overline{B}+\overline{A}\cdot B$	O=1 only when $A\neq B$
f	MUX (Multiplexer)	$A\cdot C+B\cdot\overline{C}$	O=A when C=1 O=B when C=0

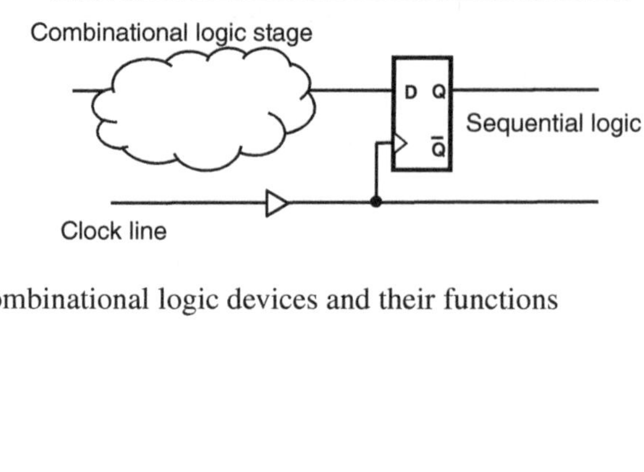

Figure 4.7 Typical combinational logic devices and their functions

4.2 Fundamentals of Electronic Systems

In this section, the fundamentals of electronic systems which will be necessary to understand Chapters 6–8 are introduced. Those who are familiar with this subject can skip this section.

4.2.1 FPGA (Field Programmable Gate Array)

Field Programmable Gate Arrays (FPGAs) are widely used in many industry fields because of their functional versatility and use of a number of different devices [6]. An FPGA consists of a matrix of logic blocks interconnected by a routing channel and track wires as illustrated in Figure 4.8.

The connection of wires can be changed by the on-offs transistors in the switch boxes.

A logic box consists of a LUT (Lookup Table) with SRAMs and a set of internal transistors and a flip-flop (FF) to retain an output. LUT can produce any kind of logic function. An example of three input AND circuits is shown in the figure of the logic block. Three inputs determine the address of SRAM whose data is transferred to the input of the FF. Only when the inputs A = B = C = 1 of a three input AND circuit is

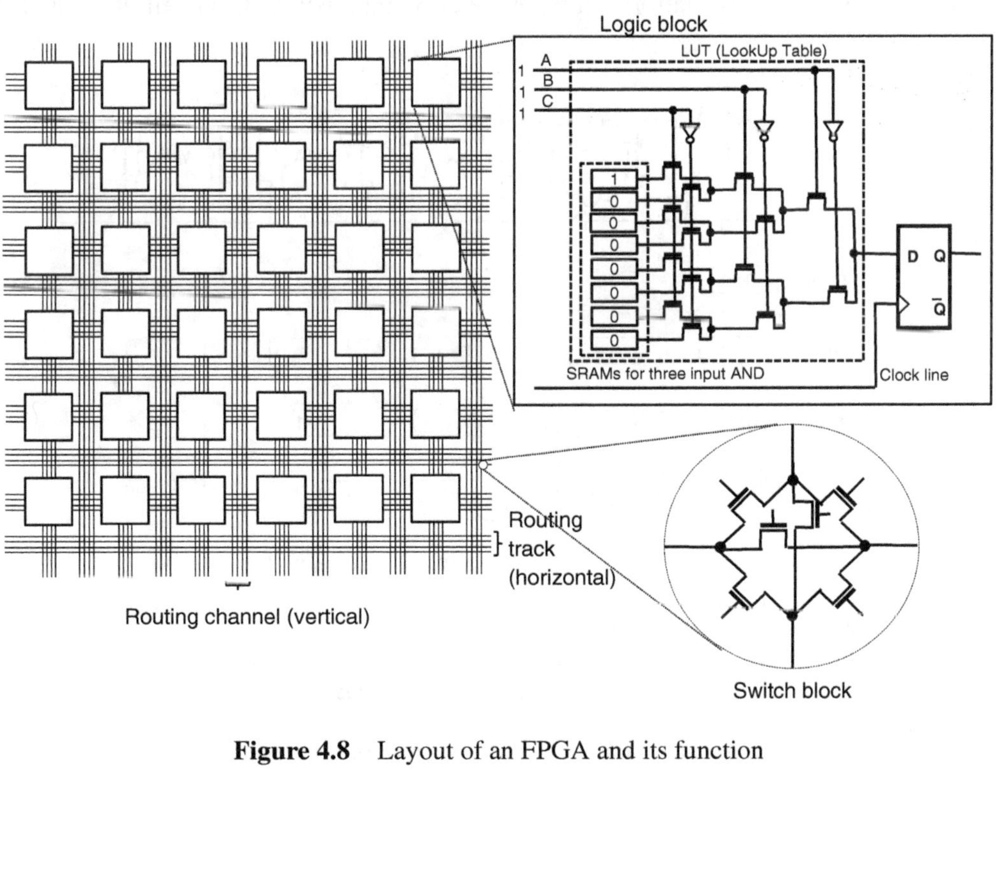

Figure 4.8 Layout of an FPGA and its function

the output '1'. By setting the data in the top SRAM at '1' and the data in other SRAMs at '0', this function is realized.

Even a complicated microprocessor can be emulated in a FPGA in this way.

4.2.2 Processor

Figure 4.9 illustrates a typical stream of a pipeline process carried out in an in-order processor [7] in which instructions are executed one by one. In general, there are five steps as follows:

1. *IF (Instruction Fetch)*: An instruction is fetched from the instruction register stored through caches from the main memory.
2. *ID (Instruction Decode/register read)*: The instruction is decoded to a control command and the register is read if necessary.
3. *EX (Execution)*: The instruction is executed by accessing data from the register and calculating the address or branching destination.
4. *MA (Memory Access)*: Load data in the memory or store data to the memory.
5. *WB (Write Back)*: Data are stored in registers.

In principle, each instruction can be treated continuously as has been illustrated in Figure 4.9.

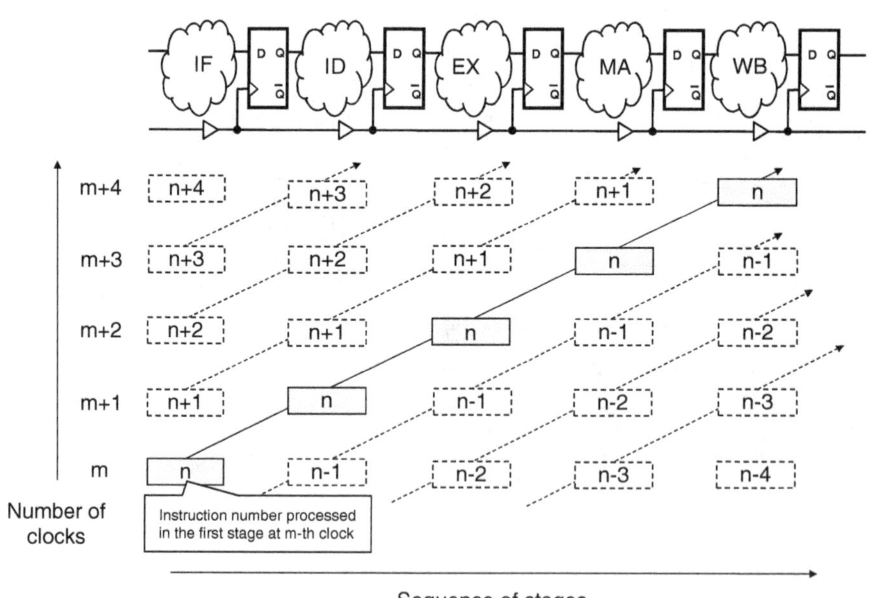

Figure 4.9 Pipeline processes in in-order processor

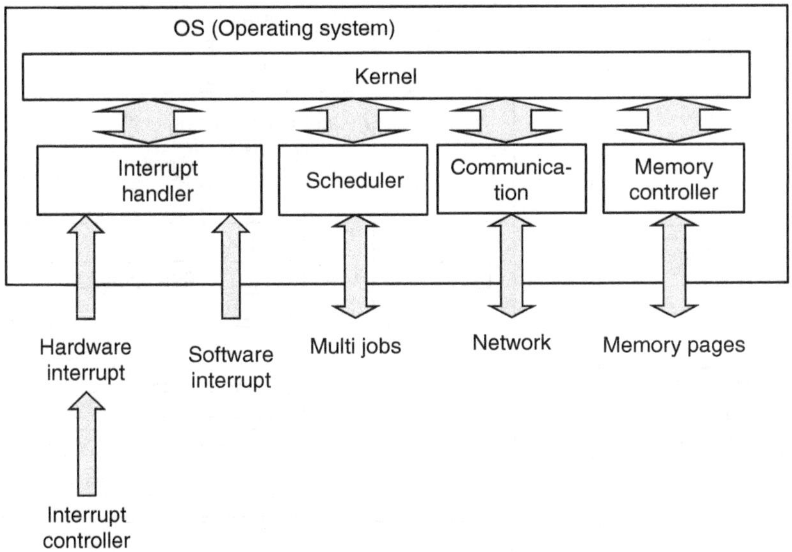

Figure 4.10 Block diagram of functions in OS

In the IF stage, for example the n-th instruction is fetched at the m-th clock. At the m + 1 clock, while the n-th instruction is decoded in the ID stage, the n + 1 instruction can be fetched in the IF stage.

When an instruction requires an execution of preceding instructions, this process may not be carried out, but the OS (operating system) can handle the sequence and timing by using data stored in the register files. The OS has this kind of function, including: scheduling, interrupt handling, memory control, communication control among relevant parts and functions as illustrated in Figure 4.10. The interrupt handler receives hardware interrupts from an interrupt controller(s) or software interrupt is made when, for example zero divide or write operation occurs in the protected area.

Figure 4.11 depicts the simplified architecture of a microprocessor. A microprocessor consists of an OS and a CPU with external IOs and a main memory. In the CPU, register files and caches are stored and referred to by a pipeline processor. In the register files, data for immediate execution of an application are stored in the data register and execution results are transferred to the main memory through caches. There are many types of registers. The following are common:

1. *Index register*: In general, the relative address of the instruction to be fetched next is stored.
2. *PC (program counter)*: The address of the instruction to be fetched next is stored.
3. *Stack pointer*: The address of the last program request in the stack (usually located in the main memory) is stored.

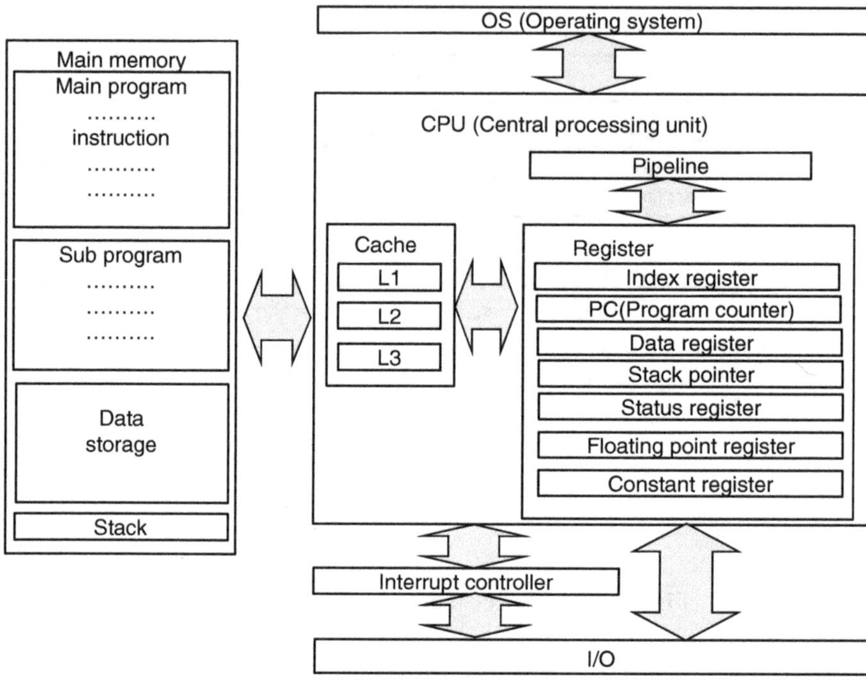

Figure 4.11 Overall architecture of a processor

4. *Status register*: A set of flag bits that express specific conditions of microprocessor is stored. The data are mainly used for branching.
5. *Floating point register*: Register to be used for floating point calculation.
6. *Constant register*: Numerical constants frequently referred to like π are stored.

4.3 Summary of Chapter 4

Fundamentals of electronic devices and systems are described to ensurea better understanding of Chapters 6–8. Structures and functions of DRAMs, inverters, SRAMs, floating gate memories, sequential and combinational logic circuits, FPGAs and processors are briefly introduced.

References

[1] Slayman, C., Ma, M. and Lindley, S. (2006) Impact of error correction code and dynamic memory reconfiguration on high-reliability/low-cost server memory. IEEE International Integrated Reliability Workshop Final Report, 2006, pp. 190–193.

[2] Baze, M., Wert, J., Clement, J. *et al.* (2006) Propagating SET characterization technique for digital CMOS libraries. *IEEE Transactions on Nuclear Science*, **53** (6), 3472–3478.

[3] Ishibashi, K. and Osada, K. (eds) (2010) *Low Power and Reliable SRAM Memory Cell and Array Design*, Springer.

[4] Cellere, G., Pellati, P., Chimenton, A. *et al.* (2001) Radiation effects on floating-gate memory cells. *IEEE Transactions on Nuclear Science*, **48** (6), 2222.

[5] Velazco, R., Foucard, G. and Peronnard, P. (2010) Combining results of accelerated radiation tests and Fault injections to predict the error rate of an application implemented in SRAM-based FPGAs. *IEEE Transactions on Nuclear Science*, **57** (6), 3500–3505.

[6] Zhang, M., Wang, N.J., Shi, Q. *et al.* (2006) Sequential element design with built-in soft error resilience. *IEEE Transactions on Very Large Scale Integration Systems*, **14** (12), 1368–1378.

[7] Kellington, J. and McBeth, R. (2007) IBM POWER6 processor soft error tolerance analysis using proton irradiation. *IEEE Workshop on Silicon Errors in Logic – System Effects* 3, Austin TX, 3 April, 4.

5

Irradiation Test Methods for Single Event Effects

5.1 Field Test

SEEs (Single Event Effects) used to be mainly measured at avionics altitudes by in-flight tests because the radiation level is $\times 100$ higher than that at ground level [1]. Even though the neutron flux level is high, adequate flight time is required to obtain statistically sound data, which results in costly experiments.

In the field test, on the other hand, DUTs (device under tests) are placed at a certain location with known latitude, longitude (and hence, geomagnetic latitude) and altitude, and the cost may be reasonable compared to in-flight tests. Since the terrestrial radiation strength is very low at sea level, it is necessary to expose thousands of DUTs to the field typically over several years. These types of tests, in principle, give straightforward and comprehensive results and, therefore, are referred to as *real time tests*. Nevertheless, care must be taken over the surrounding shielding conditions (buildings, floors), the Sun's activity and the atmospheric pressure that affects terrestrial radiation strength [2].

Attention has to be paid to space weather forecasts [3, 4] because a big solar flare may have a significant impact on the test results. GPS (Global Positioning System) data and neutron flux should be measured and recorded for an accurate analysis. In addition to the neutron flux data, the neutron energy spectrum may be measured by using the Bonner ball system [2] and the data is much more useful for data analysis. Since a real time test is carried out in remote conditions without any operators, static tests are mainly applied with long-life batteries as the power supply. Data acquisition by telecommunication from a remote office is strongly recommended.

To obtain stronger terrestrial radiation flux, HA (High Altitude), in other word, high mountain tests are regularly carried out [5–13]. Locations or facilities of such sites together with their altitudes, geographic latitudes and longitudes are summarised in

Terrestrial Radiation Effects in ULSI Devices and Electronic Systems, First Edition. Eishi H. Ibe.

Table 5.1. The relative intensity of neutron fluxes calculated by using a Flux Calculator [14] are also indicated in Table 5.1. In HA (Type HA in Table 5.1) tests, such as White Mountain (3883 m), Mauna Kea (4200, 4023 m), Pic de Bure (2552 m), Puno (3889 m), Asama (2200 m), Yungfraujoch (3570 m) and Lake Tahoe (1700, 2000 m) where neutron intensity $\times 3-16$ compared to NYC (New York City) sea level. Attention must be paid to the effects of muons and protons, since they have comparative flux at avionics altitude as previously indicated in Figure 2.3. It is also recommended that field tests near sea level (Type G in Table 5.1) are carried out to compare and validate.

Ibe *et al.*, made HA field tests of 250 nm 4 M SRAMs (static random access memories) in three locations/altitudes in Japan with a neutron dosimeter. Monte-Carlo soft-error simulation results were achieved by using CORIMS (COsmic Radiation IMpact Simulator) which concurred with the measured results [7].

Lesea *et al.* carried out HA tests in many locations such as New Mexico, White Mountain, Mauna Care, Pic de Bure together with ground tests in the Rosetta project [5]. They performed HA/G/UG (underground) field tests for Xilinx 150, 130, 90 nm FPGAs (Field Programmable Gate Arrays) together with accelerator experiments at LANSCE (neutron) and CNL (proton, UC Davis Crocker Nuclear Labs). Large discrepancies are reportedly found between the SER (soft-error report) which was estimated from simulation results based on the field neutron flux and SERs measured in the HA field tests. Estimated SERs are higher than the measured ones by a factor of 3 for 90 nm FPGAs. The root cause of the discrepancies is not identified.

Autran *et al.* construct ASTEP (Altitude Single event effects Test European Platform) that consists of facilities in HA, ground and underground associated with solid logistics and long-term technical support [8].

In order to distinguish between the contribution of alpha-ray from DUT from total SER and make corrections to SER induced by terrestrial neutron, deep UG (Type UG (under ground) in Table 5.1) tests like big caves [5] or decommissioned tunnels [11, 12] are also carried out. The effects of thermal neutrons have previously been evaluated in the field test by Kobayashi *et al.* and they demonstrated that contribution of thermal neutron to total SER must be distinguished in the field tests for DUTs that contain ^{10}B regardless of the altitude where the experiments were carried out [12]. Cadmium shielding that can attenuate thermal neutrons (See Section 5.6 for more details) can be used for the analysis. Tosaka *et al.* performed field tests of 90 nm complementary metal oxide semiconductor (CMOS) SRAM in the Mauna Kea Observatories, Hawaii, together with neutron spectrum measurements by using four Bonner balls [6]. Autran *et al.* carried out a HA field test of 40 nm SRAM at the ASTEP platform [8]. Kameyama *et al.* reported HA/G field test results of 180 and 130 nm SRAMs in California with some anomalies observed in neutron flux [13]. Torok *et al.* performed HA/G field tests of CCDs (Charge Coupled Devices) together with neutron accelerated tests in LANSCE. Collected charge spectra in CCDs were used for analyses [9].

Table 5.1 List of facilities for SEE measurements in fields

Location name	Facility name	Type[a]	Altitude	Geographic location[b]		Neuron flux normalised to NYC sea level	Note	References
			m	Latitude	Longitude	(50% maximum)		
White Mountain, CA	Barcroft Station	HA	3883	37 N	118 W	16	http://www.wmrs.edu/	Lesea [5]
Mauna Kea, Hawaii	Caltech Submillimeter Observatory (CSO)	HA	4023	19 N	155 W	8.3	http://www.cso.caltech.edu/	
	Subaru Telescope	HA	4200			9.1	http://www.naoj.org/	Tosaka [6]
Marseille, France	IM2NP[c]	G	123	43 N	5 E	0.93	http://www.im2np.fr/	Lesea [5]
Rustrel, France	Rustrel (LSBB[d])	UG	−457 (Cave)	43 N	5 E	–	http://lsbb.oca.eu/spip.php?article165	
Pic de Bure, France	–	HA	2552	44 N	5 E	7.2	–	Lesea [5], Autran [8], Alxandrescu [11] and Hubert [10]
Puno, Peru	–	HA	3889	15.5 S	70 W	8.1	–	Hubert [10]
San Jose, CA	–	G	0	37 N	122 W	0.95	–	Kameyama [13]
Kokubunji, Japan	–	G	70	35 N	139 E	0.66	–	Ibe [15]
Komoro, Japan		HA	700	36 N	138 E	1.1	–	
Asama-Mt, Japan	–	HA	2200	36 N	138 E	3.2	–	
Modane(Fejus Mt.)	LSM[e]	UG	4800 minus 1700 (Tunnel) (3100)	45 N	6 E	(9)[f]	http://www-extension-lsm.in2p3.fr/	Alxandrescu [11]
Jungfraujoch	–	HA	3570	46 N	7 E	13	–	Torok [9] and Alxandrescu [11]
Oto, Japan	Oto Cosmo Observatory	UG	845 minus 467(Tunnel) (378)	34 N	135 E	(0.83)[f]	http://wwwkm.phys.sci.osaka-u.ac.jp/info/syoukai/oto.html	Kobayashi [12]
Lake Tahoe, CA	–	HA	2000	39 N	120 W	4.7	–	
	–	HA	1700			3.8	–	Kameyama [13]
Oita, Japan	–	G	0	33 N	131 E	0.6	–	Kobayashi [12]
Yokohama, Japan	–	G	0	35 N	140 E	0.63	–	

[a]HA: high altitude; G: ground and UG: under ground.
[b]N: north; S: south, E: east and W: west.
[c]Institute Materiaux Microelecronique Nanoscience de Provence.
[d]Laboratoire Souterrain a Bas Bruit.
[e]Laboratoire Souterrain de Modane.
[f]Shielding by rock is ignored.

5.2 Alpha Ray SEE Test

Table 5.2 summarises the properties of alpha ray sources [16–21] for accelerated experiments. A sheet of foil that contains ^{241}Am is most commonly used for alpha-ray soft-error experiments. Such a ^{241}Am sheet made of aluminum is commercially available and can be used for an alpha-ray SEE (Single Event Effect) test. Spacing to the DUT surface and area to cover the DUT are defined in relevant standards like JESD89A [18], IEC60749-38 [19] and EDR4705 [20]. In some cases, accelerators are used to produce high-energy helium ion beams by using high-energy accelerators [16, 17, 22–25].

As the energy of an alpha ray is in the order of 5 MeV, as previously shown in Table 2.3, their ranges are relatively short in the DUT, in particular, in the metal layers [21]. In reality, an alpha ray source is contained or contaminated in the DUT itself so that an alpha ray can reach the sensitive portion of DUT. To evaluate the SEE by alpha rays from outside of the DUT, delidding of substrate layer may be required, in particular, from the bottom side in order to obtain a uniform penetration range of alpha particle beams.

As can be seen in Table 5.2, LET of alpha particle in Si is not so high that the main mode of soft error caused by alpha particles will be SBU (Single Bit Upset).

Roche *et al.* make alpha ray soft-error experiments in 250, 130, 90 nm SOI/bulk SRAMs by using ^{241}Am and ^{232}Th sources, demonstrating low susceptibility in SOI SRAMs compared to bulk SRAMs [16].

Gasiot and Roche further carry out ^{241}Am alpha ray soft-error tests in 130, 90, 65 nm SRAMs and show an increase in multi-cell upset (MCU) rate with 65 nm SRAM partly due to the bipolar effect [17].

KleinOsowski *et al.* [22] and Rodbell *et al.* [23] evaluate the alpha-ray soft errors in 65, 45, 32 nm SOI latches by using the IBM T. J. Watson Lab. 3MV tandem accelerator. The effects of transistor width/layout and beam angle are evaluated. The merit of using accelerators as the alpha-ray sources is that the beam can be collimated and its angle and hit location can be controlled.

Swift also utilised the TAMU (Texas A & M University) cyclotron accelerator as an alpha-ray source [24].

In order to evaluate actual alpha-ray induced SER in the field, actual alpha-ray flux from contaminated materials (mainly, ^{210}Po, ^{232}Th and ^{238}U) in devices must be measured. ^{210}Po is mainly contained in the solder bump as an impurity in Pb. ^{232}Th and ^{238}U are mainly contained in the package materials. As device scaling proceeds and Q_{crit} decreases accordingly, susceptibility of SRAMs to alpha-ray becomes so high that even alpha-ray flux of below $1 \times 10^{-3}\,h^{-1}\,cm^{-2}$ (VLA, Very Low Alpha level) may cause significant SER. The contamination level of alpha-ray emitting materials, therefore, must be controlled tightly in the manufacturing processes. Measurement of such low-level contamination, however, becomes very difficult and specific methods must be developed and used for this kind of measurements. Takasu *et al.* developed a method, in which CR-39 monomer is placed on a device in a vacuum

Table 5.2 List of methods for alpha-ray SEE measurements

Acronym	Facility name/ source	Place	Energy (MeV)	LET in Si (MeV-cm^2/mg)	Flux (α/cm^2/s)	Note	References
–	IBM TJ Watson Lab. 3MV Tandem Accelerator	Yorktown Heights, NY	4–7	0.5–4	5×10^6	65 nm SOI* latch experiment	KleinOsowski [22]
				0.6–9	–	32, 45 nm SOI latch experiment	Rodbell [23]
TAMU	Cyclotron Institute, Texas A&M University	College Station, TX	60, 99	0.11, 0.07 (Initial in vacuum) 1.5 (at Bragg peak)	–	–	Swift [24]
–	^{241}Am	–	4	0.7	–	90, 130 and 250 nm SRAM experiment	Roche [16]
						100$_{\mu}$Ci	Gasiot [17]
–	^{232}Th	–	5.4	0.6	–	90, 130 and 250 nm SRAM experiment	Roche [16]

*Silicon On Insulator

chamber. The number of etch pits in the CR-39 monomer created by alpha-particle hits is counted after chemical etching treatments, and demonstrated the detection limit $3.2 \times 10^{-5}\, h^{-1}\, cm^{-2}$ of alpha-ray flux [26], which is well below HLA (Hyper Low Alpha level; $5 \times 10^{-4}\, h^{-1}\, cm^{-2}$).

5.3 Heavy Ion Particle Irradiation Test

A number of high-energy secondary ions are produced after spallation reaction of a nucleon with a target nucleus. High-energy particles produced in an accelerator facility can be used to simulate the effects of such secondary ions or direct strikes by high-energy ions in the space environments. The effects can be single event upsets (SEUs) by a direct ionisation mechanism, TID (Total Ionisation Dose) effects and/or displacement effects as explained in Chapter 3. The high energy particles are, in general, initially supplied in a high vacuum chamber in the form of gas or vapour, and then charged up as plus or minus ions in the ionisation chamber as shown in Figure 5.1. The ions are extracted through the extraction electrodes to an accelerator chamber (Synchrotron or cyclotron or LINAC (LINear particle Accelerator)) and accelerated to a certain targeted energy and led to the irradiation chamber where the DUT or BUT (Board Under Test) is placed. The beam line is kept at a high vacuum condition, so that DUTs must be placed in a high vacuum target chamber or placed on a board in atmospheric pressure. In the latter case, accelerated ions pass through a thin film at the end of the beam line from the vacuum condition to atmospheric environment.

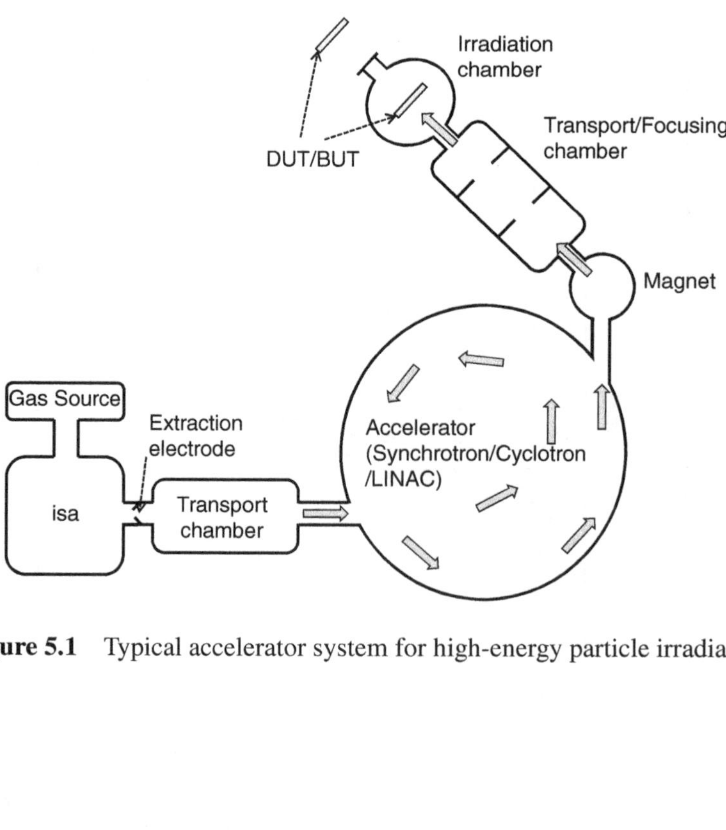

Figure 5.1 Typical accelerator system for high-energy particle irradiation

Beam strength (current) can be measured by placing a Faraday cup in the beam before irradiation or at the end of the beam in the beam dump. Care must be paid during heat-up or charge-up or activation of DUTs. DUTs or BUTs irradiated in high-energy particle facilities may be kept for some period of time in a secure room to reduce activation safely.

In many cases, DUTs have relatively thick layers or cooling fins typically made of copper and BUTs have many chips and components, which scatter penetrating ions and force them to lose their energy. The range and LET at the relevant depth in the DUT must be estimated beforehand by using, for example SRIM (Stopping and Range of Ions in Matter) [27] and other more sophisticated tools such as PHITS (Particle and Heavy Ion Transport code System) [28]. The layout and shielding of equipment and power sources should be carefully considered, since high energy X-rays, gamma rays or neutrons may be produced from the interaction of beam particles and wall materials in the beam line or the irradiation room, and they may cause malfunctions in the equipment.

Table 5.3 summarises the typical particle accelerator facilities that are used for terrestrial or cosmic radiation effect [23, 29–43]. Only *proven* data on the capabilities of the facilities are collected from the literature and summarised in Table 5.3. The same approach is taken in subsequent tables for other types of irradiation facilities.

High-energy heavy ions can be obtained by using cyclotron [23, 29–34, 38–42] or tandem accelerators [29, 37–40, 43] such as the Van de Graaf accelerator and pelletron. Tandem accelerators start with the production of minus ions as shown in Figure 5.2. Minus-charged ions are accelerated by high voltage electrodes and

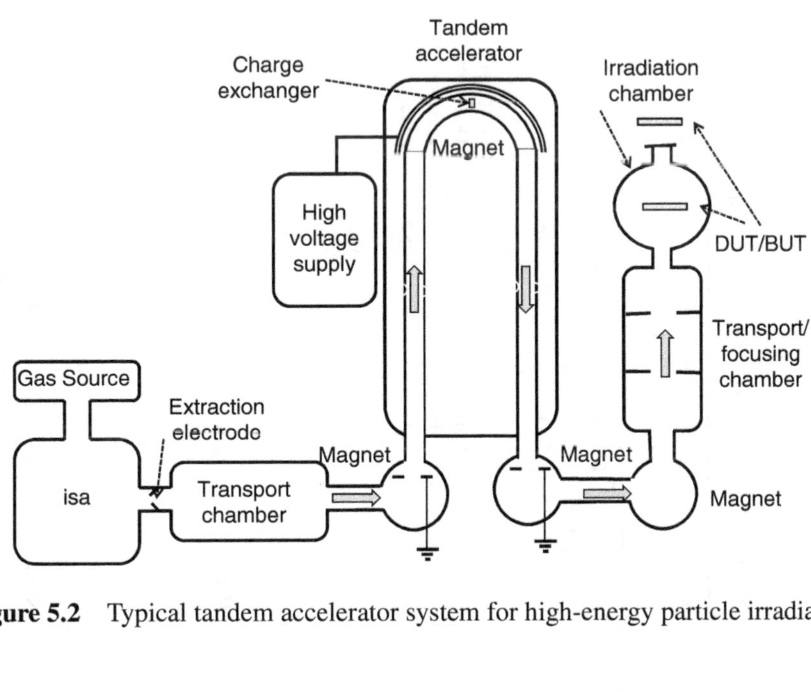

Figure 5.2 Typical tandem accelerator system for high-energy particle irradiation

Table 5.3 List of facilities for high-energy heavy-ion irradiation

Acronym	Facility name	Place	Type	Ion	Energy (MeV)	Energy (MeV/amu)	LET (MeV-cm^2/mg)	Note	References
HIF/ CYCLONE	Cyclotron of Louvain la Neuve	Louvain-la-Neuve, Belgium	Cyclotron	N, Ne, Ar, Kr, Xe	60–420	–	3.3–67.7	http://www.cyc.ucl.ac.be/HIF/HIF.html	Gerardin [29]
				He, Be,C, N, O, Ne, Ar	–	0.56–10	–	20–80 MeV neutron beam is available	Sierawski [30]
TVDG/BNL	Brookhaven National Laboratory	Upton, NY	Tandem Van De Graaff accelerator	F, Si, Ti, Br, Ag, I	–	–	3.4–59.7	http://tvdg10.phy.bnl.gov/	Irom [31]
TAMU	Cyclotron Institute, Texas A&M University	Texas, USA	Cyclotron	N, Ne, Ar, Kr, Xe	–	10	–	http://cyclotron.tamu.edu/	Rodbell [23]
				Ar, Cu, Kr, Ag, Xe, Pr, Ho, Ta	–	15, 25	5.6–130		Irom [31]
				Au, Xe	2200–2600	–	42–86		Dodd [32]
				Ne, Ar, Kr, Xe	300–1935	–	2.9–55.6		Oldham [34]
HMI	Hahn-Meitner Institute	Berlin, Germany	Cyclotron	C, O	68	–	–	–	Sonia [33]
–	Institute of Applied Physics, Nuremberg University	Erlangen, Germany	–	C, O, Fe	2	–	–	–	
NSCL	National Super-conducting Cyclotron Laboratory	Michigan State University	Cyclotron	Ne, Ar, Kr, Xe	300–14280	15, 70, 105	2.9–55.6	http://www.nscl.msu.edu/	Oldham [34]

INFN (SIRAD)	Laboratori Nazionali di Legnaro	Padova, Italy	15 MeV Tandem	O, Si, Ni, Ag	100–256	–	2.85 – 58.2	http://sirad.pd.infn.it/	Gerardin [35]
				I, Ag, Ni	239–275	–	29.3–64.2	Flux: 4×10^4 particles/cm^2/s	Larcher [36]
				Au, I	273, 286	–	85, 65	–	Cellere [37]
LBNL	Lawrence Berkeley National Laboratory	Berkeley, California	Cyclotron	O, Ne, Ar, Kr, Xe	–	10	–	http://www.lbl.gov/	Rodbell [23]
				N, Ar, Cu, Kr	234–1226	–	1.2–25		Chatterjee [38]
				N, Ne, Ar, Kr, Xe	68–612	–	2.9–64		Dodd [39]
				B, N, Ne, Ar, Cu	45–293	–	1.25–28.6		McMarr [40]
RADEF	Radiation Effects Facility	Jyväskylä, Finland	Cyclotron	Ne, Ar, Fe, Kr, Xe	186–1217	–	3.8 – 56.4	https://www.jyu.fi/fysiikka/en/research/accelerator/radef	Cellere [41]
–	California University	Berkeley, California	Cyclotron	Ne, Ar, Kr, Xe, Bi	90–950	–	5.6–190	–	Liu [42]
ALTO	Institute de Physique Nucleaire	Orsay, France	Tandem	F, Cl, Br	106–236	–	4.8–41	http://ipnweb.in2p3.fr/tandem-alto/index_E.html	FerletCavrois [43]

transferred to plus-charged ions by extracting orbital electrons typically through a metal foil then accelerated again to the ground. The ion can have an energy of $2ZV$, where Z is the atomic number of the ion and V is the terminal voltage.

Chatterjee *et al.* made heavy ion irradiation tests of 40 nm SRAM at LBNL (Lawrence Berkeley National Laboratory) [38]. It is shown that the MCU rate in a dual well SRAM is higher than a triple well SRAM for low LET (1.16 MeV-cm^2/mg), but vice versa, for high LET (24.98 MeV-cm^2/mg). This is attributed to high-density charge confinement in the triple well, which causes the affected node to be restored. This phenomena is known as 'single-event upset reversal'.

Liu *et al.* used the cyclotron at Berkeley, California University to see TID and SEE effects of 4 Mb SRAMs by irradiation of proton and heavy ions. They found that fully depleted silicon on insulator (FDSOI) SRAM show higher saturation SEU cross sections than PDSOI (Partially Depleted Silicon on Insulator) does, while threshold LET of FDSOI is much lower than that of FDSOI (Fully Depleted Silicon On Insulator) [42].

McMarr *et al.* evaluate 150 nm PDSOI SRAM by heavy ion at LBNL and 14 MeV neutron irradiation [40].

Gerardin *et al.* evaluate the effects of heavy ion irradiation at SIRAD on ESD (Electrostatic Discharge) of FDSOI MOSFET (Metal Oxide Semiconductor Field Effect Transistor) and found ESD breakdown voltage is reduced through irradiation from 5 to 3.5 V [35].

Irom and Nguyen made SEU tests on Micron 4 Gb NAND flash memory and Spansion 64 Mb NOR flash memory by using Brookhaven National Laboratory (BNL) heavy ion beam line and they report high SEU/SEFI (single event functional interruption) rates in both flash memories [31]. Catastrophic failures can be observed in NOR memory.

Oldham *et al.* evaluate SEE and TID susceptibility in 90 nm 2 Gb Micron NAND flash memory by irradiation of heavy ions at NSCL (National Superconducting Cyclotron Laboratory)/TAMU and protons at IUCF (Indiana University Cyclotron Facility) [34].

Deterioration of data retention characteristics is observed in floating gate memories by heavy ion irradiation (I, Ag and Ni ions) at SIRAD [36]. Cellere *et al.* also evaluate V_{th} shift in floating gate NOR flash memory observed in irradiation tests at SIRAD. The test results are examined in the percolation path model in the tunnel oxide where an electron-hole pair is produced by energy deposition of 17 eV and conclude that RILC (Radiation Induced Leak Current) takes place using much thicker tunnel oxide than SILC (Stress Induced Leak Current) does [37].

Cellere *et al.* also use RADEF (Radiation Effects Facility) to evaluate V_{th} shift in NOR flash memory and its dependency on the grazing angle of impinging ions [41]. The shallower the grazing angle is, the larger are the V_{th} shifts observed in NOR memories.

Dodd *et al.* utilise BNL to investigate SEB (Single Event Burnout) in lateral power MOSFETs [32].

Ferlet Cavrois *et al.* perform SET pulse analyses in long inverter chains by using laser and heavy ion irradiation at ALTO (Accélérateur Linéaire auprès du Tandem d'Orsay) and evaluate the pulse modulation mechanism, PIPB (Propagation Induced Pulse Broadening) [43]. It is pointed out that PIPB may be caused by bipolar action.

5.4 Proton Beam Test

Both protons and neutrons belong to nucleons. The only differences are isospin and mass (see Table 2.1). Proton has isospin 1/2 and neutron has isospin −1/2. Proton causes a spallation nuclear reaction as shown in Figure 3.9 with some difference because of neutron in the cross section below 50 MeV due to the Coulomb barrier effect.

Table 5.4 summarises typical proton accelerator facilities used for terrestrial or cosmic radiation effects [44–65].

Schwank *et al.* utilised TRIUMF (Tri-University Meson Facility) for space application of commercial SRAMs (500 to 140 nm). They showed that the SEU cross section of SRAM by proton increases by TID effects (10–50 krad(Si)) of proton/Co-60 gamma irradiation by a factor of 10 [51].

Dyer *et al.* measured the neutron and proton SEU cross section of 4 Mbit SRAMs by using NPL (National Physical Laboratory), TSL (The Svedberg Laboratory) and TRIUMF across a wide range of energies [46].

Effects of low-energy protons on SEU characteristics in SRAMs and latches are becoming key issues in proton irradiation tests.

Rodbell *et al.* use IBM's T. J. Watson Lab. Tandem accelerator to investigate the effects of Q_{crit} decrease on SEU characteristics 65 nm SOI SRAMs and latches [63]. They estimate Q_{crit} from the critical grazing angle of protons at which SEUs take place and obtain Q_{crit} values as low as 0.24–0.27 and 0.14–0.16 fC for SRAMs and latches, respectively. They also emphasise the effects of the metal layer structure on variation in Q_{crit}.

Sierawski *et al.* examine low-energy proton effects on SEU sensitivities in 65 nm SRAMs at GSFC (Goddard Space Flight Center), CNL, IUCF and LBNL [30]. They show that the contribution of protons with energy of below 2 MeV is much higher than those with energy of more than 100 MeV by a factor of more than 2 orders of magnitude. They attribute this increase in contribution to direct ionisation by protons with low energy. Low-energy proton has a much larger energy deposition near their Bragg peak energy compared to Q_{crit} that is decreasing because of scaling.

Lawrence *et al.* evaluated the susceptibility of 90 nm bulk CMOS SRAMs to low-energy protons at LBNL and the results are compared with high-energy proton tests at IUCF [55]. SRAMs hardened by adding capacitors and registers and non-hardened SRAMs are tested. Drastic increase in SEU sensitivity is observed for non-hardened SRAMs below 1 MeV by a factor of as high as 5–6 orders of magnitude.

Table 5.4 List of facilities for high-energy proton irradiation

Acronym	Facility name	Place	Type	Energy (MeV)	Flux $(p/cm^2/s)^a$	Note	References
PIF/TRIUMF	Canada's national laboratory for particle and nuclear physics	Vancouver, Canada	Cyclotron	34.5–498	$1.44–8.52 \times 10^{11}$	http://www.triumf.ca/	Cellere [44]
				35.4, 105, 498	–		Gerardin [29]
				20–498	–		Shaneyfelt [48]
				20–520	$1 \times 10^5 – 1 \times 10^8$		Hiemstra [49]
				50, 100, 200, 350, 500	–		Baggio [50]
				20–500	–		Schwank [51]
				105	–		Rufenscht [52]
PAULA/TSL	Uppsala University	Uppsala, Sweden	Cyclotron	24, 49, 119, 196	–	http://www.tsl.uu.se/proton-therapy/	Johansson [53]
				21, 46, 88	–		Granlund [54]
–	Northeast Proton Therapy Center	Boston,MA	Cyclotron	10–225	–	http://neurosurgery.mgh.harvard.edu/protonbeam/nptcbrochure.pdf	Liu [42]
				220	2.5×10^8		McMarr [40]
IUCF	Indiana University Cyclotron Facility	Bloomington, Indiana	Cyclotron	100, 200	–	http://www.iucf.indiana.edu/	Sonia [33]
				52, 89, 198	–		Lawrence [55]
				65, 200	–		Quinn [56]
				50, 100, 150, 200	–		Gadlage [57]
				98, 198	–		Sierawski [30]
				100, 200	–	Fluence: 1×10^{10}-$1 \times 10^{11} p/cm^2$	Oldham [34]
				27, 198	–	–	Seifert [58]
LBNL	Lawrence Berkeley National Laboratory	Berkeley, CA	Cyclotron	1.2, 6, 32.5	–	http://www.lbl.gov/	Sierawski [30]
				0.35, 3, 5	–		Skarin [59]
				0.65, 1.2	6×10^2-5×10^6		Lawrence [55]

INFN-LNL	Laboratori Nazionali di Legnaro	Legnaro, Italy	Tandem Van de Graaf accelerator	5	–	http://www.lnl.infn.it/~newweb/index.php/en/introduction-to-the-lnl-accelerators	Zanata [61]
				4.2	2.4 krad (Si)/s		Bagatin [62]
IBM Tandem	IBM T.J. Watson Research Center	Yorktown Heights, NY	Tandem Van de Graaf accelerator	0.6–6	–	–	Rodbell [23]
CNL	UC Davis Crocker Nuclear Lab.	Davis, CA	Cyclotron	19.8, 2.6	–	http://crocker.ucdavis.edu/	Sierawski [30]
				8–63	–		Schwank [51]
				1.09, 4.47	–		Seifert [58]
–	Hahn-Meitner Institute	Berlin, Germany	Cyclotron	2, 68	–	http://www.helmholtz-berlin.de/zentrum/historie/lise-meitner-campus/index_en.html	Sonia [33]
–	Francis H. Burr Proton Therapy Center	Boston, MA	Cyclotron	150	8.2×10^7	http://www.massgeneral.org/radiationoncology/BurrProtonCenter.aspx	Kellington [64] and Rao [65]
GSFC	Goddard Space Flight Center	Greenbelt, MD	Van de Graaf accelerator	< 2	–	http://www.nasa.gov/centers/goddard/home/	Sierawski [30]

[a]Otherwise indicated.

Puchner *et al.* evaluate the effects of low-energy protons on SEUs in 90 and 65 nm SRAMs using the LBNL low-energy proton beam tuned by degrader foils down to 0.35 MeV [60]. They also show an increase in the SEU cross section below 2 MeV by a factor of 2 orders of magnitude.

Seifert *et al.* examine the susceptibility of 45 and 32 nm bulk CMOS latches to low-energy protons at IUCF and CNL [58]. They use the degraders to reduce proton energies down to below 1 MeV and show no drastic increase in the SEU cross section of 45 and 32 nm latches below unless the operation voltage is reduced.

Gadlage *et al.* examine operation frequency effects on the SEU cross sections in the pipeline architecture of inverter chains with DICEs (Dual Interlocked Storage Cell) or unhardened D-flip flops (FFs) in-between [57]. As explained in Section 5.8.2, the SEU rate in D-FFs is the sum of the contribution by the direct hit of protons at sensitive nodes in FFs and the contribution by SETs (Single Event Transients) created in the combinational logic circuit (inverter chain) upstream of the FFs and captured in the FFs. The probability of capture increases as frequency increases so that the total number of SEUs in FFs starts to increase at a certain frequency (crossover frequency). They tend to approach the crossover frequency at about 1 GHz by 200 MeV proton irradiation and at about a few hundred megahertz in heavy ion irradiation.

Cellere *et al.* used TRIUMF for proton and heavy ion irradiation tests of NAND flash memories and suggest the effects of recoils from nuclear reaction of protons with tungsten [44].

Bagatin *et al.* evaluate the TID characteristics of 90 nm NOR multi-level cells and 70 nm NAND single level cell flash memories using 4.2 MeV proton beams at LNL (Laboratori Nazionali di Legnaro) [62]. They show a change in V_{th} distribution in NOR flash memories.

Gerardin *et al.* evaluate the 41 nm floating gate errors in flash memories by using heavy ion irradiation at LNL and proton beam irradiation at TRIUMF, emphasising the impact of protons on floating gate errors by direct ionisation and spallation reaction [45].

Zanata *et al.* evaluate the TID characteristics of FRAM (Ferroelectric Random Access Memory) that employs a PZT (Lead (Pb) Zirconate Titanate) ferroelectric film as capacitor and can potentially replace flash memory as non-volatile memory [61]. Radiation damage in FRAMs only appears places and appears to be strongly dependent on temperature.

Sonia *et al.* evaluate proton beam irradiation effects on AlGaN/GaN heterogeneous FETs [33].

Shaneyfelt *et al.* utilise TRIUMF, IUCF and LANSCE for proton and neutron irradiation tests of TID effects (increase in leak current) in power MOSFET [48]. Cadmium sheet is used to reduce neutron energy to a thermal level in TRIUMF.

Hiemstra *et al.* perform proton irradiation tests at TRIUMF of Pentium MMX (233 MHz, 350 nm, 4.5 M transistors) and Pentium II processors (333 MHz, 250 nm, 7.5 M transistors) with error correction code/error checking and correction (ECC) in L1 and L2 caches [49].

High-energy protons are often utilised for system level irradiation tests as well as neutrons:

Rufenacht *et al.* carry out partial irradiation tests of ESP603 single board computers and power 603r microprocessors using a TRIUMF 105 MeV proton beam [52]. They separate ESP603 into eight regions (6 cm × 6 cm each) including DRAMs (dynamic random access memories), FPGAs, clock buffers, EEPROMs (Electrically Erasable Programmable Read-Only Memories), voltage regulator and examine SEU characteristics in each region.

Kellington *et al.* evaluate derating in IBM power6 microprocessors (65 nm SOI process, dual core with on-board L2 and L3 caches) by 150 MeV proton beam tests in the Francis H. Burr Proton Therapy Center of the Massachusetts General Hospital [64]. Static and dynamic tests are carried out. The AVP (Architectural Verification Program), by which instructions are randomly created and various modes of errors can be evaluated including SDCs (Silent Data Corruptions), are applied.

Rao *et al.* also evaluated the dependency of SER on workload in IBM power6, Intel Core2 5160, Xeon Woodcrest Microprocessors by 145–150 MeV proton irradiation at the Francis H. Burr Proton Therapy Center of the Massachusetts General Hospital, and by spallation neutrons at LANSCE [65]. NOP (No Operation), operating system (OS) idle, Linpack, LDPC (Low Density Parity Check) floating point workload and Laplacian smoothing workload are applied.

Although an idle condition is commonly supposed to be less susceptible to irradiation compared to a busy condition, distinctive differences are not observed between the two conditions.

5.5 Muon Test Method

Artificial production of muons requires a very high-energy proton accelerator and the collision/beam transfer equipment so that facilities available are limited worldwide as listed in Table 5.5 [66–70].

Muons do not cause any nuclear reaction but cause direct ionisation. Their SEE can be evaluated in a similar manner to the tests with charged particles.

A simplified muon facility image is illustrated in Figure 5.3.

Muons can be produced from pions through decay reactions (Equation 5.1),

$$\begin{aligned}\pi^- &\rightarrow \mu^- + \nu_{\bar{\mu}},\\ \pi^+ &\rightarrow \mu^+ + \nu_{\mu}.\end{aligned} \tag{5.1}$$

To produce a pion π^+, a high-energy proton with kinetic energy of more than 300 MeV has to collide with a proton in a target nucleus (C, Cu, etc.) with the reaction,

$$\mathrm{p} + \mathrm{p} \rightarrow \mathrm{p} + \mathrm{n} + \pi^+. \tag{5.2}$$

Negative pion π^- can be produced by bombarding high-energy neutrons to the target

$$\mathrm{n} + \mathrm{p} \rightarrow \mathrm{p} + \mathrm{p} + \pi^-, \tag{5.3}$$

Table 5.5 List of facilities for muon irradiation

Acronym	Facility name	Place	Primary ion	Target	Energy (MeV) or momentum (MeV/c)	Flux	Note	References
MUSE/J-PARC	Japan Proton Accelerator Research Complex	Ibaraki, Japan	p	Graphite	4–55 MeV	$4 \times 10^6 (4\pi)/s$	–	http://nuclpart.kek.jp/pac/1101/pdf/JPARCPAC110113.pdf
PRISM/J-PARC			50 GeV p	–	68 MeV/c	$1 \times 10^{12} (4\pi)/s$	Under construction	http://nop.kek.jp/~arimoto/researches/prism-ffag/docs/muon_facility.pdf
MICE/ISIS	Rutherford Appleton Lab.	Oxfordshire, UK	800 MeV p	Ti hollow cylinder	140–240 MeV/c	–	Muon Ionisation cooling experiment	http://wrap.warwick.ac.uk/48594/1/WRAP_Back_1748-0221_7_05_P05009.pdf
RIKEN-RAL/ISIS			800 MeV p	–	20–120 MeV/c	–	–	http://nectar.nd.rl.ac.uk/en.html
LAMPF	Los Alamos Meson Physics Facility	Los Alamos, New Mexico	800 MeV p	Graphite	109 MeV/c	$5 \times 10^4 /s/cm^2$	Pion beam with momentum 164 MeV/c and flux $1.6 \times 10^6 /cm^2/s$ is available	Dicello [71]
M20B/TRIUMF	Tri-University Meson Facility	Vancouver, Canada	500 MeV p	Be	20–200 MeV/c	$1.4 \times 10^6 (4\pi)/s$	Six beam lines	Sierawski [72]
PSI	Paul Scherrer Institute	Villigen, Switzerland	590 MeV p	–	10-280 MeV/c	$3.0 \times 10^7 (4\pi)/s$	Six beam lines	http://lmu.web.psi.ch/facilities/facility.html

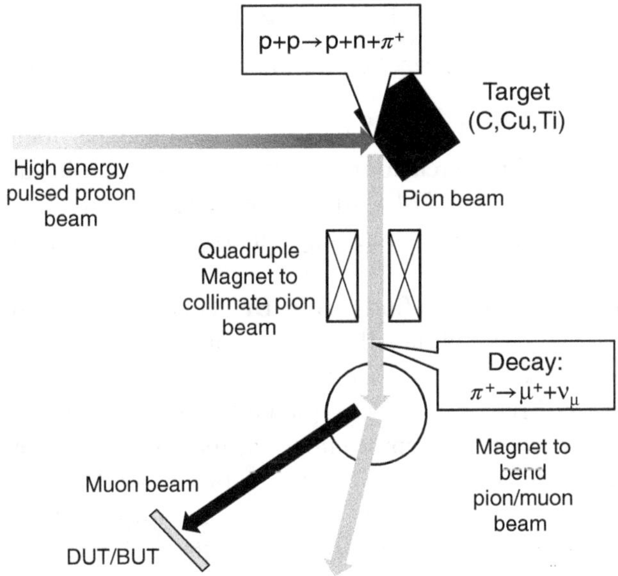

Figure 5.3 Simplified scheme of muon irradiation system

or in the vicinity of the p-p collision site where high-energy neutrons produced by the Equation (5.2) can produce π^- by the same Equation (5.3).

A muon beam is separated from the pion beam by dipole bending magnets and quadruple focussing magnets.

In most facilities in Table 5.5, pion beams may be available because pions are intermediate products of the muon beam line as illustrated in Figure 5.3, although there are only old reports by using LANSCE (formerly LAMPF, Los Alamos Meson Physics Facility) by Dicello *et al.* [71]. They report that negative pions effectively produce soft error in AM9114 EPS 4 kb SRAMs with a SEU cross section of 6×10^{-8} cm^2/chip. An SEU cross section by positive pions is less than 1×10^{-11} cm^2/chip. No error is detected by muon irradiation up to the fluence of 8×10^{10}/cm^2. The relative impact of pions on soft error is 10 000 times higher than that of muons if fluences of both particles are the same. In the actual field, however, the flux of pions is much lower than that of muons as shown in Figure 2.3 so that the impact of pions on SER should be evaluated more quantitatively.

Sierawski *et al.* demonstrate that low-energy muons, in particular μ^+, can cause 65, 45 and 40 nm SRAMs by using muon beam line M20B at TRIUMF [72] with an estimated SEU cross section of 1×10^{-13} cm^2/bit at the energy of around 0.3–0.8 MeV. They conclude that soft error induced by terrestrial muons is not substantial for current generation compared to the contribution of neutrons, but may have an impact when Q_{crit} decreases with scaling.

Ibe *et al.* evaluate the total upperbound charge collection by sea-level muons based on simulated muon spectrum and show the possible impact to be larger than neutrons when Q_{crit} decreases below 1 fC [73].

5.6 Thermal/Cold Neutron Test Methods

Irradiation facilities where thermal/epithermal neutrons are available are listed in Table 5.6 [74–80].

Thermal neutrons (0.025–1 eV) can be obtained in the experimental nuclear fission reactor where neutrons in the reactor core are thermally equilibrium (about 25 meV) with water coolant temperature [74–78]. Thermal and or epi-thermal (1 eV to 10 keV) neutrons can be obtained by attenuating high-energy neutrons through shielding materials [80].

Baumann *et al.* are the first to find ^{10}B in BPSG (Boron Phosphor Silicate Glass) film, which is used for smoothing the surface of semiconductor wafer for precise lithography and is responsible for thermal neutron induced soft error by the neutron capture reaction, $n + {}^{10}B \rightarrow \alpha + {}^{7}Li$ as explained in Section 3.6 [78]. They used cold neutrons equilibrium with 20 K environment at the NG0 beam line in CNRF (Cold Neutron Research Facility), NIST (National Institute of Standards and Technology) for SEU measurement of SRAMs and confirmed the impact of thermal neutrons by comparing the number of soft errors with and without BPSG layer. A BF_3 thermal neutron counter utilises the same capture reaction as the detection mechanism so that calibration is reportedly easy.

Cadmium and Boron (see Figure 3.11) have high capture cross sections and, therefore, they can be utilised as shielding materials for thermal neutrons. Cross-section data of cadmium to low-energy neutron is plotted in Figure 5.4 for reference.

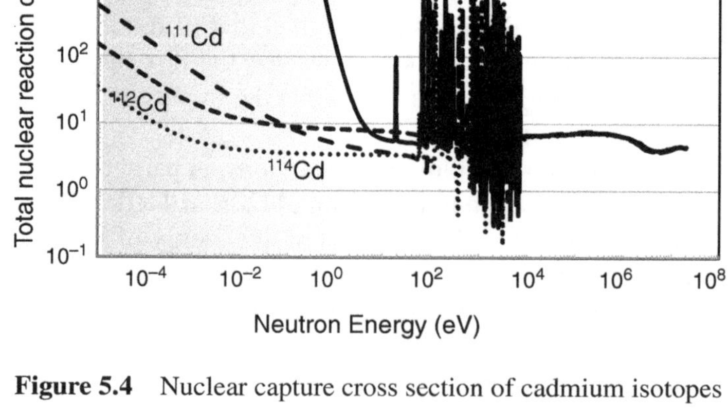

Figure 5.4 Nuclear capture cross section of cadmium isotopes

Table 5.6 List of facilities for thermal/epi-thermal neutron irradiation

Acronym	Facility name	Place	Source	Energy	Flux	Note	References
					$n/cm^2/s$		
MITR	Nuclear Reactor Laboratory	Boston, MA	Nuclear fission reactor	Thermal-epitherrmal	Epithermal 5×10^9; Thermal $<1 \times 10^{10}$	For neutron capture therapy (Boron Neutron Capture Therapy; BNCT)	http://mit.edu/nrl
Apsara	Apsara Research Reactor	Mumbai, India	Nuclear fission reactor	0.025 eV	1.5×10^4	–	http://www.nti.org/facilities/818/
KUR	Kyoto University Research Reactor Institute	Osaka, Japan	Heavy water nuclear reactor	Thermal, 24 keV	$3.9 - 28 \times 10^{12}$ 6.8×10^6(24keV)	–	http://www.rri.kyoto-u.ac.jp/en/facilities/kur, Kobayashi [12]
CNRF	Cold Neutron Research Facility (NIST)	Gaithersburg, MD	Heavy water nuclear reactor	Cold (20 K)	6.29×10^8	NG0 beam line was used	Baumann [78]
TRIUMF	Thermal Neutron Facility (TNF)	Vancouver, Canada	Cyclotron (500 MeV proton)	thermal	1×10^{12}	Epithermal beam is also available	Dyer [79]
J-PARC	Japan Proton Accelerator Research Complex	Ibaraki, Japan	Hg(p,n)	$10^{-4} - 0.1$eV (after attenuated)	$0.4–4.6 \times 10^8$	–	http://j-parc.jp/researcher/MatLife/en/instrumentation/ns3.html

Cadmium-113 is the most effective isotope to shield thermal neutrons among naturally abundant cadmium isotopes. Unlike Boron-10, Cadmium has some small resonance peaks in the energy range of about 100 eV to 10 keV. Sheets that contain such dense materials are commercially available as shielding material for thermal neutrons [81]. SEE field/accelerator tests with and without such sheets placed between the beam outlet and DUT can increase the effect of thermal neutrons [77].

Dyer *et al.* screened a wide energy range of neutrons and protons using IUCF/LENS (low-energy neutron source) (3–5 MeV neutron), TRIUMF (monoenergetic protons), NIST/NPL (National Physical Laboratory) for thermal neutrons and compared the range with high-energy neutron test data of SRAMs at LANSCE(10–800 MeV spallation neutrons), TSL(20–180 MeV quasi-monoenergetic neutrons, QMNs) and TRIUMF(10–400 MeV spallation neutrons) and concluded that some advanced SRAMs have high susceptibility to SEL (Single Event Latchup) and thermal neutron-induced soft error [79].

Zhang *et al.* evaluate thermal neutron induced soft error in 32 nm HKMG (High-k Metal Gate) SRAMs using two thermal neutron facilities [82]. They conclude that thermal neutrons do not cause MBU (Multi-Bit Upset)/SEFI/SEL of 32 nm HKMG SRAMs and pointed out the importance of the accurate measurement of flux, calculation of effective flux and the shielding effects of packaging materials.

5.7 High-Energy Neutron Test

5.7.1 Medium-Energy Neutron Source by Using Radioisotopes

Table 5.7 summarises facilities where medium-energy neutron beam is available when using radioisotopes or accelerators.

Medium-energy neutrons can be obtained by using radioisotopes such as ^{241}Am-^{9}Be [89, 90], ^{239}Pu-^{9}Be [91] and ^{252}Cf [92]. ^{241}Am-Be and ^{239}Pu-Be neutron sources containing alpha ray emitters ^{241}Am and ^{239}Pu and utilising nuclear reaction ^{9}Be(α,n)^{12}C, in principle. ^{252}Cf decays through spontaneous fission reaction to produce neutrons.

Typical neutron spectra from these neutron sources are summarised in Figure 5.5. ^{252}Cf has a single flux peak around 1 MeV, while ^{241}Am-Be and ^{239}Pu-Be sources have several peaks at similar energies around 2-10 MeV.

5.7.2 Monoenergetic Neutron Test

Monoenergetic low (1–20 MeV) energy neutrons can be produced by pT (proton-tritium), DD (deuterium-deuterium) and DT (deuterium-tritium) reactions. When an accelerated particle X collides with a target nucleus y, producing light particle ω and residual nucleus Z, this reaction is expressed in the following form,

$$X(y, \omega)Z. \tag{5.4}$$

Table 5.7 List of facilities for low-energy neutron irradiation

Acronym	Facility name	Place	Source	Energy	Flux (n/cm^2/s)	Note	References
FNL	Fast Neutron Laboratory, Tohoku Uniersity.	Sendai, Japan	T(p, n)^{3}He D(d, n)^{3}He T(d, n)^{4}He	1, 2, 5, 15 MeV	1.2–9 × 10^5 at 10 cm from the target centre	–	Nakamura [2] and Yahagi [83]
IUCF/LENS	Low Energy Neutron Source	Bloomington, IN	Be(p,n)B	3–5 MeV	–	200 MeV proton from the IU cyclotron is used.	Dyer [79]
CEA/DIF	Bruyères le châtel, France	Essonne, France	D(d, n)^{3}He	2.5, 4, 6 MeV	3 × 10^4	4 MV Van de Graaff	Baggio [84]
CEA/CVA	Valduc Centre for Nuclear Studies	Dijon, France	T(d, n)^{4}He	14 MeV	1 × 10^9 1 × 10^{11}(4π)	SAMES accelerator	Baggio [84] and Gasiot [85]
IUCF	Indiana Cyclotron Facility	Bloomington, IN	Pu-Be	1–11 MeV	–	–	Savage [86]
–	Hokkaido Univ. Japan	Sapporo, Japan	Pb(γ, n)	10–20 MeV	1 × 10^{12}(4π)	High-energy electrons from 45 MeV linac are bombarded to the Pb target	Makinaga [87]
NMIJ	National Metrology Institute of Japan	Ibaraki, Japan	^{9}Be(a, n)^{12}C^7 Li(p, n)^{7}Be	5 MeV, 8 MeV	10^{11}–10^{15}	Proton and alpha particle are from the pelletron accelerator	Harano [88]
			D(d, n)^{3}He	2.5, 14.8 MeV	10^{11}–10^{15}	Cockcroft-Walton accelerator	

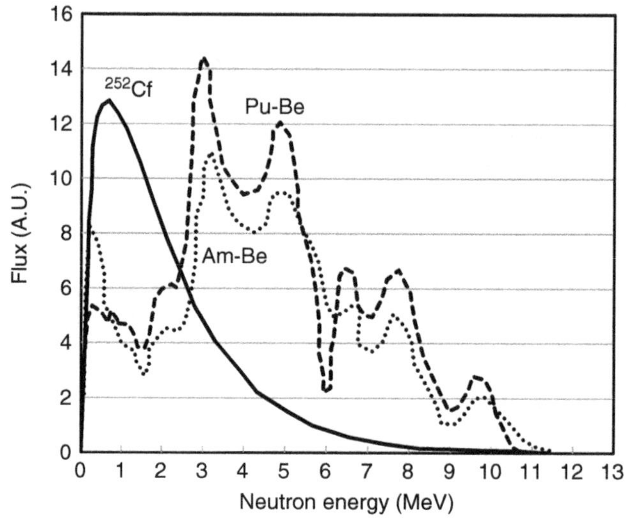

Figure 5.5 Neutron energy spectra from radioisotope neutron sources

The difference in static total mass $(m_X + m_y)c^2$ of reactants (X and y) and $(m_\omega + m_Z)c^2$ of products (ω and Z) is called Q-value and defined by,

$$Q = (m_X + m_y)c^2 - (m_\omega + m_Z)c^2. \tag{5.5}$$

When the Q value is positive, as in the case of Equations (5.7) and (5.8), the reaction is *exothermic* and only the kinetic energy of X to contact with the target is necessary to promote the reaction. When the Q value is negative as in the case of Equation (5.6) below, the reaction is *endothermic* and particle X must be accelerated to have kinetic energy greater than the Q value to trigger the reaction. There are some calculation tools for determining the Q values uploaded on the internet [93].

The pT, DD and DT reactions required to produce neutron can be expressed as follows in the case of the Fast Neutron Laboratory, Tohoku University.

$$\mathrm{T(p,n)^3He}\ (Q \text{ value} = -0.7638\ \mathrm{MeV},\ E_\mathrm{n} = 1 \text{ and } 2\ \mathrm{MeV}) \tag{5.6}$$

$$\mathrm{D(d,n)^3He}\ (3.2689\ \mathrm{MeV}, E_\mathrm{n} = 5\ \mathrm{MeV}) \tag{5.7}$$

$$\mathrm{T(d,n)^4He}\ (17.589\ \mathrm{MeV}, E_\mathrm{n} = 15\ \mathrm{MeV}) \tag{5.8}$$

Actual neutron spectra in FNL (Fast Neutron Laboratory) are monoenegetic and are shown in Figure 5.6.

Facilities where monoenergetic neutrons are available are listed in Table 5.7 along with their available locations, sources, energies and fluxes [2, 77, 83–88].

Yahagi *et al.* are the first to point out the importance of 1–5 MeV range neutrons to CMOS SRAM soft error by using FNL [83].

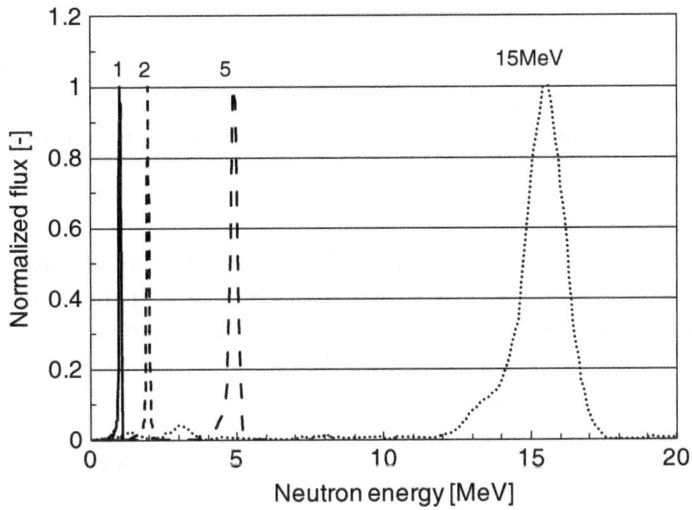

Figure 5.6 Neutron energy spectra of monoenergetic neutron beam facilities

Dyer *et al.* evaluate the sensitivity of CMOS SRAMs to soft error in a wide energy range (thermal to 800 MeV) of neutrons by using various facilities such as thermal (TRIUMF), 3–5 MeV (LENS), 20–80 MeV (TSL), 500 MeV (TRIUMF) and 800 MeV (LANSCE) as mentioned previously [79].

Gasiot *et al.* evaluate the sensitivity of 250 nm bulk and SOI SRAMs to 14 MeV neutrons at CEA (Commissariat a l'energie atomique et aux energies alternatives)/CVA (Corporate Value Associates) including angle dependency [85].

Baggio *et al.* evaluate SER in PD (Partially Depleted)/FD (Fully Depleted) SOI SRAMs for 1–10 MeV (2.5, 4.6 and 14 MeV) neutrons. It is demonstrated that body-tie reduces an SEU cross section by a factor of 100 and PD SOI has a relatively high sensitivity to low-energy neutrons [84].

Savage *et al.* used ^{239}Pu-Be neutron source at IUCF together with proton beam irradiation to evaluate Omni Wave 256 kb SRAM that is known to have very high sensitivity to soft error [86]. Angle dependency of neutron-induced soft error is reportedly observed and is attributed to knock-on collision of hydrogen (proton) contained in the device and direct ionisation by protons.

In the NMIJ (National Metrology Institute Japan), ^{9}Be(α,n)^{12}C reaction is utilised to produce 5 and 8 MeV monoenergetic neutrons in addition to ^{7}Li (p,n)^{7}Be and D(d,n) ^{3}He reactions [88].

Low-energy (10–20 MeV) neutrons can also be produced by using photonuclear reaction (γ,n).

The 45 MeV electron linac in Hokkaido University is one such facilities. The Pb target is bombarded with high-energy electrons at to produce high-energy gamma rays and hence low-energy neutrons [87]

5.7.3 Quasi-Monoenergetic Neutron Test

A QMN, whose spectrum is quasi-monoenergetic, is obtained by bombarding a thin target plate typically made of Lithium with high-energy proton. A simplified setup for the QMN facility is shown in Figure 5.7a. Neutron is produced through the following reaction,

$$p(^{7}Li,\ n)^{7}Be. \tag{5.9}$$

This reaction is a knock-on collision of an incident proton and a neutron in the target nucleus so that the kinetic energy of the neutron is almost the same as that of the impinging proton. Since the proton and target nucleus have a positive charge, the kinetic energy of proton and thus the kinetic energy of released neutron are reduced by the Coulomb barrier energy, E_c,

$$E_c = \frac{e^2}{4\pi\varepsilon_0} \times \frac{Z_I Z_T}{r_I + r_T} = 1.438 \times \frac{Z_I Z_T}{r_I + r_T} \quad (\text{MeV}), \tag{5.10}$$

where,

ε_0: dielectric constant of vacuum;
Z_I: atomic number of the incident particle;
Z_T: atomic number of the target nucleus;
r_I: radius of the incident particle (fm);
r_T: radius of the target nucleus (fm).

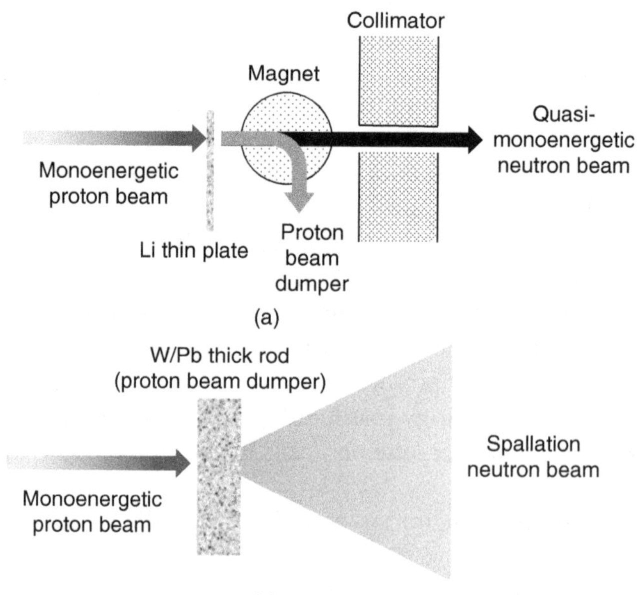

Figure 5.7 Simplified high-energy neutron beam sources. (a) Quasi-monoenergetic neutron source and (b) spallation neutron source

When the incident particle is proton ($Z_I = 1$ and $r_I = 1.2$ fm) and the target nucleus is Li ($Z_T = 3$ and $r_T = 1.55$ fm), Equation (5.10) gives $E_c = 1.57$ MeV. Thus, a neutron knocked out by a proton has a kinetic energy slightly lower than that of the incident proton.

The QMN test is applied where neutron beams with a flux peak at a specific neutron energy as exemplified in Figure 5.8 by the CYclotron and RadioIsotope Center (CYRIC), Tohokuy University and the QMN beam in TSL, Uppsala University Some of the neutron spectra have plateaus in the lower energy range, which is called as 'tail'. The SEU cross section σ_{seu} for the peak flux contribution is defined and obtained by

$$\sigma_{seu} = \frac{N_{event}^{peak}}{\Phi_{peak}} = \frac{R_{event}^{peak}}{\varphi_{peak}} = \frac{N_{event}^{total}}{\Phi_{total}} \times C_{peak} \quad (5.11)$$

where,

N_{event}^{peak}: number of events caused by neutrons in the peak flux area (errors);
Φ_{peak}: fluence in the peak flux area (n/cm^2);
R_{event}^{peak}: single event rate caused by neutrons in the peak flux area (errors/h);
ϕ_{peak}: flux in the peak flux area (n/cm^2/h);
N_{event}^{total}: number of events caused by total neutrons (errors);
Φ_{total}: total neutron fluence (n/cm^2);
C_{peak}: tail correction factor.

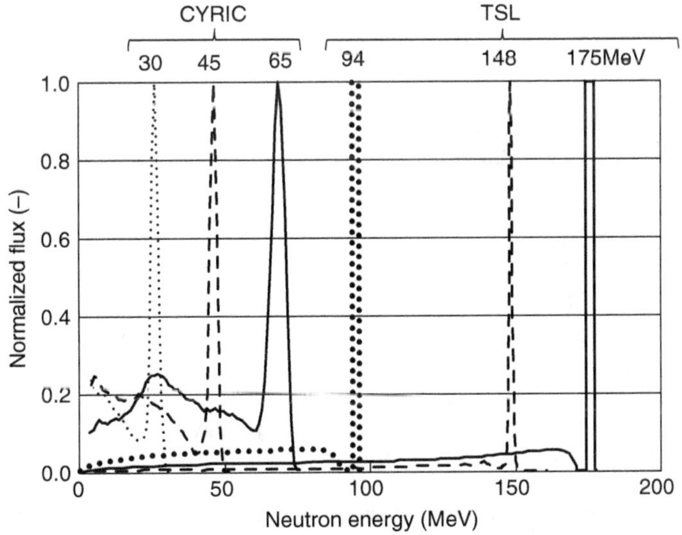

Figure 5.8 Neutron energy spectra of quasi-monoenergetic neutron beam facilities

The tail correction factor eliminates the contribution of the tail to the total number of events and can be obtained either by unfolding experimental data obtained for several (at least four) different energy peaks [53] or a soft-error simulator like CORIMS [15].

The SEU cross section $\sigma_{\text{seu}}(E_{\text{n}})$ is thus measured as a function of the neutron energy. The measured data are approximated by the Weibull Fit type excitation function $\sigma_{\text{seu}}(E_{\text{n}})$ as:

$$\sigma_{\text{seu}}(E_{\text{n}}) = \sigma_{\infty}\left[1 - \exp\left\{-\left(\frac{E_{\text{n}} - E_{\text{th}}}{W}\right)^{S}\right\}\right] \tag{5.12}$$

where,

σ_{∞}: saturation value of SEU cross section (cm^2);
E_{n}: neutron energy at the flux maximum (MeV);
E_{th}: threshold energy (MeV);
W: width factor (MeV);
S: shape factor (−).

A typical example of this type of excitation curve is described in Figure 5.9. The curve starts from E_{th} and increases gradually to the saturation value σ_{∞}.

The Weibull fit parameters σ_{∞}, E_{th} and S are intrinsic device/component/system-specific parameters and SER can be estimated for any neutron spectra including the terrestrial neutron spectrum.

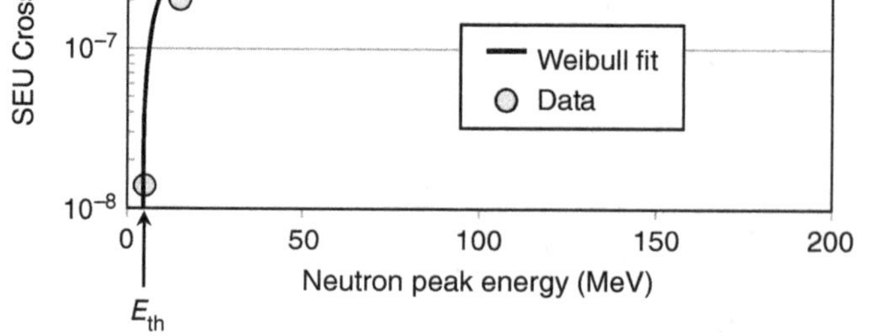

Figure 5.9 Conceptional illustration of cross section data obtained by (quasi-) monoernergetic neutron sources and fitting curve by Weibull fit

SER in any location on the Earth, for example can be obtained from integration of the Weibull Fit and differential flux over the energy range from E_{th}:

$$SER = 10^9 \times \int_{E_{\mathrm{th}}}^{\infty} \sigma_{\mathrm{seu}}(E_{\mathrm{n}}) \frac{\partial \phi(E_{\mathrm{n}})}{\partial E_{\mathrm{n}}} dE_{\mathrm{n}} \tag{5.13}$$

where,

SER: soft error rate (FIT, failure in time);
$\phi(E_{\mathrm{n}})$: neutron flux (n/cm^2/h).

Facilities where QMNs are available are listed in the upper part of Table 5.8 together with their locations, sources, energies and fluxes available [15, 94–101].

Johansson *et al.* are the first to apply the QMN beam to the SER test of memory devices and compares the test results with those of proton beam tests at TSL [53]. Substantial differences are observed between quasi-monoenergetic and proton beam tests with almost the same particle energies.

Ibe *et al.* established QMN test method as the standard test method of semiconductor memory devices by using FNL [83], CYRIC [15, 83, 94, 97] and TSL [94]. The QMN test method is accepted as the standard test method in JESD89A (JEDEC Standard) [18], IEC60749-38 [19] and EDR4705 (JEITA standard) [20].

Although high-energy proton is used for the SEE test as a substitute for high-energy neutrons [49–54], data obtained for the same SRAM device by irradiating neutrons and protons are not always consistent as exemplified in Figure 5.10 [53]. In many cases, devices show different sensitivities between protons and neutrons [53, 54].

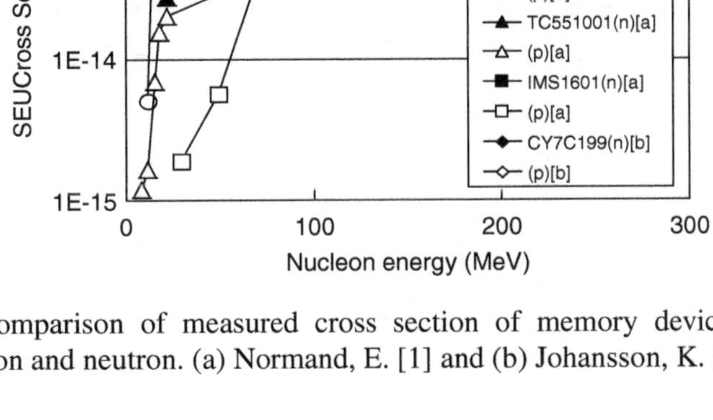

Figure 5.10 Comparison of measured cross section of memory devices irradiated by high-energy proton and neutron. (a) Normand, E. [1] and (b) Johansson, K. [53]

Table 5.8 List of facilities for high-energy neutron irradiation

Type	Acronym	Facility name	Place	Source	Energy (MeV)	Flux (n/cm²/s)	Note	References
Quasi-Mono-energetic	QMN/TSL	Quasi-Monoenergetic Neutrons/The Svedberg Lab.	Uppsala, Sweden	p → Li	21, 47, 96, 176 22, 47, 95, 144 21, 46, 88	3×10^5 –	Variable-size collimators are implemented. Beam size 1–80 cm on-line monitor of neutron flux is possible (TFBC[a])	Yahagi [83] and Ibe [94] Radaelli [95] Granlund [54]
	CYRIC	Cyclotron and RadioIsotope Center, Tohoku University	Sendai, Japan	p → Li	35, 45, 65 65	3×10^5 at 65 MeV	Beam size $10 \times 10 cm^2$	Ibe [97] and Yahagi [83] Shimbo [98] and Uezono [100]
	CYCLONE	Cyclotron of Louvain la Neuve	Louvain la Neuve	p → Li	38, 63	8×10^4	–	http://www.cyc.ucl.ac.be/Papers/hif/radecs97.pdf Berger [101]
				d → Be	20	3×10^{10} (at 14 cm from the target)		
Spallation	LANSCE	Los Alamos Neutron Science Center	Los Alamos, New Mexico	p → W	0.1–800	3×10^5 (>1MeV)	Beam size:8, 5, 2.5 cm	Eto [102] and Wender [103], http://lansce.lanl.gov/media/accelneutrontesting.pdf Dodd [104] and Gasiot [105]
	TRIUMF	Tri-University Meson Facility	Vancouver, Canada	p → Pb	1–450	3×10^6 (>10MeV)	Beam size: $5 \times 12 cm^2$	Blackmore [106]
	ANITA/TSL	Atmospheric-like Neutron from thIck Target/The Svedberg Lab.	Uppsala, Sweden	p → W	1–150	$>10^6$	Same as QMN/TSL	Prokofiev [107], Hubert [10] and Uezono [100]
	ISIS	Rutherford Appleton Laboratory	Oxfordshire, UK	p → W	0.1–800	10^5–10^7 (collimated) 10^3–10^5 (broadened)	Thermal neutron is also available collimated to $11 \times 16 cm^2$ broadened to $1–2 m^2$	Rech [108] and Frost [109]
	RCNP	Research Center for Nuclear Physics, Osaka University	Osaka, Japan	p → W	1–400	5×10^5 (> 10MeV)	Beam size: 10.8 cm in diameter	Nakauchi [110], Furuta [111] and Uemura [112]
	MLF/J-PARC	Materials and Life Science Facility	Ibaraki, Japan	p → Hg	1–400	–	–	Matsukawa [113]

[a]TFBC, thin film breakdown counter.

Baggio *et al.* obtain SEU cross section data in SRAMs (800 to 250 nm) as a function of nucleon energy by using 14 MeV neutrons and proton beams (50–500 MeV) at TRIUMF [50] and compare the SER in the field estimated from the energy dependent data and direct estimation data from the LANSCE test.

Granlund and Olsson carry 250 to 130 nm SRAM SEU tests using a PAULA proton beam and compare neutron SEU test results at QMN and LANSCE [54]. They observed that SEU cross sections by proton irradiation had much higher values by about ×1.9 than by neutrons at an energy of 46 MeV regardless of SRAM parts.

McMarr *et al.* carried out heavy ion, proton, 14 MeV neutron beam tests on 150 nm CMOS/SOI 64 kb SRAMs, essentially the equivalent number of data flips are reportedly observed between neutron and proton tests [40].

Yamamoto *et al.* utilised the Research Center for Nuclear Physics (RCNP) as neutron QMN source to measure neutron-induced SEU cross section of 180, 150, 130 nm SRAMs and FFs [99]. They fabricated a matrix of FFs similar to an SRAM matrix and demonstrated that the SEU cross sections of FFs were already comparable to those of SRAMs.

Ibe *et al.* are the first to find the bipolar-mode soft error, MCBI (Multi-Coupled Bipolar Interaction) in 130 nm CMOS SRAM by using the QMN method (see Section 5.7) [94]. They are also the first to demonstrate that soft error in memory devices can be basically overwhelmed by applying both interleaving techniques and ECC even for the aggressive MCBI, thanks to their topological natures: MCUs spread along with bit line (well) and only 1 or 2 bit width along with word line (across well). A drastic decrease in MCUs is reported near p-well taps but this is not true for SBU and Different SEU mechanisms are implicated between SBUs and MCUs. This observation is supported by Gasiot *et al.* [105] and Nakauchi *et al.* [110] for 65 nm CMOS SRAM.

Radaelli *et al.* demonstrated high MCU ratio in 150 nm SRAMs by using TSL with the peak energies 22, 48, 97 and 144 MeV [95]. The MCU ratio increases with the peak energy increases. The same observation of topological nature in MCUs as in [94] is reported. MCUs reportedly decrease only in the bits adjacent to p-well taps.

High-energy neutrons as well as protons are often utilised for system-level irradiation tests.

Shimbo *et al.* apply a partial neutron irradiation test of 90 nm router boards at CYRIC [98].

FPGA, Central Processing Unit (CPU) and SRAM chips are irradiated separately and the SRAM chip is identified as the most vulnerable part. FPGAs are also known to be significantly vulnerable to neutron-induced soft error, but the number of FPGAs in the BUT is small and thus any specific countermeasure is not applied. A part of SRAMs that do not need speed are replaced by DRAMs with ECC and total failure rate in the board turned out to be reduced by a factor of 9–10 both in the real-time tests and accelerated tests at CYRIC.

Uezono *et al.* evaluate failure behaviour of ECUs (Electronic Control Units) in automobiles with a variety of workloads using high-energy neutron irradiation tests at CYRIC (64 MeV QMN beam) and ANITA (Atmospheric-like Neutrons from thIck TArget) (Spallation neutron source) [100]. Consistent results are obtained in test results at both facilities.

5.7.4 Spallation Neutron Test

By contrast, SER at any location on the Earth can be obtained by using a different type of neutron source, spallation neutrons. A simplified setup for spallation neutron facility is also shown in Figure 5.7b. Similar to the quasi-monoenergetic source, a spallation neutron source utilises a high-energy monoenergetic proton beam to bombard a thick target typically made of tungsten [102–104, 106, 107, 114] or lead. The thick targets have a function as a proton beam dumper. The spallation neutron spectra are rather broad (less than proton energy) and are said to mimic the terrestrial neutron spectrum shown in Figure 5.11.

Since the spallation reaction of proton and target takes place only with high-energy (more than a few megaelectron volts) protons, the neutron spectrum does not have high flux in the lower energy, unlike the terrestrial neutron spectrum.

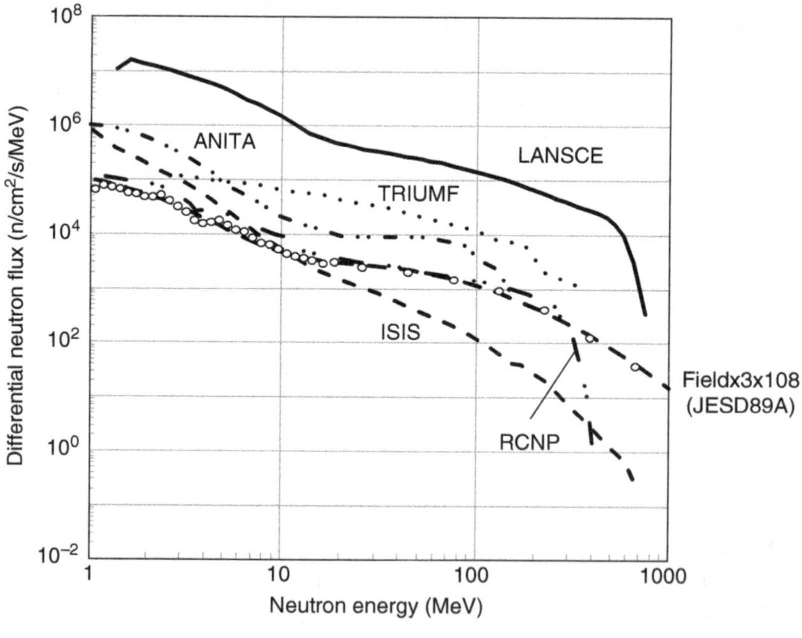

Figure 5.11 Neutron energy spectra of spallation neutron sources and terrestrial field

SER at sea level in NYC can be obtained using the following procedures:

First, define the neutron energy range E_{min} and E_{max} of the accelerator. In JESD89A, $E_{min} = 10\,\text{MeV}$ and E_{max} = maximum energy of the spallation neutron source.
Second, obtain the effective SEU cross section based on the test results:

$$\sigma_{seu}^{eff} = \frac{N_{err}}{\displaystyle\int_{E_{min}}^{E_{max}} \frac{\partial\varphi}{\partial E_n} dE_n} \tag{5.14}$$

where,

N_{err}: number of errors in the OUT for total neutron irradiation;
φ: fluence for neutron energy range between E_n and $E_n + dE_n$.

Finally, estimate RTSER (Real-Time SER) from:

$$RTSER = \sigma_{SEU}^{eff} \times \phi(E_{min}, E_{max}) \tag{5.15}$$

where, $\phi(E_{min}, E_{max})$: Flux of neutrons with an energy range between E_{min} and E_{max} at sea level in NYC. $E_{min} = 10\,\text{MeV}$ and $\phi(E_{min}, E_{max}) = 12.9\ \text{n/cm}^2\text{/h}$ are recommended in JESD89A [22].

The number of spallation neutron sources has continued to increase in the twenty-first century. Initially, the only spallation neutron sources available in the twentieth century were LANSCE [102–104, 114] and TRIUMF [106] are. In 2000–2010, ANITA [10, 107], ISIS [108, 109, 115], RCNP [99, 111, 112, 116, 117] and the J-PARC (Japan Proton Accelerator Research Complex) [113] were constructed as spallation neutron sources.

The properties of the spallation neutron sources are summarised in the lower part of Table 5.8.

Hubert *et al.* utilise ANITA to demonstrate the unique mitigation techniques of SEUs in CAMs (Content Addressable Memories) [10].

Rech *et al.* demonstrated the high susceptibility of GPUs (Graphic Processing Units) which are used in not only graphic media but also high-end computing and safety critical graphic processing in the motor industry [108]. They used ISIS for irradiation tests.

Nakauchi *et al.* evaluated the parasitic bipolar-mode soft error as introduced in [94] and [97] by using RCNP (Research Center of Nuclear Physics) and back-bias they demonstrated the mitigation technique of MCUs [110].

Furuta *et al.* evaluate the bipolar-mode soft error in 65 nm FFs by using RCNP [111]. Dependency of MCU on the tap location seems different from that of SRAMs: the MCU rate distant from tap location is high in FFs.

Uemura *et al.* also use RCNP to investigate the mitigation techniques of FFs, in particular, on MNT (Multi Node Transient)s in redundancy nodes in DICE-type hardened FFs [112].

Nishida *et al.* used LANSCE and RCNP to evaluate SEB in IGBTs (Insulated Gate Bipolar Transistors) for the motor industry [116, 117].

Carrying out neutron SER tests by using the spallation neutron sources is relatively easy, compared to the monoenergetic or QMN tests, because they do not need to test multiple energy bins. Nonetheless, the spallation neutron tests have their own setbacks:

1. The shape is not exactly the same as the terrestrial neutron spectrum as shown in Figure 5.11. The estimated SER is inevitably facility-unique [118].
2. The minimum energy E_{min} has been fixed at 10 MeV. This is acceptable only if the contribution to SEEs by neutrons with an energy rating below 10 MeV is negligibly small. Even when JESD89A was established in 2006, it was pointed out that the contribution of low-energy neutrons is substantial [83]. As scaling proceeds to levels smaller than 90 nm, there is evidence to show that low-energy (<10 MeV) protons have a much higher contribution than high-energy protons because of the decrease in critical charge Q_{crit}. As indicated in Figure 3.13, due to nuclear spallation reaction of Si in the field most portions of secondary ions consist of proton so that the contribution of low-energy neutron must be higher [119] than before and therefore E_{min} should be determined in a more reasonable manner.

5.7.5 Attenuation of Neutron Flux and Energy

In neutron SEE tests, two or more BUTs are commonly aligned along with the beam centre to save test time. Care must be paid, however, to the attenuation effects of neutron flux and energy by the BUTs themselves. Quantitative evaluation of such attenuation effects is more difficult than charged particle irradiation due to the complexity of board structure. High-energy neutrons do not attenuate greatly, but low energy neutrons do. Figure 5.12 shows the attenuation effects of high-energy neutrons by using a polyethylene block measured at LANSCE [102]. It is seen that a 1–10 MeV neutron flux is attenuated significantly by a 5–20 cm thick polyethylene block. Figure 5.13 shows the attenuation ratio of a polyethylene block calculated from the data in Figure 5.12. While a 1 MeV neutron decays by a factor of around 1000 with a 20 cm thick polyethylene block, neutrons with energy of more than 100 MeV stay almost unaffected.

A neutron beam is said to decay by around 4% [120] and can get as high as 22% [121, 122] in one BUT depending on neutron energy. It is worth noting that the energy spectrum changes simultaneously under different attenuation mechanisms. When several BUTs are aligned in the neutron beam, the low-energy part of the

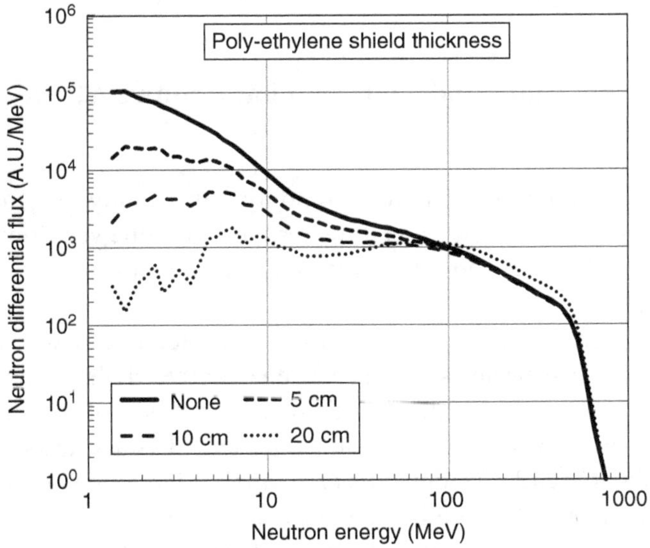

Figure 5.12 Spectrum changes by poly-ethylene shielding at LANCE

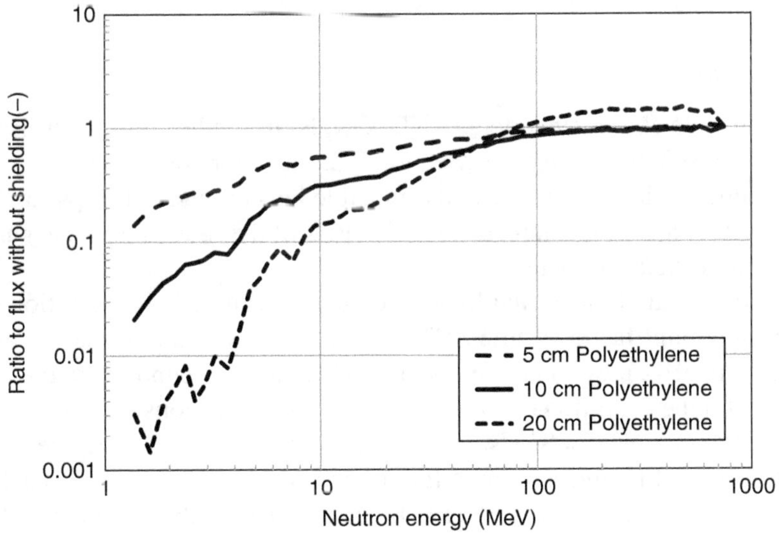

Figure 5.13 Shielding efficiency of poly-ethylene block as a function of neutron energy estimated from data in Figure 5.12

neutron spectrum may be significantly reduced compared to the front BUT, and SEU characteristics may significantly vary with the location of the BUT in the alignment.

5.8 Testing Conditions and Matters That Require Attention

5.8.1 Memories

In a field test for memories, data are commonly written only once at first, and read only once at the end of the test. Even if the test period is one year long, the frequency of the events is so low that more than two soft errors in a single bit is very rare and can be ignored.

When soft errors in memories are tested with an accelerator, a fixed '1/0' data pattern (solid/checker board/stripe and so on) is written over the whole memory matrix over a certain period of time during which less than one event is expected, and data are read at least once in each cycle. These methods are called 'static tests'. The read/write pattern may affect the SER. Tsiligiannis *et al.* apply Marching algorithms ('Dynamic test' (see Appendix A.2)) to see if the read/write pattern affects SER measurement. The stress impact of the read/write dynamic pattern turned out to accelerate SER by a maximum factor of three [123].

Temperature and operation voltage also affect SER in memories. Lowering voltage increases SER in the charge collection mode by reducing Q_{crit} but reduces SER in the bipolar mode, in general.

5.8.2 Circuits

At gate level, gate chains (inverters, AND, OR, NOR, and so on) are utilised to measure gate level SET sensitivity (see Section 7.2.1 for more details).

For sequential logic gate, scan-chains embedded in commercial chips can be utilised to measure SEU rate. A specially designed matrix of the sequential logic gate can also be used for SEU measurement.

In more complicated functional logic circuits including CPUs, operation frequency can be an important factor. While SETs take place anywhere in the circuit as illustrated in Figure 5.14, they can be SEUs only when they are captured to the FF directly downstream on the condition that the gate of the FF opens consistently with the clock timing. The total number of SEUs in the circuit is equal to the sum of the SEUs due to the direct hit of the FFs and captured SETs into the FFs and are generated in the combinational block in the upstream of the FF. Basically, the number of captured SEUs is in proportion to both the operational (clock) frequency and the intrinsic (raw) SET rate. As illustrated in Figure 5.15a, when the frequency has a fixed window during which a the SET pulse is captured in FFs and the intrinsic SET rate is high, the probability of SET capture from the input of the FFs increases and the total SEU increases at a

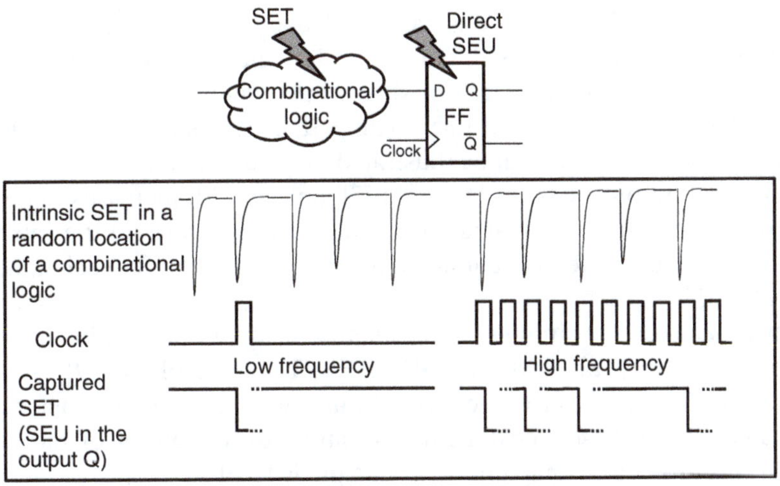

Figure 5.14 Two schemes of SEUs in flip-flop: capture of SET from combinational logic circuit and direct hit at the storage node in FF. Conceptual illustration of clock frequency effects on the capture rate of SET into FF: capture rate is in proportion to clock frequency

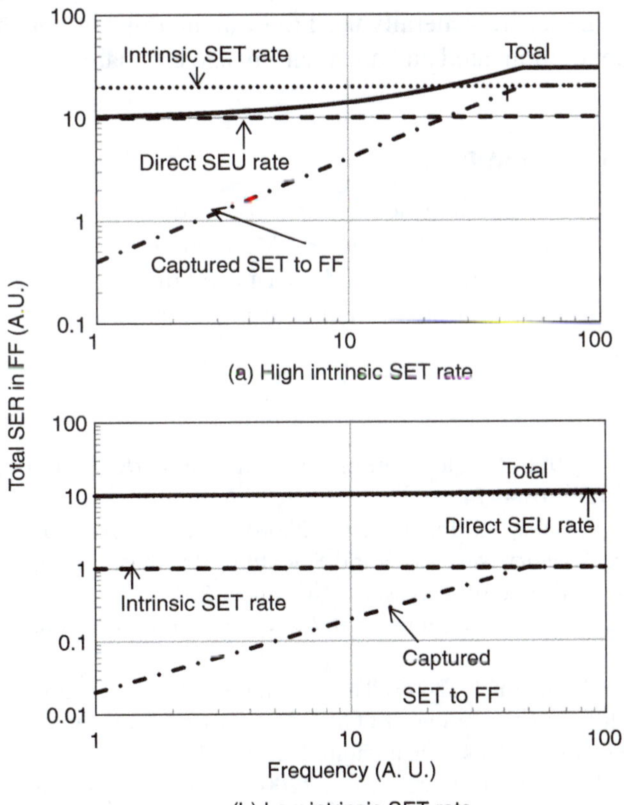

Figure 5.15 Dependency of total SEUs in a flip-flop on clock frequency. (a) High intrinsic SET rate in combinational logic and (b) Low intrinsic SET rate in combinational logic

certain frequency [124, 125]. In this case, a plateau should appear that corresponds to the intrinsic SET rate. Main SEUs are caused by the sensitive nodes in FFs receiving a direct hit of charged particles on. Meanwhile, when the frequency is low or the intrinsic SET rate in the combinational logic parts is low, such a dependency of SER on the frequency may not appear as illustrated in Figure 5.15b.

As device scaling proceeds further and SET pulse width increases to cover two or more clock pulses, the SER plateau level may increase beyond the intrinsic level. Inoue *et al.* propose a novel detection/and recovery algorithm for such multiple-cycle transients [126].

Care must be taken for circuit level irradiation testings for the global control lines including clock line [127, 128] and SET/RESET line to FFs [129]. SET on these global control lines can cause unexpected input to FFs or direct data flips in FFs. SETs-created PLL (Phase Locked Loop)s can cause distortion of the clock pulse [130, 131] resulting in mis-capture of input signals to FFs.

In real cases, a variety of applications or instructions are carried out on the same circuits. Variations in failures and their mode caused by such variations in applications or instructions have to be evaluated. A certain series of workloads like ISCAS' 85 (International Symposium on Circuits and Systems) [132, 133] or SPEC CINT2000 [134, 135] benchmarks are generally used to evaluate this kind of variation, but such workloads are not complicated and are small in size compared to other systems.

5.9 Summary of Chapter 5

Irradiation test methods and facilities for terrestrial radiation, such as fields, heavy ion, muon, proton, alpha ray, thermal to low (<1 MeV) energy neutrons, and high-energy neutrons, are reviewed and their merits, demerits and limitations are discussed. Points that must be covered by the irradiation tests are summarised.

References

[1] Normand, E. (1996) Single-event effects in avionics. *IEEE Transactions on Nuclear Science*, **43** (2), 461–474.

[2] Nakamura, T., Baba, M., Ibe, E. *et al.* (2008) *Terrestrial Neutron-Induced Soft-Errors in Advanced Memory Devices*, World Scientific, Hackensack, NJ.

[3] http://spaceweather.com/ (accessed 7 November 2013).

[4] NOAA Today's Space Weather, http://www.swpc.noaa.gov/today.html (accessed 23 May 2013)

[5] Lesea, A. and Fabura, J. (2006) The Rosetta experiment: atmospheric soft error rate testing in differing technology FPGAs. The Second Workshop on System Effects of Logic Soft Errors, Urbana-Champaign, IL, 11–12 April 2006.

[6] Tosaka, Y., Takasu, R., Uemura, T. *et al.* (2008) Simultaneous measurement of soft error rate of 90 nm CMOS SRAM and cosmic ray neutron spectra at the summit of Mauna

Kea. IEEE International Reliability Physics Symposium, Anaheim, CA, 15–19 April (SE01), pp. 727–728.

[7] Ibe, E., Yahagi, Y., Kataoka, F. *et al.* (2002) A self-consistent integrated system for terrestrial-neutron induced single event upset of semiconductor devices at the ground. 2002 International Conference on Information Technology and Application, Bathurst, Australia, 25–28 November, 2002, pp. 273–221.

[8] Autran, J.L., Serre, S., Munteanu, D. *et al.* (2012) Real-time soft-error testing of 40nm SRAMs. IEEE International Reliability Physics Symposium, Anaheim, CA, 15–19 April (3C-5).

[9] Torok, Z. and Platt, S.P. (2007) SEE-inducing effects of cosmic rays at the high-altitude research station jungfraujoch compared to accelerated test data. 9th European Conference Radiation and Its Effects on Components and System, Deauville, France, 10–14 September.

[10] Hubert, G., Velazco, R., Federico, C. *et al.* (2013) Continuous high-altitude measurements of cosmic ray neutrons and SEU/MCU at various locations:correlation and analyses based-On MUSCA SEP. *IEEE Transactions on Nuclear Science*, **60** (4), 2418–2426.

[11] Alexandrescu, D., Lhomme-Perrot, A., Schaefer, E. and Beltrando, C. (2009) Highs and lows of radiation testing. 15th IEEE International On-Line Testing Symposium, Sesimbra-Lisbon, Portugal, 24–26 June (S3.1).

[12] Kobayashi, H., Shiraishi, K., Tsuchiya, H. *et al.* (2002) Soft errors in SRAM devices induced by high energy neutrons, thermal neutrons and alpha particles. International Electron Devices Meeting, San Francisco, CA, 9–11 December, 2002, pp. 337–340.

[13] Kameyama, H., Yahagi, Y. and Ibe, E. (2007) A quantitative analysis of neutron-induced multi-cell upset in deep sub micron SRAMs and of the impact due to anomalous noise. 45th Internationall Reliability Physics Symposium, Phoenix, AZ, 15–19 April 2007, pp. 678–679.

[14] Flux Calculation http://www.seutest.com/cgi-bin/FluxCalculator.cgi (accessed 25 April 2013)

[15] Ibe, E., Yahagi, Y., Kataoka, F. *et al.* (2002) A self-consistent integrated system for terrestrial-neutron induced sungle event upset of semiconductor devices at the ground. 2002 International Conference on Information Technology and Application, Bathurst, Australia, 25–28 November 2002, pp. 273–321.

[16] Roche, P., Gasiot, G., Forbes, K. *et al.* (2003) Comparisons of soft error rate for srams in commercial SOI and bulk below the 130-nm technology node. *IEEE Transactions on Nuclear Science*, **50** (6), 2046–2054.

[17] Gasiot, G., Giot, D. and Roche, P. (2006) Alpha-induced multiple cell upsets in standard and radiation hardened SRAMs manufactured in a 65 nm CMOS technology. *IEEE Transactions on Nuclear Science*, **53** (6), 3479–3486.

[18] JEDEC JESD89A. (2006) *Measurement and Reporting of Alpha Particle and Terrestrial Cosmic Ray Induced Soft Errors in Semicondyctor Devices*, JEDEC, pp. 1–94.

[19] IEC IEC60749-38. (2008) *IEC60749 Part 38: Soft Error Test Method for Semiconductor Devices with Memory. Semiconductor Devices. Mechanical and Climatic Test Methods (Edition 1.0)*, IEC, pp. 1–9.

[20] JEITA (2005) JEITA SER Testing Guideline. EIAJ EDR-4705, pp. 1–62.

[21] Tang, H.H.K., Murray, C.E., Fiorenza, G. and Rodbell, K.P. (2009) Modeling of alpha-induced single event upsets for 45 nm node SOI devices using realistic C4 And 3D BEOL geometries. *IEEE Transactions on Nuclear Science*, **56** (6), 3093–3097.

[22] KleinOsowski, A.J., Cannon, E.H., Gordon, M.S. *et al.* (2007) Latch design techniques for mitigating single event upsets in 65 nm SOI device technology. *IEEE Transactions on Nuclear Science*, **54** (6), 2021–2027.

[23] Rodbell, K.P., Heidel, D.F., Gordon, M.S. *et al.* (2011) 32 and 45 nm Radiation Hardened by Design (RHBD) SOI latches. *IEEE Transactions on Nuclear Science*, **58** (6), 2702–2710.

[24] Swift, G. (2011) An improved approach to characterizing alpha – induced upset susceptibility. 2011 IEEE Workshop on Silicon Errors in Logic – System Effects, Champaign, IL, 29–30 March.

[25] Radiation Effects Facility http://cyclotron.tamu.edu/ref/beams.php (accessed 7 November 2013)

[26] Takasu, R., Tosaka, Y., Fukuda, H. and Kataoka, Y. (2005) A novel method for accurately estimating alpha-induced soft error rates. 2005 IEEE International Reliability Physics Symposium Proceedings, 17–21 April, San Jose, CA, 17–21 April, 2005, pp. 230–233.

[27] SRIM Particle Interactions with Matteer, http://www.srim.org (acceced 17 April 2013)

[28] Wakasa, T., Hirao, T., Sanami, T. *et al.* (2004) Measurement and analysis of single event transient current induced in si devices by quasi-monoenergetic neutrons. The 6th International Workshop on Radiation Effects on Semiconductor Devices for Space Application, Tsukuba, Japan, 6–8 October 2004, pp. 213–216.

[29] Gerardin, S., Bagatin, M., Paccagnella, A. *et al.* (2011) Single event effects in 90-nm phase change memories. *IEEE Transactions on Nuclear Science*, **58** (6), 2755–2760.

[30] Sierawski, B.D., Reed, R.A., Schrimpf, R.D. *et al.* (2009) Impact of low-energy proton induced upsets on test methods and rate predictions. *IEEE Transactions on Nuclear Science*, **56** (6), 3085–3092.

[31] Irom, F. and Nguyen, D.N. (2007) Single event effect characterization of high density commercial NAND and NOR nonvolatile flash memories. *IEEE Transactions on Nuclear Science*, **54** (6), 2547–2553.

[32] Dodd, P.E., Shaneyfelt, M.R., Draper, B.L. *et al.* (2009) Development of a radiation hardened lateral power MOSFET for POL applications. *IEEE Transactions on Nuclear Science*, **56** (6), 3456–3462.

[33] Sonia, G., Mai, F., Brunner, A. *et al.* (2006) Proton and heavy ion irradiation effects onAlGaN/GaN HFET devices. *IEEE Transactions on Nuclear Science*, **53** (6), 3661–3666.

[34] Oldham, T.R., Ladbury, R.L., Friendlich, M. *et al.* (2006) SEE and TID characterization of an advanced commercial 2Gbit NAND flash nonvolatile memory. *IEEE Transactions on Nuclear Science*, **53** (6), 3217–3222.

[35] Gerardin, S., Cester, A., Tazzoli, A. *et al.* (2007) Electrostatic discharge effects in irradiated fully depleted SOI MOSFETs with ultra-thin gate oxide. *IEEE Transactions on Nuclear Science*, **54** (6), 2204–2209.

[36] Larcher, L., Cellerle, G., Paccagnella, A. *et al.* (2003) Data retention after heavy ion exposure of floating gate memories:analysia and simulation. *IEEE Transactions on Nuclear Science*, **50** (6), 2176–2183.

[37] Cellere, G., Paccagnella, A., Visconti, A. and Bonanomi, M. (2006) Variability in FG memories performance after irradiation. *IEEE Transactions on Nuclear Science*, **53** (6), 3349–3355.

[38] Chatterjee, I., Mahatme, N., Bhuva, B. *et al.* (2011) Single-event charge collection and upset in 40 nm dual- andtriple-well bulk CMOS SRAMs. *IEEE Transactions on Nuclear Science*, **58** (6), 2761–2767.

[39] Dodd, P.E., Shaneyfelt, M.R., Draper, B.L., Young, R.W., Savington, D., Witcher, J.B., Vizkelethy, G., Schwank, J.R., Shen, Z.J., Shea, P., Landowski, M., Dalton, S.M. (2009) Development of a Radiation Hardened Lateral Power MOSFET for POL Applications. *Trans. Nucl. Sci.*, **56** (6), 3456–3462.

[40] McMarr, P., Nelso, M.E., Liu, S.T. *et al.* (2005) Asymmetric SEU in SOI SRAM. *IEEE Transactions on Nuclear Science*, **52** (6), 2481–2486.

[41] Cellere, G., Paccagnella, A., Visconti, A. *et al.* (2007) Traces of errors due to single ion in floating gate memories. International Conference on IC Design and Technology, Austin, TX, 18–20 May.

[42] Liu, S.T., Heikkila, W.W., Golke, K.W. *et al.* (2003) Single event effects in PDSOI 4 M SRAM fabricated in UNIBOND. *IEEE Transactions on Nuclear Science*, **50** (6), 2095–2100.

[43] Ferlet Cavrois, V., Pouget, V., McMorrow, D. *et al.* (2008) Investigation of the Propagation Induced PulseBroadening (PIPB) effect on single event transients in SOI and bulk inverter chains. *IEEE Transactions on Nuclear Science*, **55** (6), 2842–2853.

[44] Cellere, G., Paccagnella, A., Visconti, A. *et al.* (2008) Direct evidence of secondary recoiled nuclei from high energy protons. *IEEE Transactions on Nuclear Science*, **55** (6), 2904–2913.

[45] Gerardin, S., Bagatin, M., Paccagnella, A. *et al.* (2011) Proton-induced Upsets in 41-nm NAND floating gate cells. The Conference on Radiation Effects on Components and Systems, Sevilla, Spain, 19–23 September (C-2).

[46] Dyer, C.S., Clucas, S.N., Sanderson, C. *et al.* (2004) An experimental study of single-event effects induced in commercial srams by neutrons and protons from thermal energies to 500 MeV. *IEEE Transactions on Nuclear Science*, **51** (5), 2817–2824.

[47] Truscott, P., Dyer, C., Frydland, A. *et al.* (2005) Neutron energy-deposition spectra and seu measurements and comparisons with predictions. 2005 Radiation and Its Effects on Components and Systems, 19–23 September, Palais des Congres, Cap d'Agde, France (LN-11).

[48] Shaneyfelt, M.R., Felix, J.A., Dodd, P.E. *et al.* (2008) Enhanced proton and neutron induced degradationand its impact on hardness assurance testing. *IEEE Transactions on Nuclear Science*, **55** (6), 3096–3105.

[49] Hiemstra, D.H., Baril, A. and Dettwiler, M. (1999) Single event upset characterization of the pentium MMX and pentium II microprocessors using proton irradiation. *IEEE Transactions on Nuclear Science*, **46** (6), 1453–1460.

[50] Baggio, J., Ferlet-Cavrois, V. and Flament, O. (2004) Analysis of proton/neutron SEU sensitivity of commercial SRAMs-application to the terrestrial environment test method. *IEEE Transactions on Nuclear Science*, **51** (6), 3420–3426.

[51] Schwank, J.R., Dodd, P.E., Shaneyfelt, M.R. *et al.* (2004) Issues for single-event proton testing of SRAMs. *IEEE Transactions on Nuclear Science*, **51** (6), 3692–3700.

[52] Rufenacht, H., Le, K.A., Gazdewich, J. *et al.* (2005) Single event upset characterization of the ESP603 single board spacecomputer with the power PC603r processor using proton irradiation. Radiation Effects Data workshop, Seattle, Washington, 11–15 July 2005 (W-11), pp. 65–69.

[53] Johansson, K., Dyreklev, P., Granbom, B. *et al.* (1998) Energy-resolved neutron SEU measurements from 22 to 160 MeV. *IEEE Transactions on Nuclear Science*, **45** (6), 2519–2526.

[54] Granlund, T. and Olsson, N. (2005) A comparative study between proton and neutron induced SEUs in SRAMs. *IEEE Transactions on Nuclear Science*, **53** (4), 1871–1875.

[55] Lawrence, R.K., Ross, J.F., Haddad, N. *et al.* (2009) Soft error sensitivities in 90 nm bulk CMOS SRAMs. 2009 IEEE Radiation Effects Data Workshop, Quebac, Canada, 20–24 July, pp. 123–126.

[56] Quinn, H., Morgan, K., Graham, P. *et al.* (2007) Static proton and heavy ion testing of the Xilinx Virtex-5 device. Radiation Effects Data Workshop, Honolulu, HI, 23–27 July (W-31), pp. 177–184.

[57] Gadlage, M.J., Turflinger, T., Eaton, P.H. and Benedetto, J.M. (2005) Comparison of heavy ion and proton induced combinatorial and sequential logic error rates in a deep submicron process. *IEEE Transactions on Nuclear Science*, **52** (6), 2120–2124.

[58] Seifert, N., Gill, B., Pellish, J.A. *et al.* (2011) The susceptibility of 45 and 32 nm bulk CMOS latches to low-energy protons. *IEEE Transactions on Nuclear Science*, **58** (6), 2711–2718.

[59] Skarin, D., Karlsson, J. (2008) Software Implemented Detection and Recovery of Soft Errors in Brake-by-Wire System, IEEE Seventh European Dependable Computing Conference, Kaunas, Lithuania, May 7–9, 2008, 145–154.

[60] Puchner, H., Tausch, J. and Koga, R. (2011) Proton-induced single event upsets in 90 nm technology high performance sram memories. Radiation Effects Data Workshop, Seattle, Washington, 11–15 July 2005 (W-26), pp. 1–3

[61] Zanata, M., Wrachien, N. and Cester, A. (2008) Ionizing radiation effect on ferroelectric nonvolatile memories and its dependence on the irradiation temperature. *IEEE Transactions on Nuclear Science*, **55** (6), 3237–3245.

[62] Bagatin, M., Gerardin, S., Cellere, G. *et al.* (2009) Error instability in floating gate flash memories exposed to TID. *IEEE Transactions on Nuclear Science*, **56** (6), 3267–3273.

[63] Rodbell, K.P., Gordon, M.S., Heidel, D.F. *et al.* (2007) Low energy proton induced single event upsets in 65 nm silicon-on-insulator latches and memory cells. *IEEE Transactions on Nuclear Science*, **54** (6), 2474–2479.

[64] Kellington, J. and McBeth, R. (2007) IBM POWER6 processor soft error tolerance analysis using proton irradiation. IEEE Workshop on Silicon Errors in Logic – System Effects 3, Austin Texas, 3 and 4 April.

[65] Rao, S., Hong, T., Sanda, P. *et al.* (2008) Examining workload dependence of soft error rates. IEEE Workshop on Silicon Errors in Logic – System Effects, University of Texas at Austin, 26 and 27 March.

[66] Miyake, Y. Current Status of J-PARC MUSE, http://nuclpart.kek.jp/pac/1101/pdf/JPARCPAC110113.pdf (accessed 6 November 2013).

[67] TRIUMF http://nop.kek.jp/~arimoto/researches/prism-ffag/docs/muon_facility.pdf (accessed 7 November 2013).

[68] Bogomilov, M., Karadzhov, Y., Kolev, D. *et al.* The MICE Muon Beam on ISIS and the Beam-Line Instrumentation of the Muon Ionization Cooling Experiment, http://wrap.warwick.ac.uk/48594/1/WRAP_Back_1748-0221_7_05_P05009.pdf (accessed 7 November 2013)

[69] TRIUMF http://www.triumf.ca/ (accessed 7 November 2013).

[70] http://www.psi.ch/lmu/ (accessed August 22, 2014).

[71] Dicello, J.F., McCabe, C.W., Doss, J.D. and Paciotti, M. (1983) The relative efficiency of soft-error induction in 4 K static RAMs by Muons and Pions. *IEEE Transactions on Nuclear Science*, **30** (6), 4613–4615.

[72] Sierawski, B.D., Mendenhall, M.H., Reed, R.A. *et al.* (2010) Muon-induced single event upsets in deep-submicron technology. *IEEE Transactions on Nuclear Science*, **57** (6), 3273–3278.

[73] Ibe, E., Toba, T., Shimbo, K. and Taniguchi, H. (2012) Fault-based reliable design-on-upper-bound of electronic systems for terrestrial radiation including muons, electrons, protons and low energy neutrons. IEEE International On-Line Testing Symposium, Sitges, Spain, 27–29 June 2012 (3.2).

[74] MIT Nuclear Reactor Laboratory http://mit.edu/nrl (accessed 7 November 2013).

[75] NTI http://www.nti.org/facilities/818/ (accessed 7 November 2013).

[76] Kyoto University Research Reactor Institute http://www.rri.kyoto-u.ac.jp/en/facilities/kur (accessed 7 November 2013).

[77] Kobayashi, H., Usuki, H., Shiraishi, K. *et al.* (2004) Comparison between neutron-induced system-SER and accelerated-SER in SRAMs. 2004 IEEE International Reliability Physics Symposium, Phoenix, AZ, 25–29 April, pp. 288–293.

[78] Baumann, R.C. and Smith, E.B. (2000) Neutron-induced boron fission as a major source of soft errors in deep submicron SRAM devices. 2000 IEEE International Reliability Physics Symposium Proceedings, San Jose, CA, 10–13 April, pp. 152–157.

[79] Dyer, C., Hands, A., Hunter, K. *et al.* (2006) Neutron-induced single event effects testing across a wide range of energies and facilities and implications for standards. *IEEE Transactions on Nuclear Science*, **53** (6), 3596–3601.

[80] J-PARC http://j-parc.jp/researcher/MatLife/en/instrumentation/ns3.html (accessed 7 November 2013).

[81] indiamart http://www.indiamart.com/shah-marketingco/radiation-shielding-products.html (accessed 7 November 2013).

[82] Zhang, M., Park, J., Kim, G. *et al.* (2011) Thermal neutron SER testing and analysis: findings from a 32nm HKMG SRAM case study. 2011 IEEE Workshop on Silicon Errors in Logic – System Effects, Champaign, IL, 29–30 March, pp. 7–10.

[83] Yahagi, Y., Ibe, E., Takahashi, Y. *et al.* (2004) Threshold energy of neutron-induced single event upset as a critical factor. 2004 IEEE International Reliability Physics Symposium, 25–29 April, Phoenix, AZ, pp. 669–670.

[84] Baggio, J., Lambert, D., Ferlet-Cavrois, V. *et al.* (2007) Single event upsets induced by 1-10 MeV neutrons in static-RAMs using mono-energetic neutron sources. *IEEE Transactions on Nuclear Science*, **54** (6), 2149–2155.

[85] Gasiot, G., Ferlet-Cavrois, V., Baggio, j. *et al.* (2002) SEU sensitivity of bulk and SOI technologies to 14-MeV neutrons. *IEEE Transactions on Nuclear Science*, **52** (6), 2433–2437.

[86] Savage, M.W., McNulty, P.J., Roth, D.R. and Foster, C.C. (1998) Possible role for secondary particles in proton-induced single event upsets of modern devices. *IEEE Transactions on Nuclear Science*, **45** (6), 2745–2751.

[87] Makinaga, A., Akimune, H., Kim, G.N. *et al.* (2012) Activities on the Neutron and Photon Experiments at Hokkaido University, http://anrdw-3.knu.ac.kr/resources/PDF/17-IV_2_3rdAASPP-makinaga.pdf (accessed 18 April 2014).

[88] Harano, H., Matsumoto, T., Taniimura, Y. *et al.* (2010) Monoenergetic and quasi-monoenergetic neutron reference fields in Japan. *Radiation Measurements*, **45** (10), 1076–1082.

[89] Chabane, H., Vaille, J-R., Barelaud, B. *et al.* (2005) Comparison of experimental and simulated Am/Be neutron source energy spectra obtained in silicon detector. 8th European Conference on Radiation Effects on Components and Sysatems, Cap d'Agde, France, 19–23 September, pp. LN2-1–LN2-4.

[90] Marsh, J.W., Thomas, D.J. and Burke, M. (1995) High resolution measurements of neutron energy spectra from Am-Be and Am-B neutron sources. *Nuclear Insrrumcnts and Methods in Physics Research A*, **366**, 340–348.

[91] Héctor René Vega, C. and Celia Torres, M. (0000) Low Energy Neutrons From a ^{239}PuBe Isotopic Neutron Source Inserted in Moderating Media, http://www.ejournal.unam.mx/rmf/no485/RMF48503.pdf (accessed 13 November 2013).

[92] Reinig, W.C. Californium-252: A New Isotopic Source for Neutron Radiography, http://www.orau.org/ptp/PTP%20Library/library/Subject/Neutrons/dpms6848.pdf (accessed 13 November 2013).

[93] http://nrv.jinr.ru/nrv/webnrv/qcalc/ (accessed 13 November 2013).

[94] Ibe, E., Chung, S., Wen, S. *et al.* (2006) Spreading diversity in multi-cell neutron-induced upsets with device scaling. The 2006 IEEE Custom Integrated Circuits Conference, San Jose, CA, 10–13 September, 2006, pp. 437–444.

[95] Radaelli, D., Puchner, H., Chia, P. *et al.* (2005) Investigation of multi-bit upsets in a 150 nm technology SRAM device. *IEEE Transactions on Nuclear Science*, **49** (6), 3032–3037.

[96] Tsiligiannis, G., Dilillo, L., Bosio, A. *et al.* (2012) Evaluation of test algorithms stress effect on SRAMs under neutron radiation. IEEE International On-Line Testing Symposium, Sitges, Spain, 27–29 June 2012 (7.4), pp. 121–122.

[97] Ibe, E., Kameyama, H., Yahagi, Y. *et al.* (2004) Distinctive asymmetry in neutron-induced multiple error patterns of 0.13umocess SRAM. The 6th International Workshop on Radiation Effects on Semiconductor Devices for Space Application, Tsukuba, 6–8 October 2004, pp. 19–23.

[98] Shimbo, K., Toba, T., Nishii, K. *et al.* (2011) Quantification & mitigation techniques of soft-error rates in routers validated in accelerated neutron irradiation test and field test. 2011 IEEE Workshop on Silicon Errors in Logic – System Effects, Champaign, IL, 29–30March, pp. 11–15.

[99] Yamamoto, S., Kokuryou, K., Okada, Y. *et al.* (2004) Neutron-induced Soft-error in logic device using quasi-monoenergetic neutron beam. 2004 IEEE International Reliability Physics Symposium, Phoenix, AZ, 25–29 April, pp. 305–309.

[100] Uezono, T., Yoneki, S., Toba, T. *et al.* (2013) Evaluation of neutron-induced soft Error effects on CPU in automotive microcontrollers. 2013 Convention on Radiation Effects on Components and Systems, Oxford, UK, 23–27 September.

[101] Berger, G. *et al.* (1997) CYCLONE-A Multipurpose Heavy Ion, Proton, and Neutron SEE Test Site, RADECS Workshop, September 15–29, 1997, Cannes, France, 51–55.

[102] Eto, A., Hidaka, M., Okuyama, Y. *et al.* (1998) Impact of neutron flux on soft errors in MOS memories. International Electron Devices Meeting, San Francisco, CA, 6–9 December 1998, pp. 367–370.

[103] Wender, S. (2003) Neutron single-event effects testing at LANSCE. IEEE International Reliability Physics Symposium, Tutorials, Dallas, Texas, 30 March – 4 April (242.2).

[104] Dodd, P.E., Shaneyfelt, M.R., Schwank, J.R. and Hash, G.L. (2002) Neutron-induced softerrors, latchup, and comparison of SER test methods for SRAM technologies. International Electron Devices Meeting, San Francisco, CA, 9–11 December 2002, pp. 333–336.

[105] Gasiot, G., Giot, D. and Roche, P. (2007) Multiple cell upsets as the key contribution to the total SER of 65 nm CMOS SRAMs and its dependence on well engineering. *IEEE Transactions on Nuclear Science*, **54** (6), 2468–2473.

[106] Blackmore, E.W. (2009) Development of a large area neutron beam for system testing at TRIUMF. 2009 IEEE Radiation Effects Data Workshop, Quebec City, Canada, 20–24 July, pp. 157–160.

[107] Prokofiev, A.V., Blomgren, J., Nolte, R. *et al.* (2009) ANITA – a new neutron facility for accelerated SEE testing at the svedberg laboratory. IEEE International Reliability Physics Symposium 2009, Montreal, Quebec, Canada, 28–30 April (SE2), pp. 929–935.

[108] Rech, P. and Carro, L. (2013) Experimental evaluation of neutron-induced effects in graphic processing units. The 9th Workshop on Silicon Errors in Logic – System Effects, Palo Alto, CA, 26–27 March (5.3).

[109] Frost, C.D., Ansell, S. and Gorini, G. (2009) A new dedicated neutron facility for accelerated SEE testing at the ISIS facility. IEEE International Reliability Physics Symposium 2009, Montreal, Quebec,Canada, 28–30 April (SE6), pp. 952–954.

[110] Nakauchi, T., Mikami, N., Oyama, A. *et al.* (2008) A novel technique for mitigating neutron-induced multi-cell upset by means of back bias. IEEE International Reliability Physics Symposium, Anaheim, CA, 15–19 April (2 F.2), pp. 187–191.

[111] Furuta, J., Kobayashi, K. and Onodera, H. (2010) Measurement results of multiple cell upsets on a 65nm tapless flip-flop array. IEEE Workshop on System Efects of Logic Soft Errors, Stanford University, 23 and 24 March.

[112] Uemura, T., Sakoda, T. and Matsuyama, H. (2011) Layout optimization to maximize tolerance in SEILA: soft error immune latch. International Conference on IC Design and Technology, Kaohsiung, Taiwan, 2–4 May 2011.

[113] Matsukawa, F., Sato, T., Sato, K. *et al.* (2012) Neutron dose rate measurements in J-PARC MLF. *Progress in Nuclear Science and Technology*, **3**, 76–78.

[114] LANSCE http://lansce.lanl.gov/media/accelneutrontesting.pdf (accessed 7 November 2013).

[115] Platt, S.P. and Torok, Z. (2007) Charge-Collection and single-event upset measurements at the isis neutron source. 9th European Conference Radiation and Its Effects on Components and Systems, Deauville, France, 10–14 September (F-2).

[116] Nishida, S., Shoji, T., Ohnishi, T. *et al.* (2010) Cosmic ray ruggedness of IGBTs for hybrid vehicles. The 22nd International Symposium on Power Semiconductor Devices & ICs, Hiroshima, Japan, 6–10 June, pp. 129–132.

[117] Shoji, T., Nishida, S., Ohnishi, T. *et al.* (2010) Neutron induced single-event burnout of IGBT. The 2010 International Power Electronics Conference, Sapporo, Hokkaido, Japan, 21–24 June, pp. 142–148.

[118] Slayman, C.W. (2010) Theoretical correlation of broad spectrum neutron sources for accelerated soft error testing. *IEEE Transactions on Nuclear Science*, **57** (6), 3163–3168.

[119] Ibe, E., Taniguchi, H., Yahagi, Y. *et al.* (2010) Impact of scaling on neutron-induced soft error in SRAMs from a 250 nm to a 22 nm design rule. *IEEE Transactions on Electron Devices*, **57** (7), 1527–1538.

[120] Hazucha, P., Karnik, T., Maiz, J. *et al.* (2003) Neutron soft error rate measurements in a 90-nm CMOS process and scaling trends in SRAM from 0.25-micron to 90-nm generation. 2003 IEEE International Electron Devices Meeting, Washington, DC, 7–10 December, 2003 (21.5).

[121] Ando, H. and Hatanaka, S. (2007) Accelerated testing of a 90nm SPARC64 V microprocessor for neutron SER. IEEE Workshop on Silicon Errors in Logic – System Effects 3, Austin Texas, April 3, 4.

[122] Tsiligiannis, G., Vatajelu, E., DiLillo, L. *et al.* (2013) SRAM soft error rate evaluation under atmospheric neutron radiation and PVT variations. 19th IEEE International On-Line Testing Symposium, Chania, Crete, July 8–10 (8.2), pp. 145–150.

[123] Buchner, S., Baze, M., Brown, D. *et al.* (1997) Comparison of error rates in combinational and sequential logic. *IEEE Transactions on Nuclear Science*, **44** (6), 2209–2216.

[124] Jagannathan, S., Loveless, T.D., Bhuva, B.L. *et al.* (2012) Frequency dependence of alpha-particle induced soft error rates of flip-flops in40-nm CMOS technology. *IEEE Transactions on Nuclear Science*, **59** (6), 2796–2802.

[125] Grando, C., Lisboa, C., Moreira, A. and Carro, L. (2009) Invariant checkers: an efficient low cost technique for run-time transient errors detection. 15th IEEE International On-Line Testing Symposium, Sesimbra-Lisbon, Portugal, 24–26 June (2.2).

[126] Inoue, T., Henmi, H., Yoshikawa, Y. and Ichihara, H. (2011) High-level synthesis for multi-cycle transient fault tolerant datapaths. 17th IEEE International On-Line Testing Symposium, Athens, Greece, 13–15 July (1.3), pp. 13–18.

[127] Seifert, N., Gill, B., Zhang, M. *et al.* (2007) On the scalability of redundancy based SER mitigation schemes. International Conference on IC Design and Technology, Austin, Texas, 18–20 May (G2), pp. 197–205.

[128] Uemura, T., Tosaka, Y. and Matsuyama, H. (2010) SEILA: soft error immune latch for mitigating multi-node-SEU and local-clock-SET. IEEE International Reliability Physics Symposium 2010, Anaheim, CA, 2–6 May, pp. 218–223.

[129] Cabanas-Holmen, M., Cannon, E.H., Kleinosowski, A. *et al.* (2009) Clock and reset transients in a 90 nm RHBD single-core tilera processor. *IEEE Transactions on Nuclear Science*, **56** (6), 3505–3510.

[130] Seifert, N., Shipley, P., Pant, M.D. *et al.* (2005) Radiation-induced clock jitter and race. 2005 IEEE International Reliability Physics Symposium Proceedings, April 17–21, San Jose, CA, 17–21 April, 2005, pp. 215–222.

[131] Kim, S., Tsuchiya, A. and Onodera, H. (2013) Perturbation-immune radiation-hardened PLL with a switchable DMR structure. 19th IEEE On-Line Testing Symposium, Chania, Crete, 8–10 July.

[132] Chen, L. and Tahoori, M. (2012) An efficient probability framework for error propagation and correlation estimation. IEEE International On-Line Testing Symposium, Sitges, Spain, 27–29 June, 2012 (9.1).

[133] George, N., Lach, J. and Brown, C.L. (2011) Characterization of logical masking and error propagation in combinational circuits and effects on system vulnerability. 2011 International Conference on Dependable Systems and Networks, Hong Kong, China, 28–30 June.

[134] Cook, J. and Zilles, C. (2008) Characterizing instruction-level error derating. IEEE Workshop on Silicon Errors in Logic – System Effects, University of Texas at Austin, 26 and 27 March.

[135] Mizan, E. (2008) Reliability improvements enabled by self-imposed temporal redundancy. IEEE Workshop on Silicon Errors in Logic – System Effects, University of Texas at Austin, 26 and 27 March.

6

Integrated Device Level Simulation Techniques

6.1 Overall Multi-scale and Multi-physics Soft-Error Analysis System

Figure 6.1 illustrates the overall macroscopic soft-error analysis system *SECIS* (SElf Consistent Integrated System for terrestrial neutron soft error) [1–4] introduced in this chapter, which consists of the following activities:

(i) SEE (Single Event Effect) -MC (Monte-Carlo) simulators, CORIMS (COsmic Radiation Impact Simulator and SEALER (Single Event Adverse and Local Effects Reliever)
(ii) Programming language Visual Basic™ is used for many reasons. (see Appendices A.3 and A.4),
(iii) TCAD (Technology Computer-Aided Design) device simulators,
(iv) an automatic device model builder based on the device layout database GDS (Graphic Data System)-II
(v) database handler based on SQL (see Appendices A.5 and A.6),
(vi) databases of LET (Linear Energy Transfer), isotopes, nuclear reactions, materials, geophysics and terrestrial radiation spectra.
(vii) the Sun's activity data,
(viii) field neutron data,
(ix) field soft-error data,
(x) GPS (Global Positioning System) data of the fields and
(xi) irradiation data for single event effect (SEE) cross sections at the accelerators.

Databases referred to above enable realistic simulations and validate simulation results. As a typical device, SRAM (Static Random Access Memory) models are used for simulations.

A SRAM consists of two n^+ storage nodes in the p-well and two p^+ storage nodes in the n-well as shown in Figure 6.2. Typically, two sets of adjacent n^+ and p^+ nodes are

Terrestrial Radiation Effects in ULSI Devices and Electronic Systems, First Edition. Eishi H. Ibe.

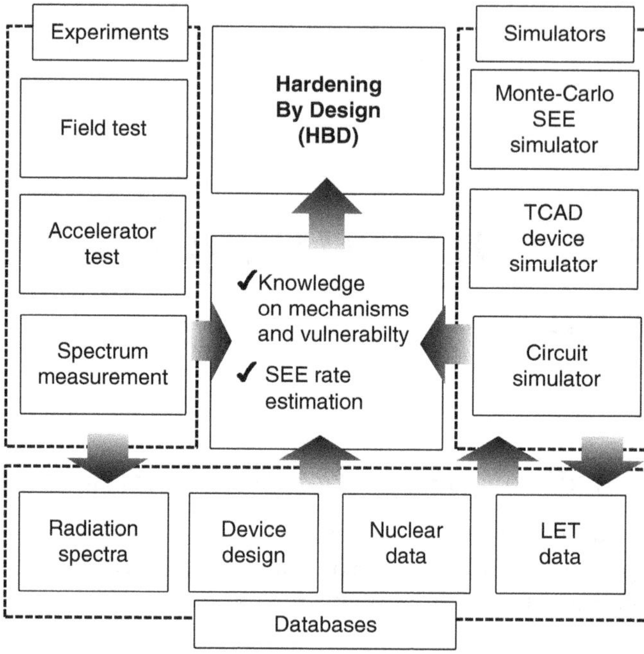

Figure 6.1 Overall scheme of SEE simulation/evaluation/hardening system

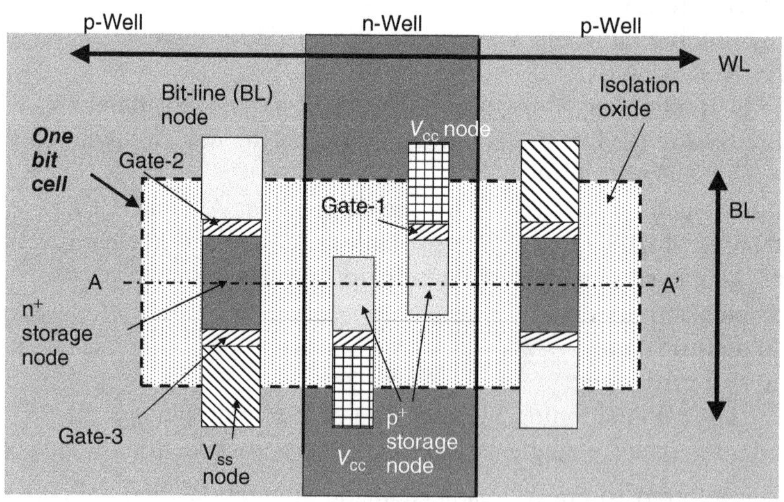

Figure 6.2 Top view of a SRAM model

connected by a metal wire above the transistor, and correspond to either 'high or low' potential states. The memory data '1' or '0' is assigned to either side (right or left in the figure) that has highest potential. The p^+ nodes are connected to V_{cc} when the gate-1 potential is low, while the n^+ nodes are connected to the ground or V_{ss} when the gate-2 potential is high. When the gate-3 potential is high, the data in the bit-line node is written to the set of n^+ and p^+ nodes or the data in the set of the nodes is written through the bit-line node. Each set of the bit-line node, the gate-2, the n^+ node, the gate-3 and the ground node has the common p-well (n^+ active area). Each set of the V_{cc} node, the gate-1 and the p^+ storage node has the common n-well (p^+ active area) as well. The n^+ active areas and p^+ active area are isolated by the STI (Shallow Trench Isolation) (SiO_2) as shown in Figure 6.3. Figure 6.4 depicts the equivalent circuit of the SRAM two nodes Q and $\overline{Q}$ corresponding to two sets of p^+ and n^+ storage nodes in Figure 7.2, for example. Once a ballistic neutron penetrates into the SRAM, nuclear spallation reaction may take place between the neutron and the nucleus (mostly Si) in the device as illustrated in Figure 6.5. As a prompt reaction, nucleons (protons and neutrons) collide with each other in the nucleus. Some of the nucleons may escape from the nucleus when they have enough kinetic energies. This process is known as INC (Intra Nuclear Cascade) [5]. After this rapid process, light nuclei may be 'evaporated" from the residual excited nucleus [6]. As a consequence, nucleons, light nuclei and the residual nucleus run inside the SRAM cell producing electron-hole pairs along with the ion track. Energy necessary to produce one pair of electrons and holes is 3.6 eV in Si. When one such secondary ion hits the storage nodes, some of the charges are collected to the storage node mainly through the funnelling effect [7] and drift/diffusion process. In the drift/diffusion process, a charged particle does not have to pass through the depletion layer but passes under the vicinity of the depletion layer and deposited charge is collected to the diffusion layer by diffusion. If the amount of the charges exceeds the critical charge, Q_{crit} can flips the logical state of the SRAM, a soft error takes place in the SRAM.

Figure 6.6 shows the overall simulation flow. First, an energy spectrum is selected for a particle and it is separated into multi energy bins. Particles with energy in a

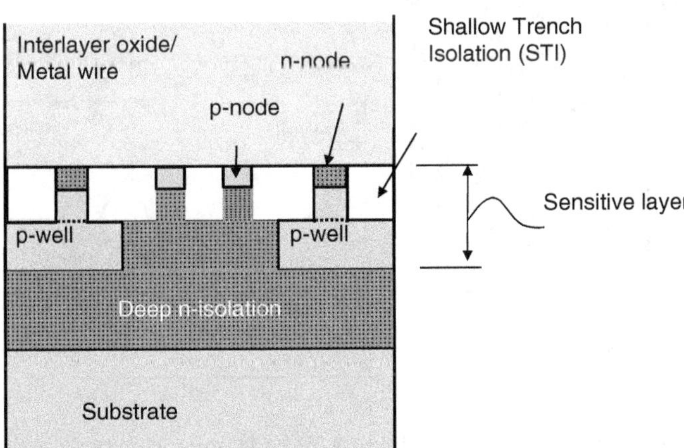

Figure 6.3 Cross section of a SRAM model (A-A′ cross section in Figure 6.2)

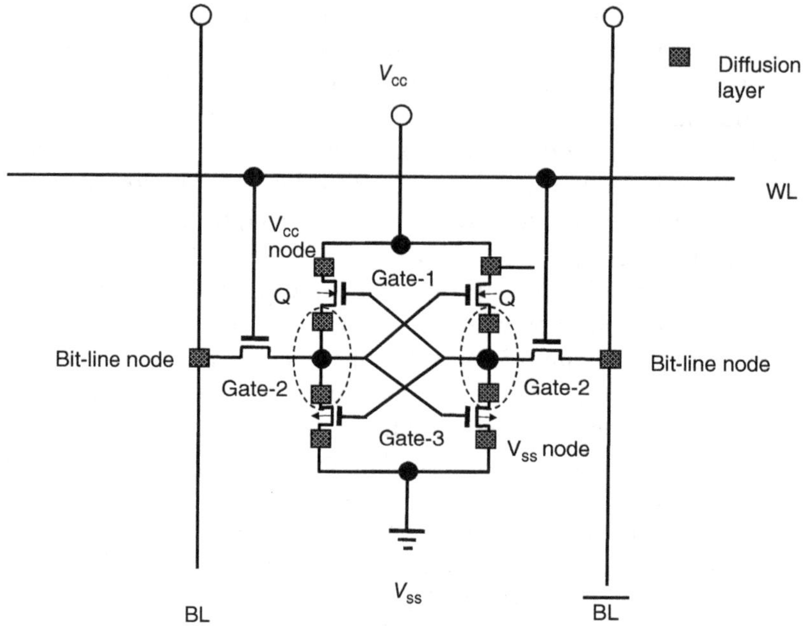

Figure 6.4 Relation of physical layouts in Figures 6.2 and 6.3 to SRAM circuits

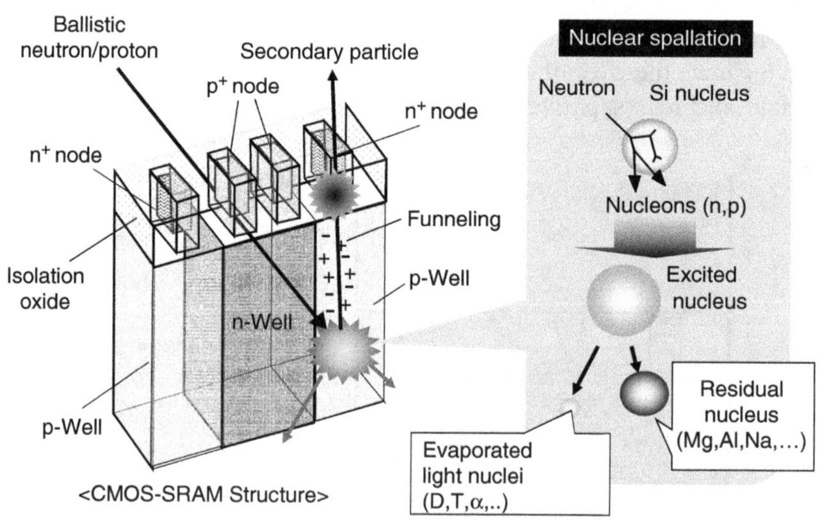

Figure 6.5 Simplified picture of intra-nuclear cascade and evaporation models in spallation nuclear reaction in a Si SRAM device (re-display of Figure 3.10)

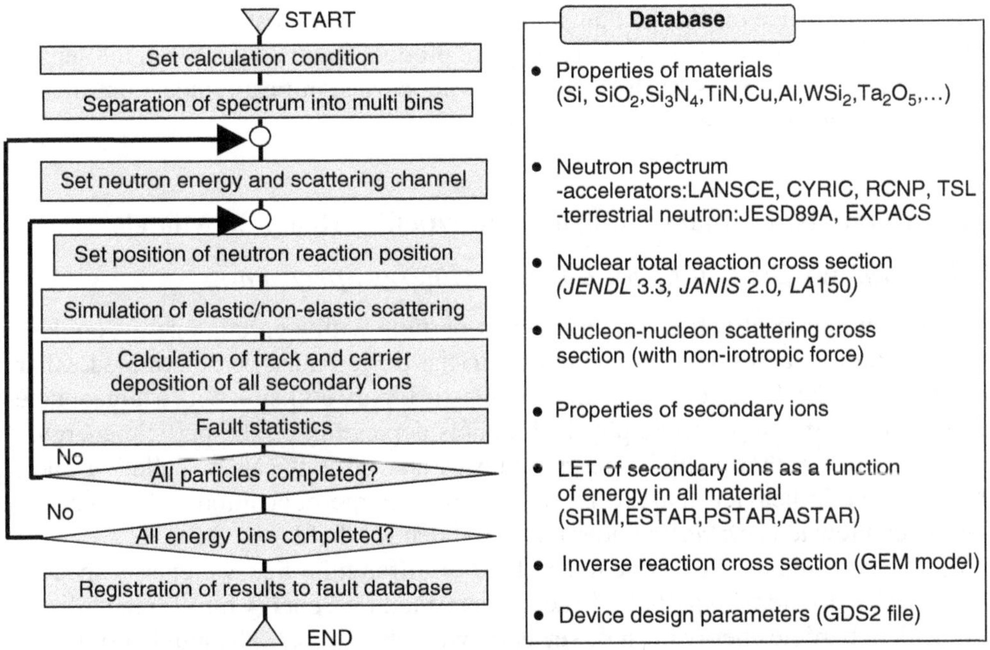

Figure 6.6 Overall simulation flow and databases for single event effects in semiconductor devices

certain energy bin are injected into a device model in random locations. Energy and charge deposited along with the track of primary or secondary particles from a nuclear reaction site are calculated. Charge collected to a storage node is calculated based on certain physical models and stored in an interim database as a fault. SEU (Single Event Upset) is analysed, classified and stored in the final database. Databases for materials, particle spectra, nuclear reaction data, LET, and so on, are referred to when necessary. A database is easily handled with Visual Basic™. A sample code for choosing data of a specific variable name is shown in Appendix A.5. Of all semiconductor materials, only Si is dealt with in this chapter, because the effects of other heavier material such as Cu, W only make a slight difference in SEU simulation during full modelling of materials [8]. At first sight, heavier elements seem to have a significant impact on SEEs because they have larger numbers of orbital electrons and copper wire and tungsten used in them are commonly located near sources and drains of transistors. The facts revealed by conducting a full simulation using SEALER are that:

(i) Since most momentum of secondary particles after spallation reaction is given to light particles like proton, heavy ions cannot have a large kinetic energy due to the law of conservation of energy and momentum.
(ii) Heavy secondary ions are produced in heavy materials that have strong stopping power in them, thus their ranges are inevitably very short.

As a consequence of the mechanisms (i) and (ii) above, most heavy secondary ions remain in the heavy materials where they are produced and do not reach the sensitive part of the nearby transistor, thereby, they do not have significant impact on soft-error in semiconductor devices.

6.2 Relativistic Binary Collision and Nuclear Reaction Models

6.2.1 Energy Bin Setting for a Particle Energy Spectrum

In MC simulation of soft error, an incident particle with a certain energy, angle and position on the target device penetrates into the device must be determined. Charge deposition in the device or elastic and non-elastic binary collision with a target nucleus is analysed using appropriate physical models depending on the particle energy. As explained in Chapter 3, high-energy neutrons and protons cause spallation nuclear reaction, while heavier charged particles cause charge deposition with energies of less than a few tens of MeVs under terrestrial neutron fields.

In order to simulate SEEs due to any kind of differential flux-vs.-energy spectra of terrestrial or accelerated particles in accelerator facilities, energy bins for the full range of incident particle energy must be set as shown in Figure 6.7. The number of incident particles in an energy bin is given in proportion to the flux in the corresponding energy bin in the simulation. As the differential flux generally decreases drastically as the particle energy increases as illustrated in Figure 2.3, a wide variation in the number of incident particles to be analysed in the simulation may take place. This would cause wide variation in statistical confidence by energy bin to bin. The following energy bin

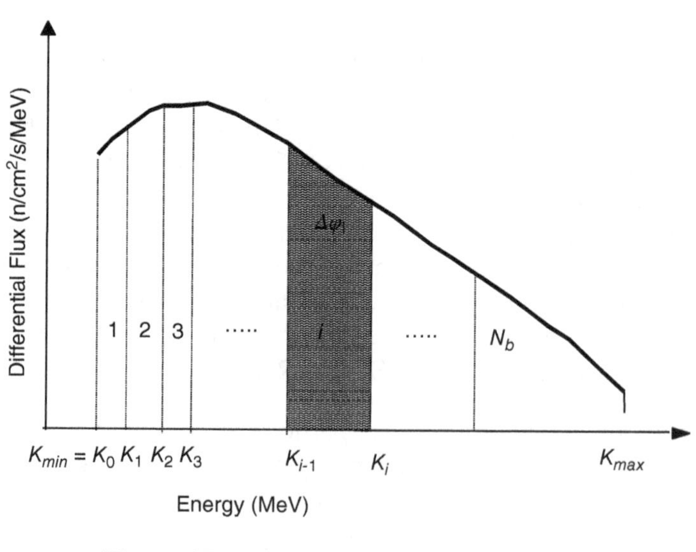

Figure 6.7 Schematic of energy bin setting

optimisation technique equalises the number of incident particles in the simulation and its use is recommended to avoid this problem.

First, we give the minimum and maximum boundary energies K_{i-1} and K_i. Then, we define the representative (average) energy K_i^*,

$$K_i^* = (K_{i-1} + K_i)/2. \tag{6.1}$$

With the pre-determined number of energy bins N_b and the total incident particle N_{Total} between minimum and maximum particle energy, E_{min} and E_{max}, the actual total flux ϕ_{TOT} can be obtained by,

$$\phi_{TOT} = \int_{E_{min}}^{E_{max}} \frac{d\phi_p}{dE_p} dE_p. \tag{6.2}$$

Each of the boundary energies K_{i-1} and K_i can be obtained in order to satisfy the following equation,

$$\int_{K_{i-1}}^{K_i} \frac{d\varphi_p}{dE_p} dE_p = \phi_{TOT}/N_b, \tag{6.3}$$

starting upward from $K_{min}(i = 1)$ or downward from $K_{max}(i = N_b)$, one by one.

In each energy bin, a simulation is carried out for incident particles of energy K_i^*. The number of particles is almost equally set to be Int(N_{Total}/N_b). (Int(x) means the maximum integer that is not larger than the real number x).

After setting the number of particles in the set of bins, the simulation is carried out by the bin to bin and all faults are registered in the interim database as shown in the flowchart of Figure 6.8.

6.2.2 Relativistic Binary Collision Model

As mentioned before, the initial physical stage of nuclear spallation reaction is regarded as a cascade of a binary elastic scattering problem of an incident neutron or proton with a nucleon in a nucleus. First, the details of the relativistic elastic scattering model are introduced as the basis of the simulation model for nuclear spallation reaction and low-energy neutron capture reaction by ^{10}B [9, 10].

Particles produced as a consequence of nuclear reaction are scattered, following the law of conservation of energy and momentum. As for nuclear spallation reaction, the scattering problem must be dealt with based on the special theory of relativity, because the particle energy can be very high, thus its energy exceeds simple kinetic energy $mv^2/2$ where m is particle mass and v is particle velocity. A scattering process between two particles is illustrated in Figure 6.9.

$$\mathrm{I} + \mathrm{T}(\rightarrow\ \mathrm{C}) \rightarrow\ \mathrm{X} + \mathrm{R}, \tag{6.4}$$

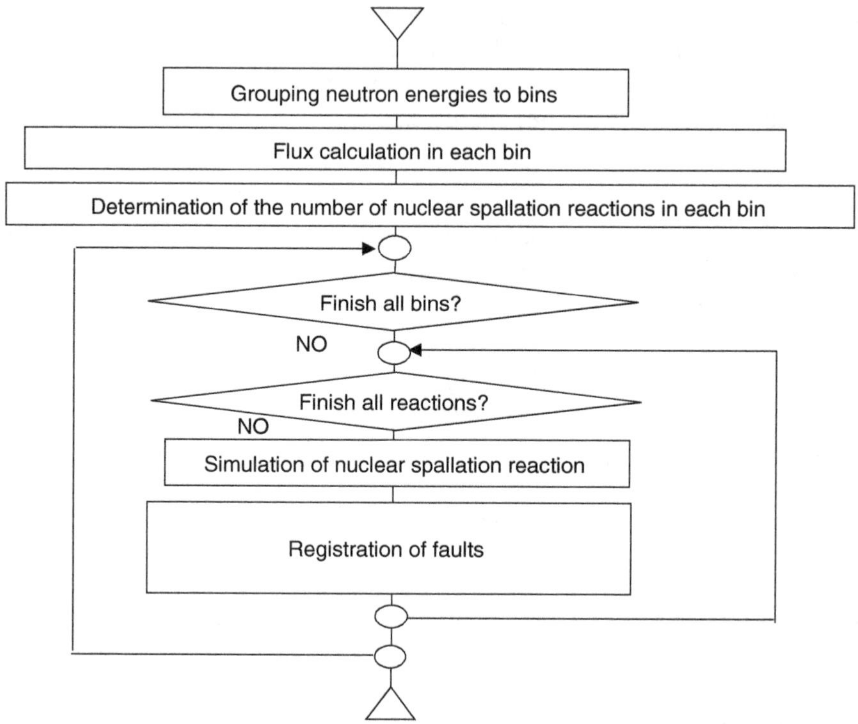

Figure 6.8 Simulation flow of particle induced faults in energy bins

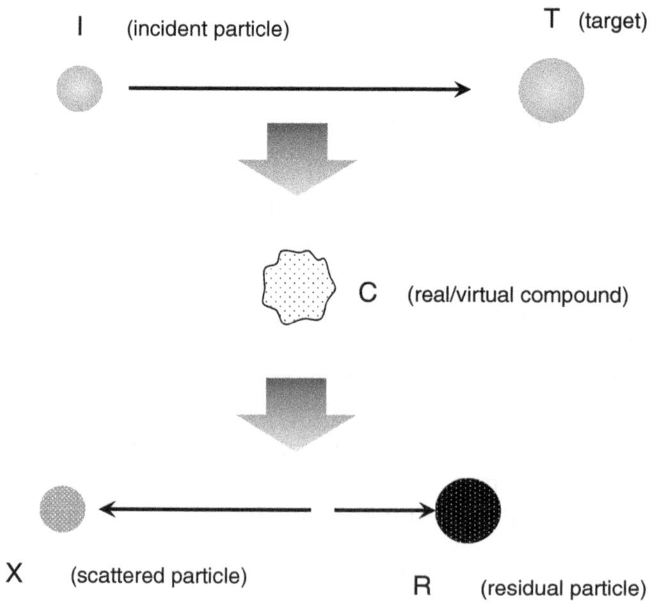

Figure 6.9 Concepts of scattering of ions in a component. Scattering of ions: I (secondary ion) + T (target nucleus) → C (compound nucleus) → X (released light nucleus) + R (residual nucleus)

where,

I: incident particle
T: target particle
C: combined nucleus (can be virtual)
X: light particle produced from C
R: residual particle from C that released X.

6.2.3 ALS (Absolute Laboratory System) and ALLS (Aligned Laboratory System)

Let us define the laboratory system fixed to a silicon wafer that is the target subject to irradiation, ALS (Absolute Laboratory System). For the particle q with relativistic total energy E_q, and momentum p_q, rest mass m_q and kinetic energy K_q is given by,

$$E_q = \sqrt{{m_q}^2c^4 + c^2{p_q}^2} = m_qc^2 + K_q. \tag{6.5}$$

Thus, the velocity $\beta_c = (v_c/c)$ of CCM (Classical Center of Mass) of the particles I, T relative to the ALS is given by,

$$\beta_c = \frac{cp_c^{lab}}{E_c^{lab}} = \frac{\sqrt{2m_Ic^2K_I + K_I^2}}{K_I^{lab} + m_Ic^2 + m_Tc^2} \tag{6.6}$$

The direction of the incident particle, in other words, the direction of origin of CCM to ALS is not equal to the direction of the movement of the ALS origin. In a special relativistic theory, however, Lorentz transformation that can be used for calculation of energy and momentum can only be applied to the movement direction of the origin of CCM. Therefore, a coordinate transformation that follows the CCM direction must be assigned to one of three coordinates, where we take z-axis, to apply the Lorentz transformation. This transformation is also illustrated in Figure 6.10. The arrow indicates the direction of the particle movement to the axis X_0, Y_0, Z_0 in the ALS. Let us define the zenith angle θ as the angle between the axis Z_0 and the arrow, and azimuth angle φ as the angle between the projection of the arrow to the X_0-Y_0 plane and the X_0 axis. The new coordinate system, with which the Z-axis is parallel to the direction of the particle movement, can be obtained by the following mathematical processes (i) to (iv) below. We call the new coordinate system *Aligned Laboratory System* (*ALLS*). We define the positive rotation of coordinates as clockwise rotation looking at the positive direction of the axis of rotation without exception.

(i) Rotate with angle φ around Z_0 axis to make the coordinate system X_1-Y_1-Z_1 in ALS. The Z_0 and Z_1 axis is the same.
(ii) Then, rotate with angle θ around the Y_1 axis to make the coordinate system X_2-Y_2-Z_2 in ALLS. Here, the Y_1 axis and Y_2 axis are the same.

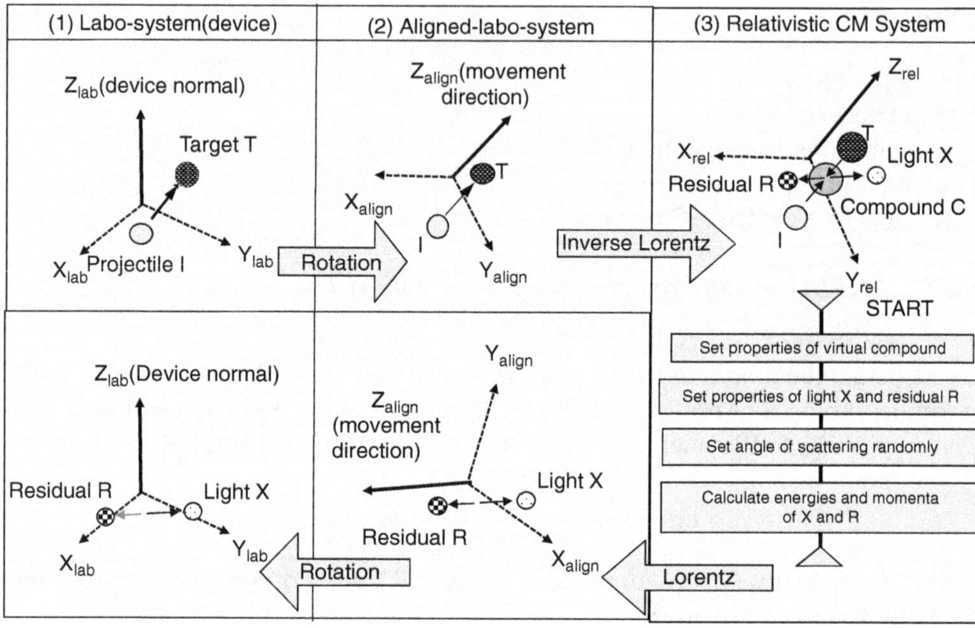

Figure 6.10 Rotation of the axes from the labo system to aligned labo system (ALS) and to relativistic centre of mass system and reverse rotations

The vector (x_0, y_0, z_0) in the X_0-Y_0-Z_0 system can be transformed by affine transformation to the vector in X_1-Y_1-Z_1 and X_2-Y_2-Z_2 system by,

$$\vec{\mathbf{r}}_1 = \begin{pmatrix} x_1 \\ y_1 \\ z_1 \end{pmatrix} = \begin{pmatrix} \cos\phi & -\sin\phi & 0 \\ \sin\phi & \cos\phi & 0 \\ 0 & 0 & 1 \end{pmatrix} \begin{pmatrix} x_0 \\ y_0 \\ z_0 \end{pmatrix} = \mathbf{R}_\phi \vec{\mathbf{r}}_0, \tag{6.7}$$

and,

$$\vec{\mathbf{r}}_2 = \begin{pmatrix} x_2 \\ y_2 \\ z_2 \end{pmatrix} = \begin{pmatrix} \cos\theta & 0 & \sin\theta \\ 0 & 1 & 0 \\ -\sin\theta & 0 & \cos\theta \end{pmatrix} \begin{pmatrix} x_1 \\ y_1 \\ z_1 \end{pmatrix} = \mathbf{R}_\theta \vec{\mathbf{r}}_1 = \mathbf{R}_\theta \mathbf{R}_\phi \vec{\mathbf{r}}_0. \tag{6.8}$$

By using time derivative of Equation (6.8), we obtain the velocity vector in ALLS,

$$\vec{\mathbf{v}}_2 = \begin{pmatrix} v_{2,x} \\ v_{2,y} \\ v_{2,z} \end{pmatrix} = \mathbf{R}_\theta \mathbf{R}_\phi \vec{\mathbf{v}}_0, \tag{6.9}$$

and, therefore, we obtain the momentum vector by simply multiplying the mass of the particle and velocity in Equation (6.9),

$$\vec{\mathbf{p}}_2 = \mathbf{R}_\theta \mathbf{R}_\phi \vec{\mathbf{p}}_0 \tag{6.10}$$

In the Cartesian coordinate system X_2-Y_2-Z_2, the positive direction on Z-axis is set to the incident particle direction.

(iii) Calculation of momenta in RCM (Relativistic Centre of Mass system) system.

The Lorentz transformation matrix **L** can be given for the Z-axis only by,

$$\mathbf{L} = \begin{pmatrix} \gamma_\mathrm{c} & 0 & 0 & \beta_\mathrm{c}\gamma_\mathrm{c} \\ 0 & 1 & 0 & 0 \\ 0 & 0 & 1 & 0 \\ \beta_\mathrm{c}\gamma_\mathrm{c} & 0 & 0 & \gamma_\mathrm{c} \end{pmatrix}, \tag{6.11}$$

or the inverse Lorentz transformation matrix $\mathbf{L}^{-1}$,

$$\mathbf{L}^{-1} = \begin{pmatrix} \gamma_\mathrm{c} & 0 & 0 & -\beta_\mathrm{c}\gamma_\mathrm{c} \\ 0 & 1 & 0 & 0 \\ 0 & 0 & 1 & 0 \\ -\beta_\mathrm{c}\gamma_\mathrm{c} & 0 & 0 & \gamma_\mathrm{c} \end{pmatrix} \tag{6.12}$$

where,

$$\gamma_c = \frac{1}{\sqrt{1-\beta_c{}^2}}.$$

In the special theory of relativity, four-dimensional momenta of the particle I and T are given as $(E_\mathrm{I}, 0, 0, cp_{x,\mathrm{I}})$, $(E_\mathrm{T}, 0, 0, 0)$, respectively.

Relativistic four-dimensional momenta can be calculated by the following inverse Lorentz transformation in the RCM,

$$\begin{pmatrix} E_\mathrm{I}^* \\ cp_{x,\mathrm{I}}^* \\ cp_{y,\mathrm{I}}^* \\ cp_{z,\mathrm{I}}^* \end{pmatrix} = \mathbf{L}^{-1} \begin{pmatrix} E_\mathrm{I} \\ 0 \\ 0 \\ cp_{z,\mathrm{I}} \end{pmatrix} \tag{6.13}$$

and,

$$\begin{pmatrix} E_\mathrm{T}^* \\ cp_{x,\mathrm{T}}^* \\ cp_{y,\mathrm{T}}^* \\ cp_{z,\mathrm{T}}^* \end{pmatrix} = \mathbf{L}^{-1} \begin{pmatrix} E_\mathrm{T} \\ 0 \\ 0 \\ 0 \end{pmatrix}, \tag{6.14}$$

respectively.

All binary scattering parameters in RCM can be obtained by the processes above.

(iv) The (virtual) compound particle C of I and T is at rest at the origin of the system with total energy $U = E_\mathrm{I}{}^* + E_\mathrm{T}{}^* + G_\mathrm{I} + G_\mathrm{T}$.

Next, let us suppose that the compound particle C splits into two particles X and R. Considering that the momenta of the particles X and R are the same in the

RCM, the total energies and momenta of the particles X and R in the RCM can be given by,

$$E_X^* = \frac{U^2 + m_X^2 c^4 - m_R^2 c^4}{2U} \tag{6.15a}$$

$$E_R^* = \frac{U^2 + m_R^2 c^4 - m_X^2 c^4}{2U} \tag{6.15b}$$

$$p^* = \frac{\sqrt{\{U^2 - (m_X - m_R)^2 c^4\}\{U^2 - (m_X + m_R)^2 c^4\}}}{2U} \tag{6.15c}$$

The directions of the two particle movements must be strictly opposite in the RCM system, and can be set isotropically. Let us define the zenith and azimuth angles of the particles X, θ^* and φ^*, respectively, and take the direction of the particle X to be the new Z-axis. The momenta of the two particles in the new coordinate system can be given by,

$$\vec{\mathbf{p}}_X = {}^t(0, 0, p^*), \vec{\mathbf{p}}_R = {}^t(0, 0, -p^*).$$

The momenta and total energy in the RCM system for the particle q (= X, R) can be given by,

$$\vec{\mathbf{p}}_q^{CM} = \mathbf{R}_{\phi^*}^{-1} \mathbf{R}_{\theta^*}^{-1} \vec{\mathbf{p}}_q, \tag{6.16}$$

and,

$$E_q^{CM} = \sqrt{m_q^2 c^4 + c^2 |\vec{\mathbf{p}}_q^{CM}|}, \tag{6.17}$$

respectively.

The total energies and momenta of the particles in the ALLS, therefore, can be given with superscript 'lab' by,

$${}^t(E_q^{lab'}, cp_{q,x}^{lab'}, cp_{q,y}^{lab'}, cp_{q,z}^{lab'}) = L\,{}^t(E_q^{CM}, cp_{q,x}^{CM}, cp_{q,y}^{CM}, cp_{q,z}^{CM}). \tag{6.18}$$

In the ALS,

$$E_q^{lab} = E_q^{lab'} \tag{6.19}$$

$$\vec{\mathbf{p}}_q^{lab} = \mathbf{R}_\phi^{-1} \mathbf{R}_\theta^{-1} \vec{\mathbf{p}}_q^{lab'} \tag{6.20}$$

with kinetic energy,

$$K_q^{lab} = E_q^{lab} - m_q c^2, \tag{6.21}$$

and with angles of rupture,

$$\varphi_q^{lab} = \tan^{-1}\left(\frac{p_{q,y}^{lab}}{p_{q,x}^{lab}}\right) \qquad (p_{q,x}^{lab} \geq 0); \tag{6.22}$$

$$\varphi_{\mathrm{q}}^{\mathrm{lab}} = \pi + \tan^{-1}\left(\frac{p_{\mathrm{q},y}^{\mathrm{lab}}}{p_{\mathrm{q},x}^{\mathrm{lab}}}\right) \qquad (p_{\mathrm{q},x}^{\mathrm{lab}} < 0); \tag{6.23}$$

$$\theta_{\mathrm{q}}^{\mathrm{lab}} = \tan^{-1}\left(\frac{\sqrt{\left(p_{\mathrm{q},x}^{\mathrm{lab}}\right)^2 + (p_{\mathrm{q},x}^{\mathrm{lab}})^2}}{|p_{\mathrm{q},z}^{\mathrm{lab}}|}\right) \qquad (p_{\mathrm{q},z}^{\mathrm{lab}} \geq 0); \tag{6.24}$$

$$\theta_{\mathrm{q}}^{\mathrm{lab}} = \frac{\pi}{2} + \tan^{-1}\left(\frac{\sqrt{\left(p_{\mathrm{q},x}^{\mathrm{lab}}\right)^2 + (p_{\mathrm{q},x}^{\mathrm{lab}})^2}}{|p_{\mathrm{q},z}^{\mathrm{lab}}|}\right) \qquad (p_{\mathrm{q},z}^{\mathrm{lab}} < 0). \tag{6.25}$$

Thus, all the scattering parameters in ALS can be obtained.

6.3 Intra-nuclear Cascade (INC) Model for High-Energy Neutrons and Protons

Figure 6.11 illustrates the simulation procedures under which INC is regarded as a sequence of binary elastic collisions between two nucleons in the target nucleus. Starting with an incident nucleon, the kinematics of the nucleon are traced until they are stopped or released from the target nucleus. The kinematics of nucleons excited by the incident nucleons (first generation) are recorded sequentially and the kinematics of the nucleons are traced after collision in a similar way. Likewise, second or further generation nucleons are traced until evolution of the next generation is not possible. All kinematics of released nucleons are recorded to set the initial conditions of the evaporation process that takes place after the INC process. Energies of nucleons that are not released out of the nucleus are added up to obtain elevated energy of excited residual nucleus.

Details of the procedures consist of the following steps:

6.3.1 *Penetration of a Nucleon into a Target Nucleus*

Effective reaction radius R_{eff} is determined by,

$$R_{\mathrm{eff}} = r_I + r_{\mathrm{T}} + R_{\mathrm{nuc}} \quad \text{(fm)}, \tag{6.26}$$

where,

r_{I}: radius of the incident nucleon
r_{T}: radius of the target nucleus
R_{nuc}: effective distance of nuclear force (=1.4 fm).

A sphere with a radius of R_{eff} is set as the target in the simulation model.

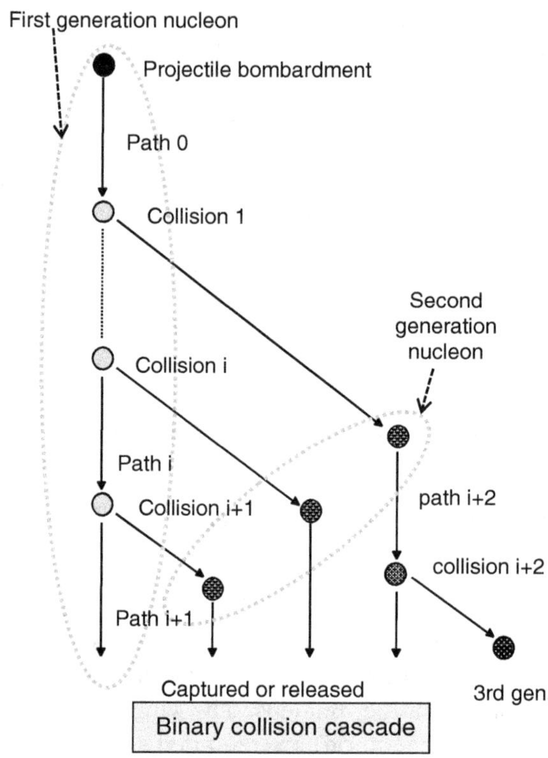

Figure 6.11 Concept of the intra-nuclear cascade (INC) model

Radius of nucleus can be calculated by,

$$r_q = r_0 \sqrt[3]{A_q} \quad (r_0 = 1.2\ \text{fm}). \tag{6.27}$$

As the probability that an incident particle has scattering parameter b_c is proportional to $b_c{}^2 db$, the scattering parameter of an incident nucleon is set randomly by using random number *Rnd*,

$$b_c = \sqrt{R_{\text{eff}}^2 \times Rnd}. \tag{6.28}$$

All nucleons are set to penetrate to the target within a distance of R_{eff} from the centre of the target. When the nucleon is proton, protons with energy lower than the Coulomb barrier,

$$V_{\text{coul}} = \frac{Z_I Z_T e^2}{R_{\text{eff}}} = 1.44 \frac{Z_I Z_T}{R_{\text{eff}}} \quad (\text{When } R_{\text{eff}} \text{ is in the unit of fm, the unit of } V_{\text{coul}} \text{ is MeV}) \tag{6.29}$$

cannot penetrate into the target nucleus, thus causing no reaction.

6.3.2 Calculation of Probability of Binary Collision between Two Nucleons in the Target Nucleus

Incident nucleon or nucleons contained in the target nucleus cannot always react in the target nucleus. They almost certainly react with each other. Here, we assume that the number density ρ_N of nucleons in a nucleus is homogeneous and 0.18/fm^3.

The mean free path $\lambda_{n,p}$ of a nucleon in a nucleus is given by,

$$\lambda_{n,p} = \frac{1}{\rho_N \overline{\sigma}_{n,p}} \tag{6.30}$$

where, $\overline{\sigma}_{n,p}$: average collision cross section of a neutron or proton, and is given by,

$$\overline{\sigma}_n = \frac{N_n \sigma_{nn}(E_q) + N_p \sigma_{np}(E_q)}{N_n + N_p} \tag{6.31}$$

or

$$\overline{\sigma}_p = \frac{N_n \sigma_{pn}(E_q) + N_p \sigma_{pp}(E_q)}{N_n + N_p}; \tag{6.32}$$

N_n, N_p: the number of neutron and proton in the nucleus.

Let us define L_{path} as the distance of an excited nucleon that has an energy of more than Fermi energy at the surface of the target nucleus. The probability P_{out} that the nucleon can move with the distance L_{path} without collision with other nucleons is given by,

$$P_{out} = \exp\left(-\frac{L_{path}}{\lambda_{n,p}}\right) \tag{6.33}$$

so that when the chosen random number *Rnd* is less than P_{out}, the nucleon is not set to react without collisions at the surface of the target nucleus. Otherwise, the position of collision is set randomly within the distance L_{path}. Determining whether the nucleon reached the surface of the target nucleus and then escaped from the target nucleus or not is explained later.

Whether the target nucleon is a neutron or proton is determined by using the following equation,

$$Iso_n = \frac{N_n \sigma_{nn}(E_n)}{N_n \sigma_{nn}(E_n) + N_p \sigma_{np}(E_n)}. \tag{6.34}$$

When the incident nucleon is a neutron and the chosen random number *Rnd* is smaller than Iso_n, the target nucleon is set to neutron, otherwise to proton.

6.3.3 Determination of Condition in Nucleon-Nucleon Collision

In order to analyse one nucleon-nucleon collision, the energy of the target nucleon has to be set first and the Fermi gas model is applied [11]. The potential of nuclear force in the target nucleus is assumed to be the square-well potential and the volume Ω_{nuc} of the target nucleus that is necessary for calculation of the Fermi level introduced later in Equation (6.36) is assumed to be,

$$\Omega_{nuc} = \frac{4\pi}{3} R_{eff}^3. \tag{6.35}$$

The Fermi level E_{qF} is given by,

$$E_{qF} = \frac{3^{2/3}\pi^4\hbar^2}{2m_q}\left(\frac{N_q}{\Omega_{nuc}}\right). \tag{6.36}$$

The values of E_{qF} for neutron and proton are, in general, very similar and in the ^{28}Si case, the simple average $E_F = (E_{nF} + E_{pF})/2$ is used. The binding energy of nucleon at the Fermi level is set to the average binding energy B_{av} of nucleon, which is obtained from the database.

Effective kinetic energies K^*_q of nucleons in the nucleus beyond the Fermi level below the potential barrier are given by,

$$K^*_q < \begin{cases} V_0 \\ V_0 + V_{coul} \end{cases} \quad \text{(The upper is neutron and the lower is proton),} \tag{6.37}$$

where, V_o: energy difference between vacuum and the Fermi level.

When any nucleon cannot escape from the nucleus by itself nor give enough energy to other nucleon to escape, further collision analysis of the nucleon is aborted and their kinetic energy is added up to the excitation energy of the residual nucleus when all nucleons have energies below the potential barrier. On the other hand, nucleons that have energies beyond the barrier at the surface of the target nucleus are set to escape. The kinematics of residual excited nucleus are determined from the kinematics of these escaped nucleons.

6.4 Evaporation Model for High-Energy Neutrons and Protons

After the INC process, light particles **X** like n, p, D (deuterium), T (tritium), ^3He, ^4He are 'evaporated' from the excited nucleus leaving a residual excited nucleus. This process is regarded as a sequence of 'inverse' binary collision as shown in Figure 6.12 conceptually. In the evaporation process, the reaction *channel* that determines the number of each isotope stochastically is defined.

Properties of the initial excited residual nucleus after the INC process are calculated as follows:

(i) *Rest mass and mass defect*: Atom number and mass number of the residual nucleus are calculated based on the number of neutrons n_n and protons n_p

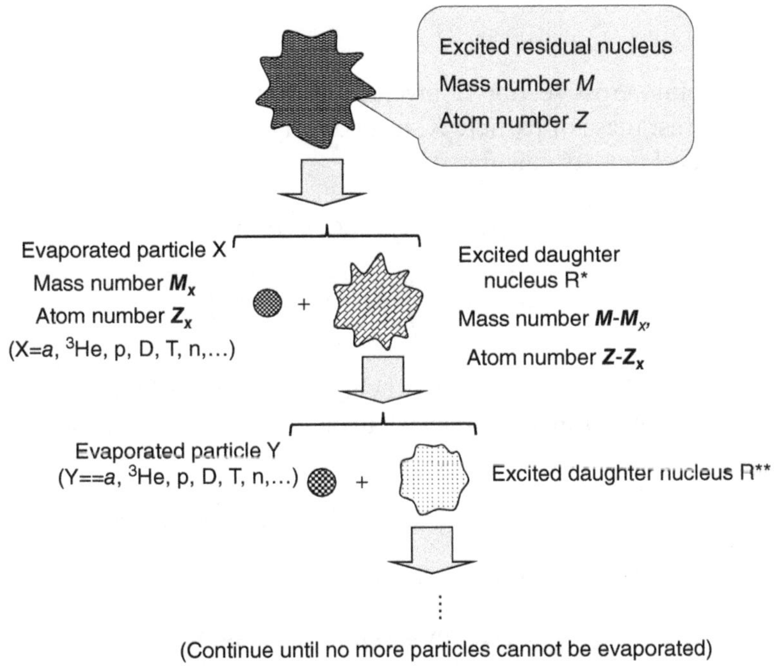

Figure 6.12 Concept of the evaporation model in a nuclear spallation reaction

released from the target nucleus with mass m_T. The rest mass of the residual nucleus m_{resid} is determined from the database of isotopes. Mass defect Δm is calculated by,

$$\Delta m = m_T - n_n m_n - n_p m_p - m_{resid}. \tag{6.38}$$

(ii) *Excitation energy*: The sum of energies above the Fermi level of nucleons that are not released from the target nucleus and Δm.
(iii) *Momentum vector*: The initial momentum of the target nucleus plus the incident nucleon minus the sum of momenta of nucleons released from the target nucleus.
(iv) *Kinetic energy*: calculated by using Equation (6.15) with momenta and mass of the residual nucleus.

Analysis of the evaporation process in which ground-state particles are 'evaporated' from a nucleus T* with excitation energy $E_T{}^*$ producing a residual excited nucleus is performed as follows:

First, the probability of the evaporation of particle X can be calculated by using the Weisskopf-Ewing equation,

$$P_x(E_T^*; \varepsilon) = \frac{(2S_x + 1)m_x}{\pi^2 \hbar^3} \varepsilon \sigma_{RX}(E_T^*; \varepsilon) \frac{\omega_R(E_R^*)}{\omega_T(E_T^*)}, \tag{6.39}$$

where,

σ_{RX}: inverse capture cross section of inverse reaction $R^* + X \rightarrow T^*$
S_x, m_x: spin and rest mass of particle X, respectively
$\omega(E^*)$: energy level density calculated by

$$\omega(E^*) = \omega_0 \exp(2\sqrt{aE^*})$$
$$a = \frac{A}{7.3}\left(1 + 1.5\frac{\Delta^2}{A^2}\right) \tag{6.40}$$

$E_R{}^*$ can be obtained by solving the following Equation (6.41) self-consistently (see Appendix A.7) starting from the value calculated from the first and second terms.

$$E_R^* = E_T^* - B(X + R \rightarrow T) - \delta_R - \varepsilon\left(1 - \frac{m_X}{m_X + m_R + E_R^*/c^2}\right) \tag{6.41}$$

where, $B(X + R \rightarrow T)$ is the mass defect and,

$$B(X + R \rightarrow T) = (m_X + m_R - m_T)c^2. \tag{6.42}$$

δ_R is the correction factor based on the odd-even combination of the number of neutrons and protons and expressed as follows:

$$\delta_R = \begin{cases} {}^{11}/\sqrt{A_R} \\ -{}^{11}/\sqrt{A_R} \\ 0 \end{cases} \quad \text{(MeV) (even-even, even-odd, odd-odd from the top to bottom)} \tag{6.43}$$

X has maximum energy,

$$\varepsilon_{max} = \left(1 + \frac{m_X}{m_R}\right)(E_T^* - B - \delta_R) \tag{6.44}$$

when $E_R{}^*$ is 0.

The evaporation channel for a combination of released particle and residual nucleus is determined by the following procedures:

(i) Production probability W_X of particle X is obtained by integration of $P_X(E_T{}^*; \varepsilon)$ over the energy from the Coulomb barrier V_c of $X + R^*$ to the maximum energy ε_{max},

$$W_X = \int_{V_C}^{\varepsilon_{max}} d\varepsilon P_X(E_T^*; \varepsilon) = \frac{(2S_X + 1)m_X}{N_0}\int_{V_C}^{\varepsilon_{max}} d\varepsilon\varepsilon\sigma_{RX}(E_T^*; \varepsilon)\omega_R(E_R^*) \tag{6.45}$$

(ii) Calculate the sum of overall channels to produce j-th particle (j-1,2, ...) and the probability of particle q production can be calculated by,

$$P_q = W_q / \sum_j W_j. \tag{6.46}$$

In order to obtain the kinetic energy of the particle produced, the following distribution function,

$$f(\varepsilon) = \frac{(\varepsilon - V_c)}{N_\varepsilon} \exp\left[-\left(\frac{\varepsilon - V_c}{\varepsilon_0}\right)\right] \qquad (V_c < \varepsilon < \varepsilon_{max}) \tag{6.47}$$

$$\varepsilon_0 = \frac{1}{2} < \varepsilon - V_c > \frac{1 - (1 + \overline{\varepsilon}_{max}/\varepsilon_0)\exp(-\overline{\varepsilon}_{max}/\varepsilon_0)}{1 - \left[1 + \overline{\varepsilon}_{max}/\varepsilon_0 + \frac{1}{2}\left(\overline{\varepsilon}_{max}/\varepsilon_0\right)^2\right]\exp(-\overline{\varepsilon}_{max}/\varepsilon_0)} \tag{6.48}$$

$<\varepsilon - V_c>$: the average of $\varepsilon - V_c$;

$$\overline{\varepsilon}_{max} = \varepsilon_{max} - V_c.$$

Equations (6.47) and (6.48) can be solved for ε_0 self-consistently, starting with $<\varepsilon - V_c>/2$.

The kinetic energy of X obtained here is the value for the case that the residual nucleus at rest in labo-system. Actually, both T* and R* has momenta and are not at rest. Thus, the kinematics of R* and X are calculated on the RCM. The absolute momenta of two particles are the same and the direction is opposite. The total energy of $E_R{}^* + E_X$ does not depend on the coordinate system. Thereforc, the following relationship can be obtained.

$$f(p_{CM}) = \sqrt{c^2 p_{CM}{}^2 + m_X^2 c^4} + \sqrt{c^2 p_{CM}{}^2 + m_R^2 c^4} - \varepsilon + m_X c^2 + m_R c^2 = 0 \tag{6.49}$$

From Equation (6.49), p_{CM} can be obtained numerically by using the Newton iteration method, for example. By setting the zenith and azimuth angles of rupture, θ_{CM} and φ_{CM} of the particle X on relativistic CM randomly, and momenta of X and R*, $(0, 0, p_{CM})$, $(0, 0, -p_{CM})$ on the Z-axis in parallel with the movement of X, momenta in the RCM can be calculated. The kinematics of two particles in the ALS can be obtained using the same procedures. The procedure for evaporation is continued until the excitation energy reaches zero.

6.5 Gencralised Evaporation Model (GEM) for Inverse Reaction Cross Sections

Inverse reaction cross section σ_{RX} in Equation (6.39) can be calculated based on the GEM (Generalized Evaporation Model) [12] for any combination of R* + X → T* by,

$$\sigma_{RX}(\varepsilon) = \pi R_b^2 c_j \left(1 - \frac{k_j V}{\varepsilon}\right) \tag{6.50}$$

$$V = \frac{Z_j Z_T e^2}{R_c} \tag{6.51}$$

For a charged evaporated particle,
where,

Z_j: atomic number of evaporated particles;
Z_T: atomic number of excited nucleus;
R_b: nuclear distance;
R_c: nuclear distance for Coulomb barrier;
c_j: Dostrovsky's parameter;
k_j: Coulomb barrier transmission coefficient.

R_b and R_c can be calculated by

$$R_b = 1.5{A_T}^{1/3} \text{ (for neutron and proton)}$$
$$R_b = 1.5({A_T}^{1/3} + {A_j}^{1/3}) \text{ (for other particles)}$$
$$R_c = 1.7{A_T}^{1/3} \text{ (for neutron and proton)}$$
$$R_c = 1.7{A_T}^{1/3} + 1.2 \text{ (for other particle).}$$

where,

A_T: mass number of excited nucleus
A_j: mass number of evaporated particles.

When X is neutron,

$$\sigma_{RX}(\varepsilon) = \pi R_b^2 c_n \left(1 + \frac{b}{\varepsilon}\right) \tag{6.52}$$

$$c_n = 0.76 + 1.93{A_T}^{-1/3} \tag{6.53}$$

$$b = \frac{1.66A_T^{-2/3} - 0.050}{0.76 + 1.93A_T^{-1/3}}. \tag{6.54}$$

With the parameters in Table 6.1, c_j and k_j can be calculated as follows:

$$c_p = 1 + c$$
$$c_d = 1 + c/2$$
$$c_t = 1 + c/3$$
$$c_{3He} = c_\alpha = 0$$

Table 6.1 Parameters for calculation of inverse reaction cross section based on generalised evaporation model (GEM)

Z_T	k	$k\alpha$	c
<= 20	0.51	0.81	0
30	0.6	0.85	−0.06
40	0.66	0.89	−0.1
>= 50	0.68	0.93	−0.1

and,

$$\begin{aligned} k_p &= k \\ k_d &= k + 0.06, \\ k_t &= k + 0.12 \\ k_{3He} &= k_\alpha - 0.06. \end{aligned}$$

Figure 6.13 shows the consistency of the calculated results by using GEM with measured the inverse reaction cross section [13].

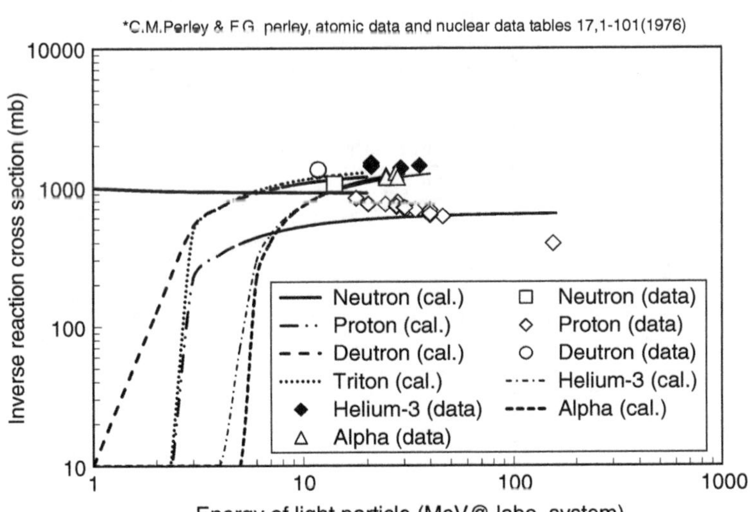

Figure 6.13 Comparison of measured inverse reaction cross sections for binary sets of Si and light particles and calculated results from generalised evaporation model (GEM)

6.6 Neutron Capture Reaction Model

Figure 6.14 shows the cumulative flux of a low-energy (<1 MeV) neutron at NYC (New York City) sea level.

The energy spectra of low-energy neutrons, neutron capture cross section data in Figure 3.11 and the neutron spectrum data are used for simulation. The amount of ^{10}B contained in a SRAM cell is calculated on the base of the assumption that 10^{20} Boron-10 atoms/cm^3 is contained in a metal layer 5 μm thick [9].

The number of nuclear reactions in CORIMS is determined based on the product of the capture cross section and differential neutron flux for each energy bin.

In the case of low-energy neutrons, the reaction channel is only one ($n + {}^{10}B \rightarrow Li + \alpha$) so that the momenta of two particles in the RCM system should be exactly the same.

The directions of the two particles are strictly opposite in the RCM system and can be set randomly. Directions of incoming neutrons are randomly set.

Energies and directions of Li and α in the labo system can be calculated based on these assumptions.

Calculated energy spectra are shown in Figure 6.15. When the kinetic energy of an incident neutron is low enough, it can be neglected compared to the Q-value of the reaction and therefore the energies of ^7Li and ^4He are almost exactly 0.84 MeV and 1.47 MeV, respectively. In reality, incident neutron has a higher or lower energy than an epi-thermal neutron so that the energy spectra of ^4He and ^7Li have a certain width.

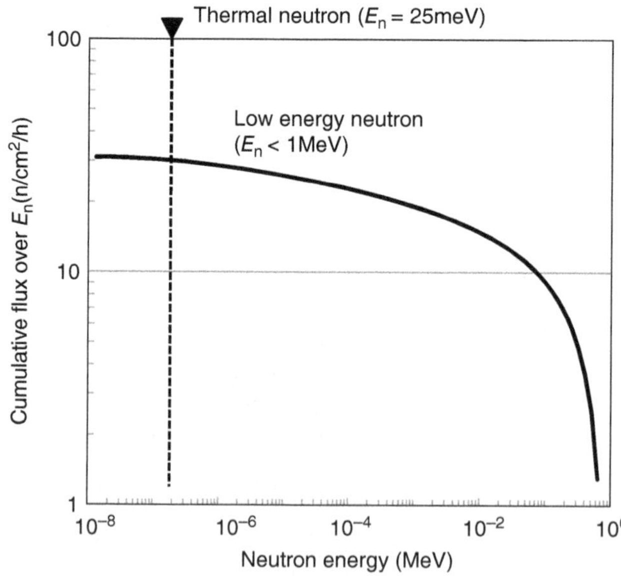

Figure 6.14 Cumulative flux of low-energy (<1 MeV) neutrons at NYC sea level

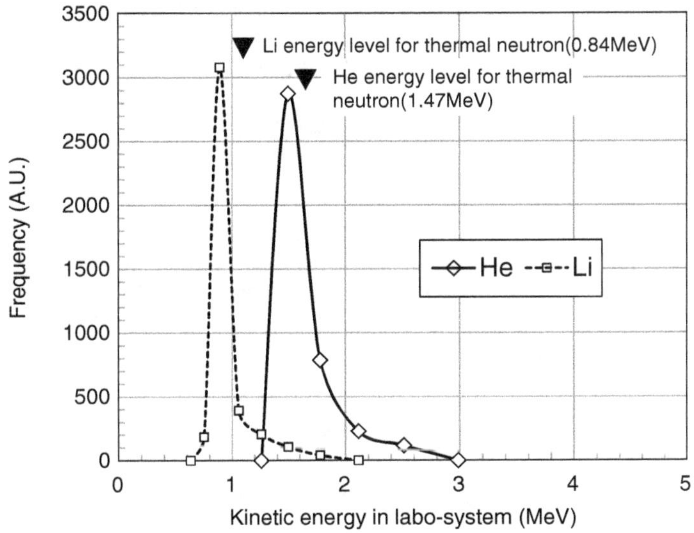

Figure 6.15 Energy spectra of α and ^{7}Li produced by neutron capture reaction of ^{10}B with low-energy neutrons at NYC sea level

6.7 Automated Device Modelling

Two-dimensional layout information of semiconductor devices can be obtained through a GDS-II file. With depth profile information, a three-dimensional device model can be constructed semi-automatically by using computer geometry techniques.

First, semiconductor devices, which consist of layers and components, like substrate, isolation oxide, channels, diffusion layer, depletion layer, metal wires, contacts, and so on, are regarded as a complex of layers and components.

Components have several geometrical layers or parts such as points, lines, tiles, unit (cell), block, device as shown in Figure 6.16.

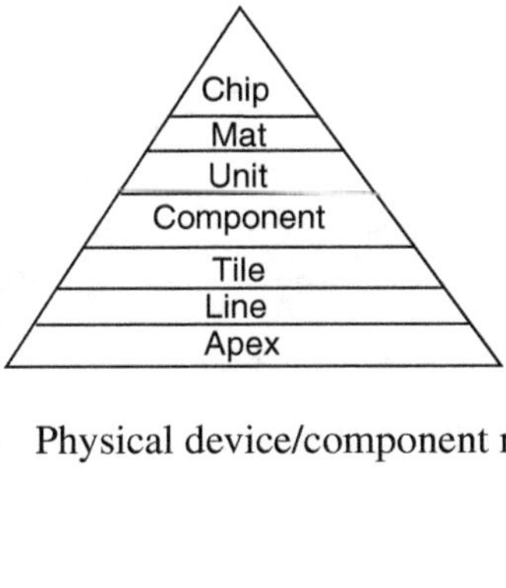

Figure 6.16 Physical device/component model layers

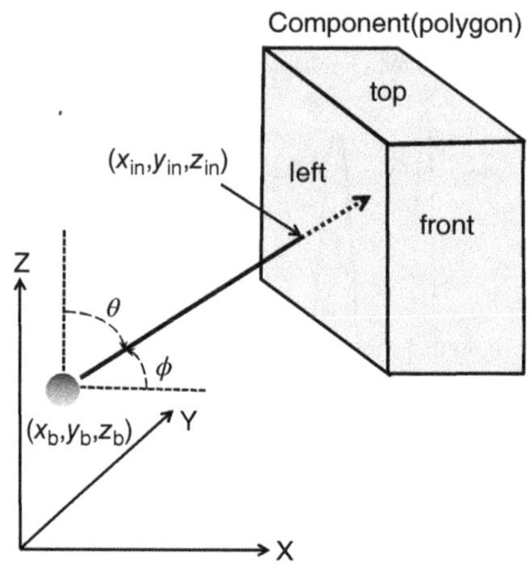

Figure 6.17 Outline of particle tracking

Points have three-dimensional coordinate values. Lines have point IDs at both ends. Tiles have the number and a sequence of points that makes a tile. A tile can also be a set of lines but this is not applicable here.

A rectangular parallelepiped has six tiles, top, bottom, right, left, front and back tiles as defined in Figure 6.17. Particle movement is defined by azimuth (ϕ) and zenith (θ) angles with kinetic energy. The inlet point (x_{in}, y_{in}, z_{in}) and outlet point (x_{out}, y_{out}, z_{out}) of a component can be calculated as a cross point of the path of the particle and tiles as shown in Section 6.9.

Sometimes, components in semiconductor devices can be modelled as a rectangular parallelepiped. Nevertheless, sometimes they are not, like the honeycomb structure of dynamic random access memory (DRAM) cells. A more general definition of a tile is given as a polygon:

As illustrated in Figure 6.18, as for a polygon ABCD ... X on the top tile, the coordinates of the first adjacent point B to the anchor point A can be calculated from the length and azimuth angle of the first edge of the polygon. The coordinates of the next point C can be calculated by using the length of edge L_2 and angle ϕ_1. All coordinates of points in a polygon can be calculated by these procedures.

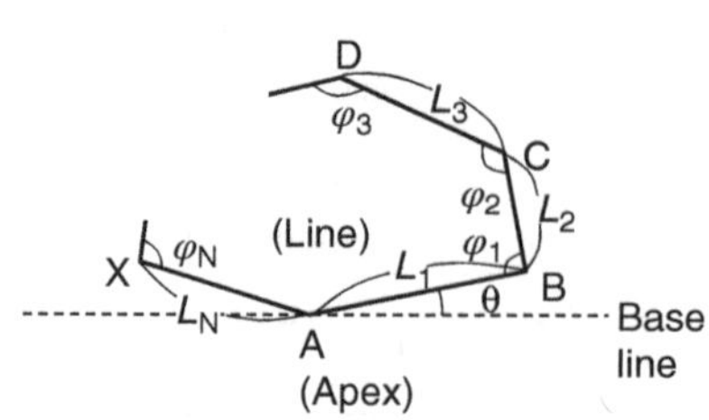

Figure 6.18 Schematic of polygon definition

Front surface of a tile is defined as the direction of right-hand thread movement when the thread is rotated in the sequence of the points within the tile. This information can be used to see whether an ion is near to a tile or not and whether an ion is inside a component or not.

The Unit (an SRAM memory cell) has unique properties compared to other components in order to have the overall necessary properties for assembling a cell. Critical charge Q_{crit} is the amount of charge collected to the storage node necessary to cause a flip of the state of SRAM or a soft error. The critical charge Q_{crit} is one of the most important properties of the Unit. As mentioned before, an SRAM has two storage nodes that are connected to a set of a p^+ node in an n-well and an n^+ node in a p-well. We call the set of nodes 'logical nodes' and call individual p^+ and n^+ nodes 'physical nodes'. There are three other diffusion layers that are connected to V_{cc},V_{ss} and data line. No visible impact is expected when the layers connected to V_{cc} and V_{ss} are hit by an ion, because they are not physical nodes. When some charge collected to the storage node which is connected to the data line and the transistors is 'ON' to read the data, erroneous data may be read depending on the sensitivity of the sense amps. We call this error mechanism 'bit-line mode'. In this case, the diffusion layer connected to the data line may be registered as a component. Moreover, the drift-diffusion process is thought to be the cause of soft error and a certain volume below the diffusion layer can be registered as a component.

When the upper structure has an extra source of secondary ion, relevant parts have to be registered as components of a unit. The possible cases are:

(i) When the parts contain high concentration of ^{10}B, low-energy neutrons may cause soft error through the neutron capture reaction $^{10}\text{B} + \text{n} \rightarrow {}^{7}\text{Li} + \alpha$. The parts emit ^{7}Li and α in a random direction from a random location inside the parts.

(ii) When the parts are contaminated by a α-ray source, α-ray-induced soft error may take place. The parts emit α-rays in random directions from a random location inside the parts.

6.8 Setting of Random Position of Spallation Reaction Point in a Component

In principle, a nuclear reaction location has to be set randomly (homogeneously) in the volume of one unit. This is simple, if the Unit shape is rectangular parallelepiped: a random location can be given by choosing random coordinates x, y, z on the X-Y-Z axes separately, as follows.

$$
\begin{aligned}
x &= rnd \times L_{\text{x}} \\
y &= rnd \times L_{\text{y}} \\
z &= rnd \times L_{\text{z}}
\end{aligned}
\qquad (6.55)
$$

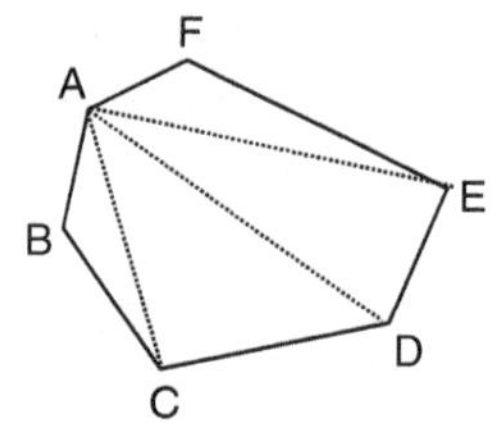

Figure 6.19 Triangle decomposition of a polygon

where,

L_x, L_y, L_z: length of edges of the Unit in parallel with the X, Y, Z axis, respectively;
rnd: random number generated independently for each edge.

However, in other cases of polygons, the above procedure cannot be applied. The following procedures are introduced as a method that is more versatile:

First, a polygon of the top tile of the Unit or component can be divided into triangles as shown in Figure 6.19. The area of each triangle can be calculated from the exterior product of two vectors made from two edges of a triangle ΔABC.

$$S_{\Delta ABC} = \frac{1}{2}|\overrightarrow{AB} \times \overrightarrow{AC}| \tag{6.56}$$

Next, a triangular space where spallation reaction takes place is chosen based on the probability P_i in proportion to the area S_i of i-th triangle,

$$P_i = \frac{S_i}{\sum_j S_j} \tag{6.57}$$

For each random number *rnd* generated, an i-th triangle is chosen when

$$P_{i-1} < rnd \leq P_i \qquad (P_0 = 0) \tag{6.58}$$

XY coordinates on the triangular top tile where spallation reaction takes place can be determined by the following procedures:

From the original triangle ΔABC, a triangle ΔABD is made by making a symmetry point D of point A through median point O as illustrated in Figure 6.20. Now, ABCD makes a parallelogram. With the properties of the parallelogram L (width), h (height, θ (slope), random x, y coordinates of point P can be determined homogeneously in the parallelogram:

$$x = rnd \times L + h/\tan\theta$$
$$y = rnd \times h \tag{6.59}$$

When the x, y coordinates of P are located in the triangular ΔABD, new x and y coordinates of P' can be given from symmetry point to O in the original ΔABC. In order

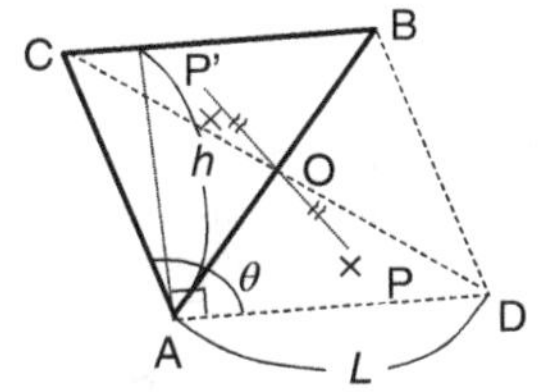

Figure 6.20 Formation of a parallelogram from a triangle to choose a random reaction point

to judge whether a point is located within a polygon, the following procedures may be applied.

First, make vectors from the point to all apexes of the polygon in the apex sequence and define the positive direction (clockwise, for example) of rotation around the point.

Second, make the exterior product of each set of two sequential vectors, for example point P in a polygon ABCD.

$$\overrightarrow{PA} \times \overrightarrow{PB} = (0, 0, (a_x - p_x)(b_y - p_y) - (a_y - p_y)(b_x - p_x))$$
$$|\overrightarrow{PA} \times \overrightarrow{PB}| = |\overrightarrow{PA}||\overrightarrow{PB}||\sin \angle APB| \tag{6.60}$$

The Z-coordinate of the exterior products gives direction of rotation from apex A to B.

An absolute rotation angle can be obtained from the second equation above. If the point P is within the polygon, the sum of the rotation angles (some angles are negative depending on the direction of the rotation) gives 2π.

Finally, the depth of the spallation reaction position can be determined by choice of random position within the thickness of the Unit or component.

A sample coder in Visual Basic™ for random location selection in a triangle is shown in Appendix A.8.

6.9 Algorithms for Ion Tracking

First, the starting point of ions or nuclear reaction point has to be set as explained in the previous section in the specific matrix cell randomly. After setting the point, ion tracking starts as shown in Figure 6.17.

The track of terrestrial charged particles, secondary ions produced from nuclear reaction and α, β-rays from radioisotopes can be traced by using CAD (Computer Aided Design) algorithms:

By the term 'tracking', we mean the process by which the path, energy of the charged particle and charge deposited by the charged particle are determined. A basic flow chart of ion tracking is summarised in Figure 6.21.

The volume that may cause soft error is limited in the semiconductor device. As for the 6 T-SRAM case, the layer that contains wells is only the part where a fault may be produced. We call this layer the susceptive layer. If the initial location of the charged

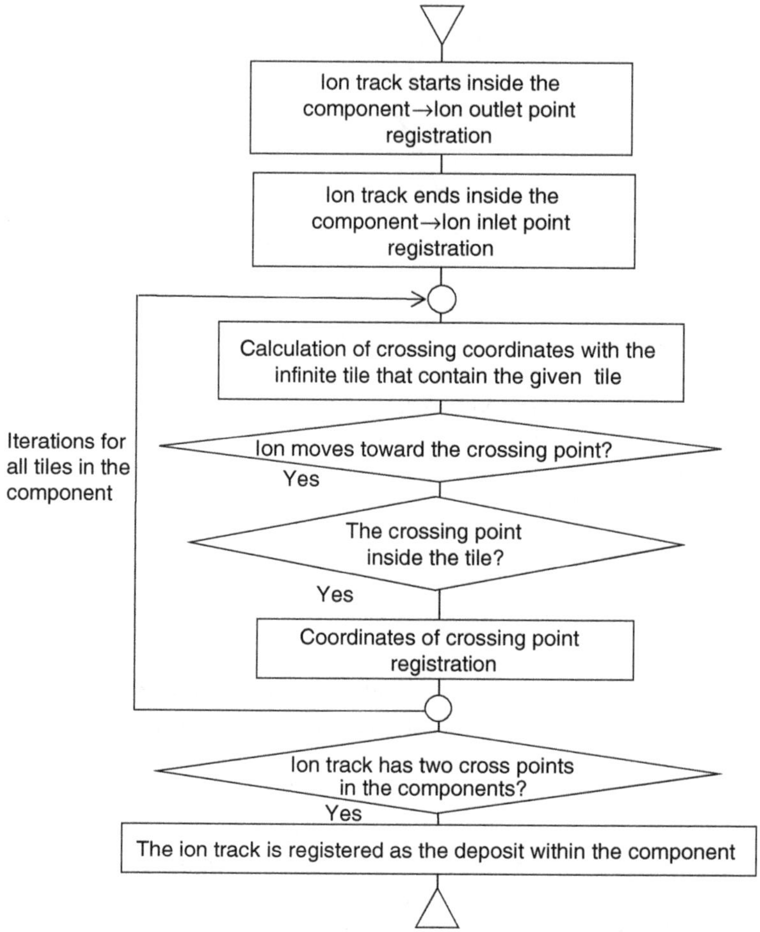

Figure 6.21 General flow of particle tracking in a component

particle is out of the susceptive layer and the ion move away from the susceptive layer, there is no possibility for the ion to cause a fault, and it can be excluded for further tracking.

Coordinates x, y, z of the charged particle are expressed by,

$$
\begin{aligned}
x &= x_0 + l_p \sin\theta \cos\phi \\
y &= y_0 + l_p \sin\theta \sin\phi \\
z &= z_0 + l_p \cos\theta
\end{aligned}
\tag{6.61}
$$

where,

x_0, y_0, z_0: initial coordinates
l_p: distance from the initial coordinate

θ: zenith angle of the ion track
ψ: azimuth angle of the ion track.

When coordinates on a layer satisfy the following equation,

$$ax + by + cz + d = 0, \tag{6.62}$$

the length l_{p} to the cross point of the track and the layer is given by,

$$l_{\mathrm{p}} = -\frac{ax_0 + by_0 + cz_0 + d}{a \sin\theta \cos\varphi + b \sin\theta \sin\varphi + c \cos\theta}. \tag{6.63}$$

When l_{p} has negative value, it means that the ion moves away from the layer. Coordinates of the cross point can be calculated from Equation (6.61). When two cross points of the upper and lower planes of the susceptive layer are obtained, cells through which the ion passes are analysed. Let us define $l_{\mathrm{p,near}}$ and $l_{\mathrm{p,far}}$ for l_{p} of the cross point closest to the starting point and then the furthest cross point. When the starting point or stopping point of the ion is within the susceptive layer, these points are at one end of the track. If $l_{\mathrm{p,near}}$ is shorter than the range of the ion, the ion cannot cause fault production and can be excluded for further calculation.

Now, cells for further analysis are determined in the device. The next step is to determine whether the ion causes fault production within the cell.

As for 6T-SRAM, components where faults can be produced are physical nodes (depletion layer), channel, drift volume and wells. Drift volume contains physical nodes and depletion layers so it is important to judge whether the ion will pass through the drift volume 'high' and well. A specific flow chart for ion tracking in a storage node is summarised in Figure 6.22. The judgement is made by the cross point calculation for all tiles that make the components.

When the ion passes through the drift volume, further judgement on whether the ion passes through the depletion layer in the drift volume is made. When the ion passes through the depletion layer, a funnelling model is applied to calculate the charge collected to the physical node. When the ion does not pass through the depletion layer a part of deposited charge in the drift volume is supposed to be collected to the physical node.

If the ion passes through two pn-junctions in p-wells, funnelling takes place on both surfaces of pn-junctions and half of the deposited charge flows out of the p-well at both sides.

6.10 Fault Mode Models

As indicated in Figure 6.23a, a funnelling model proposed by Hu *et al.* is applied [7] when the charged particles penetrate into the depletion layers under the p^+ and n^+ storage nodes. No charge collection is assumed to take place when the charged

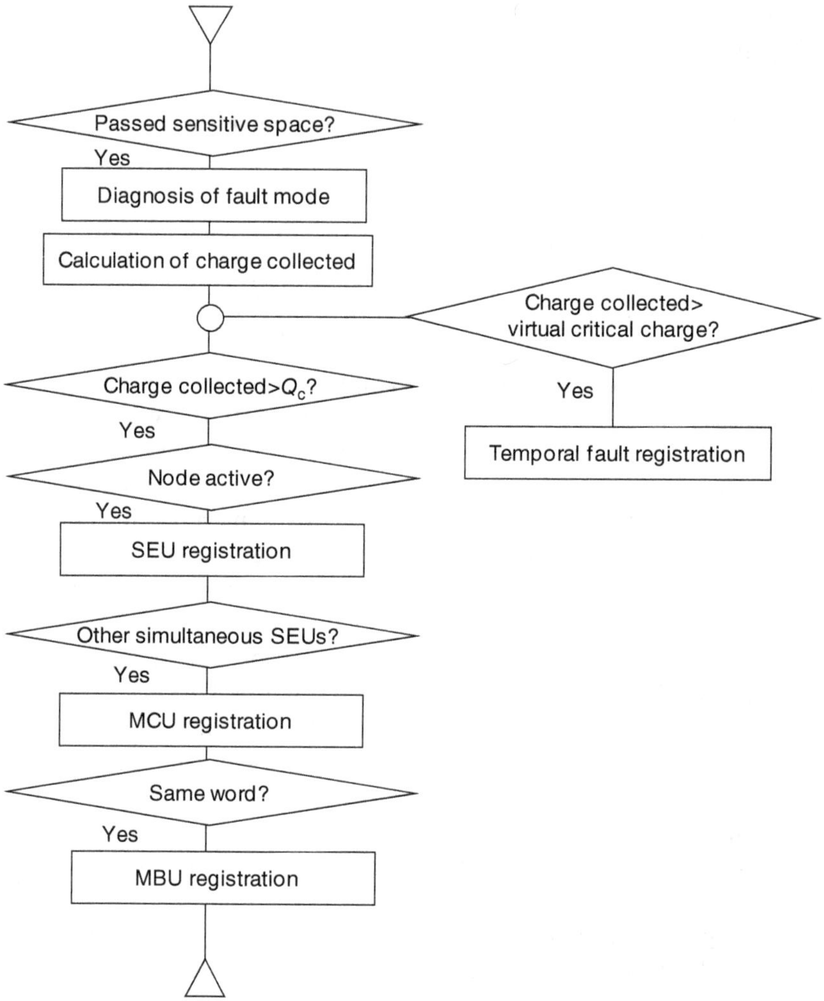

Figure 6.22 Simulation flow of secondary particle tracking in a sensitive node

particle penetrates into the STIs, storage nodes themselves and the upper regions (5 μm thickness Si layer is given in a positive z direction (z=0 at the top of the diffusion layer), but not shown in Figure 6.23) where recombination of electrons and holes are dominant.

The funnelling length x_c, within which a deposited charge is collected to the storage node from the bottom of the storage node is expressed as follows:

$$x_c = \left(1 + \frac{\mu_n}{\mu_e}\right) \frac{W}{|\cos \omega|}, \tag{6.64}$$

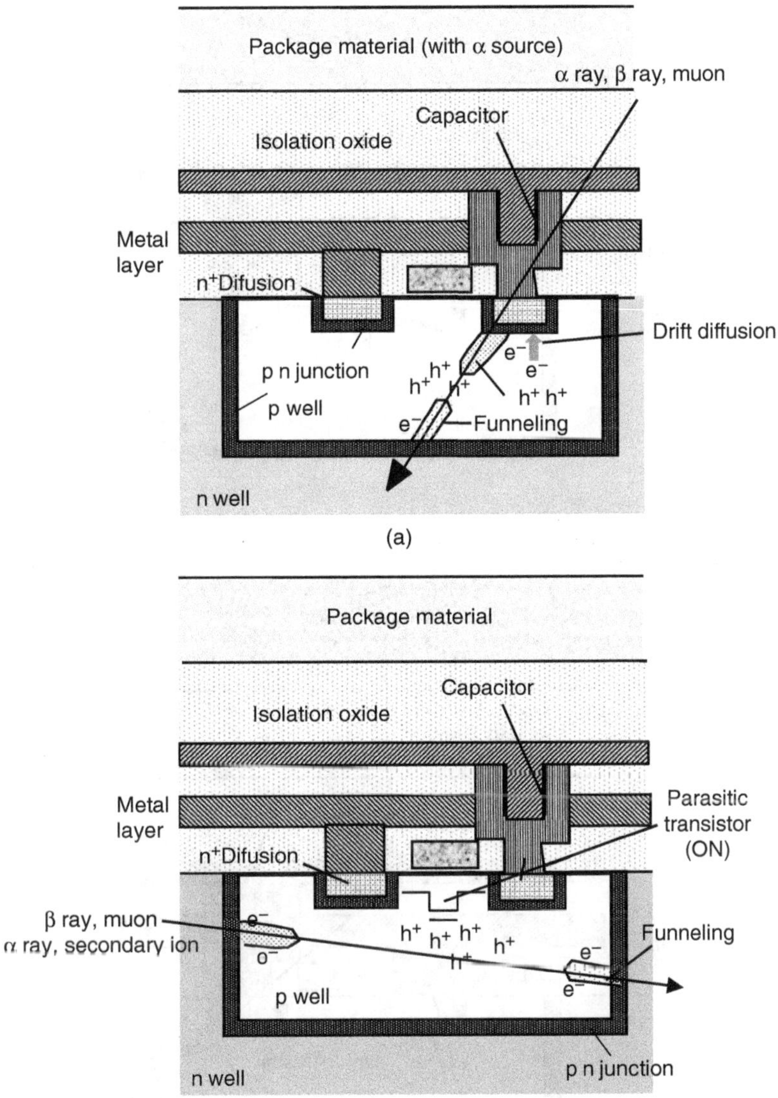

Figure 6.23 (a) Error modes in a memory cell with triple well: direct ionisation. (b) Error modes in a memory cell with triple well: bipolar mode. (c) Error modes in a memory cell with triple well: ALPEN (ALpha Particle source/drain PENtration) mode [14]. (d) Error modes in a memory cell with triple well: low/thermal neutron capture reaction with ^{10}B

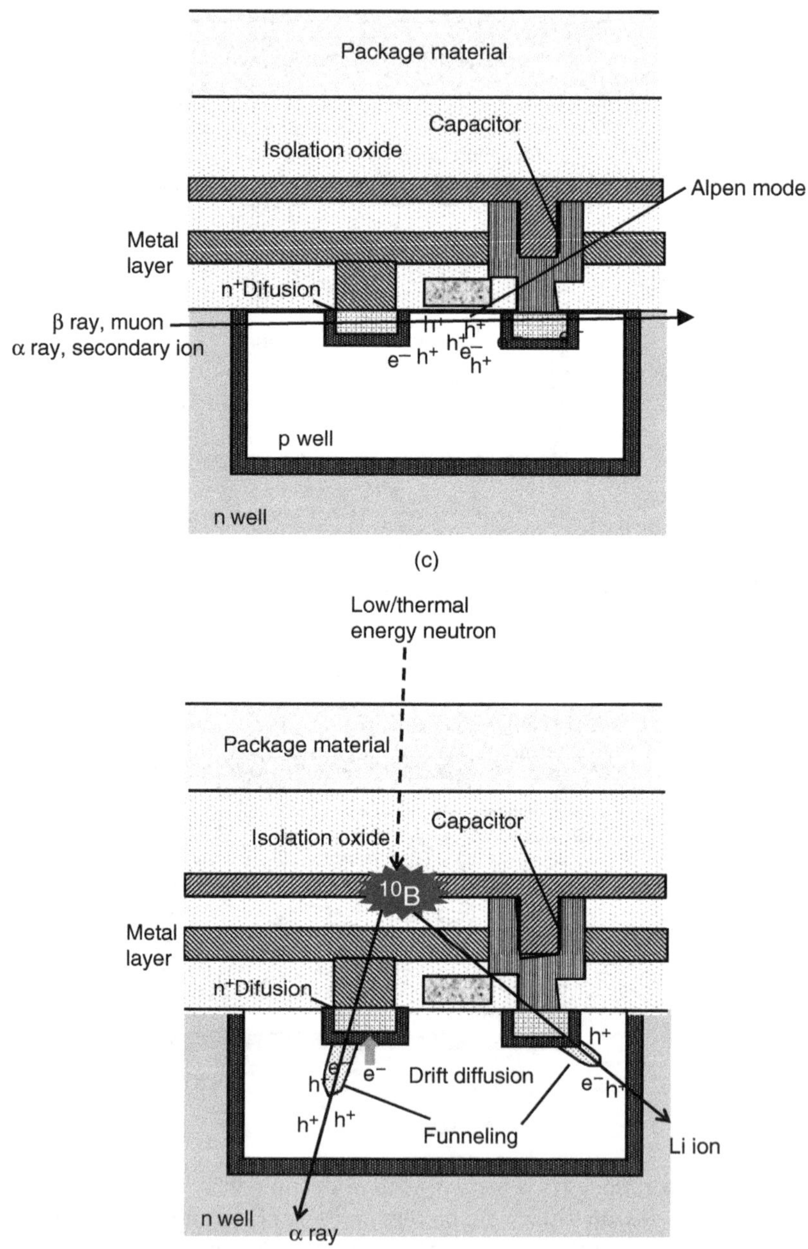

Figure 6.23 *(continued)*

where,

μ_n: mobility of hole
μ_e: mobility of electron
W: thickness of depletion layer (μm)
ω: zenith angle.

When the charged particle penetrates two p-n junctions in the bottom of the n$^+$ diffusion layer and the wall of the p-well, the funnelling grows towards both p-n junctions and it is assumed that eventually only half of the deposited charge is collected to the n$^+$ node.

Equation (6.63) may give very large maximum funnelling length when ω approaches $\pi/2$. A more realistic picture is that the ratio of collected charge to the n$^+$ node would be decreased as the distance from the n$^+$ node becomes longer along with the funnelling path due to a recombination effect of electrons and holes. Assuming that such decrease is an exponential function, the following equation is given to correct the collected charge $Q_{collect}$.

$$Q_{collect} = Q_{all} \exp\left(-\frac{x_c}{L_{max}}\right), \tag{6.65}$$

where,

Q_{all}: all deposited charge in the funnelling length x_c calculated from Equation (6.65).
L_{max}: effective maximum funnelling length (assumed to be 4 μm to give a better fit to experimental data).

Together with the funnelling model, a drift-diffusion model is also applied when the secondary ion does not penetrate the n$^+$ node but passes very close to the n$^+$ node. The drift diffusion box with a thickness of 0.1 μm is assumed below the n$^+$ node. All deposited charge is assumed to be collected to the n$^+$ node.

The sum of charge collected by the funnelling and drift diffusion is a metric of this fault mode – *cell mode*.

When the charge collection takes place in the storage nodes connected to bit line (BL), a similar charge collection mechanism takes place. This fault mode is known as *bit mode* and the primary metric of this type of fault is set to the charge collected.

In the p-well of a triple well, electrons are collected through the pn-junctions to the n-well or n-substrate as shown in Figure 6.23b, leaving holes in the p-well that cause a bipolar action. Whether bipolar action causes error or not depends primarily on the amount of charge left in the p-well. Therefore, deposited charge in the p-well is regarded as a primary metric for a bipolar-type fault.

As for the p$^+$ node in the n-well, only charge in the depletion layer is assumed to be collected to the node. Neither funnelling nor drift diffusion effect is applied to pMOSFET (p-channel Metal Oxide Semiconductor Field Effect Transistor) because of low mobility of holes.

This type of physical device model for *bipolar mode* may be over simplified compared to full TCAD models where three-dimensional electromagnetic responses of the semiconductor transistor structure to an injected charged particle are fully analysed [12], but such simplification is unavoidable for MC simulations where hundreds of thousands of nuclear reactions are incorporated in order to obtain satisfactory statistics.

When an ion traverses the side walls of source drain and the channel in-between, the channel can be 'ON' and cause soft error as shown in Figure 6.23c. This mechanism is known as 'ALPEN effect' [14] and channels are regarded as components to analyse this effect. The primary metric of the *ALPEN-mode* fault is set to the amount of charge deposited in the channel. In the upper portion of the active volume, metal or poly-Si gates, vias, metal wires are made in the oxide layer. In the soft-error simulation, gates, vias and metal wires are not important so they are neglected in the simulation model.

When thermal or epi-thermal neutrons react with ^{10}B in the device, secondary ^{7}Li and alpha particles traverse the device as shown in Figure 6.23d and all the same modes can take place due to the two secondary particles.

All fault data are stored in an interim database with relevant charges and fault types based on the flow chart in Figure 6.24.

By using the fault simulation interim database, SEUs including multi cell upsets (MCUs) can be extracted as shown in the flow chart in Figure 6.25 by comparing deposited charge in each fault by Q_{crit} for each mode. Since the event number is stored

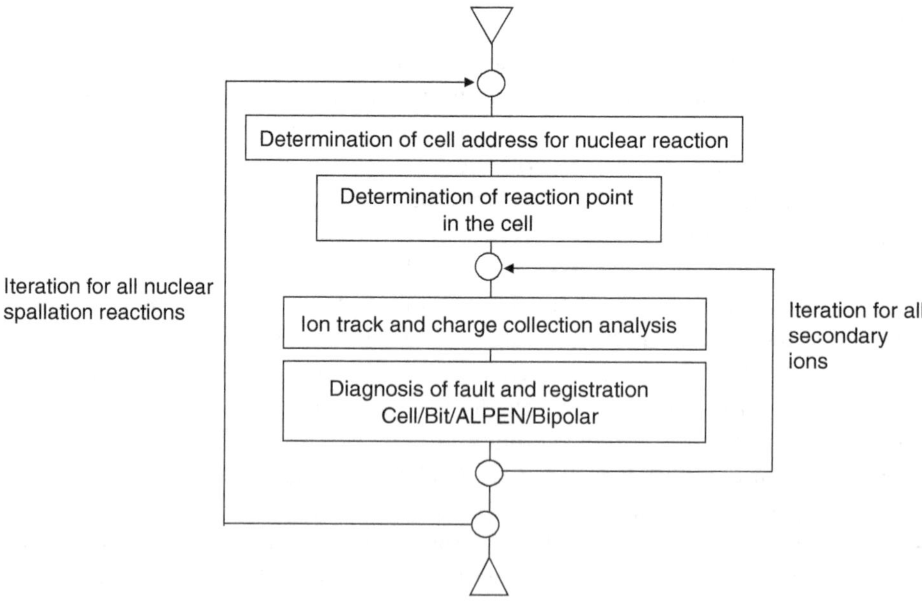

Figure 6.24 Classification flow of faults

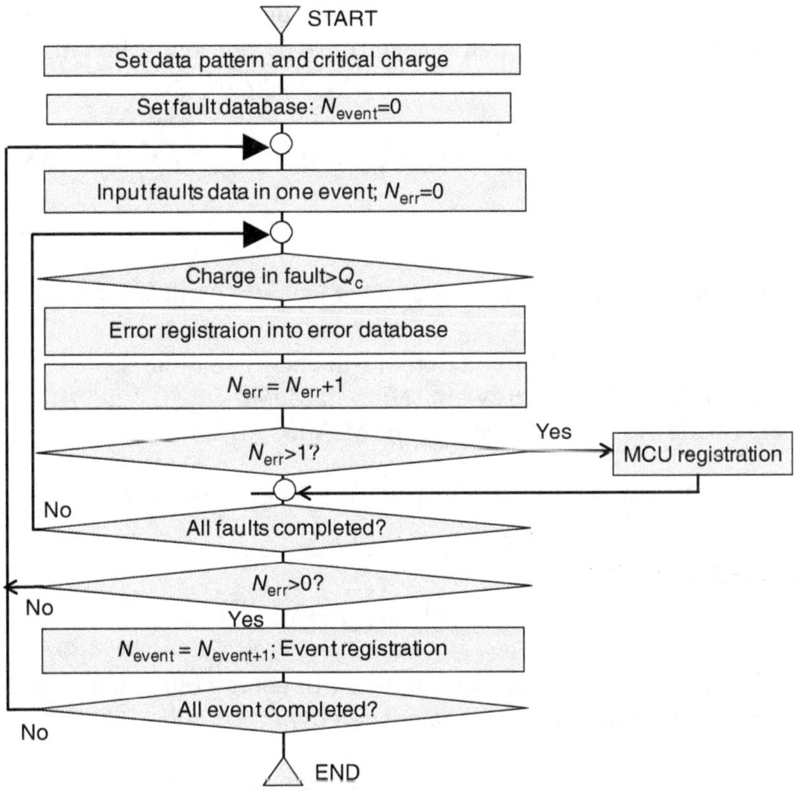

Figure 6.25 Classification flow of particle induced errors by event-base

in each fault in the database, MCU can be registered if two or more errors have the same event number. Once the interim database is constructed, a powerful parameter survey for evaluation of variations in SBU (Single Bit Upset) and MCU rate with many design parameters such as Q_{crit}, data pattern for each mode can be possible.

6.11 Calculation of Cross Section

The number density of target atoms N_{tar}(/cm^3) is given by,

$$N_{tar} = \rho_{tar} N_A / M_{tar}, \tag{6.66}$$

where ρ_{tar}: density of the target (g/cm^3);

N_A: Avogadro number
M_{tar}: atomic weight of the target.

In the simulation of nuclear spallation reaction, the total number of target particle N_{total} in the target volume $V_{\mathrm{sens}}(\mathrm{cm}^3)$ is set initially and the following relationship holds with the differential flux of incident particle $\partial\phi/\partial K_{\mathrm{n}}(/\mathrm{cm}^2/\mathrm{s})$,

$$N_{\mathrm{total}} = N_{\mathrm{tar}} V_{\mathrm{sens}} T_{\mathrm{irrad}} \int_{E_{\mathrm{min}}}^{E_{\mathrm{max}}} \frac{\partial\phi}{\partial K_{\mathrm{n}}} \sigma(K_{\mathrm{n}}) dK_{\mathrm{n}} = N_{\mathrm{tar}} V_{\mathrm{sens}} T_{\mathrm{irrad}} \sum_{\mathrm{i}=1}^{N_{\mathrm{b}}} \Delta\phi_{\mathrm{i}} \sigma_{\mathrm{i}} \Delta K_{\mathrm{n,i}}, \quad (6.67)$$

where,

$\sigma(K_{\mathrm{n}})$: total cross section of nuclear reaction at incident particle kinetic energy of K_{n} (cm^2)
σ_{i}: total cross section of nuclear reaction of i-th energy bin (cm^2)
$\Delta\phi_{\mathrm{i}}$: differential flux of i-th energy bin (n/cm^2/s/MeV)
$\Delta K_{\mathrm{n,i}}$: energy bandwidth in i-th energy bin (MeV).

Actual duration $T_{\mathrm{irrad}}(\mathrm{s})$ corresponding to N_{total} can be obtained by,

$$T_{\mathrm{irrad}} = \frac{N_{\mathrm{total}}}{N_{\mathrm{tar}} V_{\mathrm{sens}} \sum_{\mathrm{i}=1}^{N_{\mathrm{b}}} \Delta\phi_{\mathrm{i}} \sigma_{\mathrm{i}} \Delta K_{\mathrm{i}}}. \quad (6.68)$$

Differential cross section of production in a j-th energy bin when n_{p} p-th particles are produced for the duration can be calculated by,

$$\frac{\partial\sigma_{\mathrm{pj}}}{\partial K_{\mathrm{p}}} = \frac{n_{\mathrm{p}}/T_{\mathrm{irrad}}}{\Delta K_{\mathrm{pj}} \int_{K_{\mathrm{min}}}^{K_{\mathrm{max}}} \frac{\partial\phi}{\partial K_{\mathrm{n}}} dK_{\mathrm{n}}}. \quad (6.69)$$

When the number of events N_{event} during this duration is obtained, the error rate soft error rate (SER) and SEE cross section is given by,

$$SER = \frac{N_{\mathrm{event}}}{T_{\mathrm{irrad}}} \times 3600 \times 10^9 \ \ (\mathrm{FIT}), \quad (6.70)$$

where failure in time (FIT) is the unit of error rate in events/(1×10^9 hour·device)

$$\sigma_{\mathrm{SEE}} = \frac{N_{\mathrm{event}}/T_{\mathrm{irrad}}}{\int_{K_{\mathrm{min}}}^{K_{\mathrm{max}}} \frac{\partial\phi}{\partial K_{\mathrm{n}}} dK_{\mathrm{n}}} \quad (6.71)$$

6.12 Prediction for Scaling Effects of Soft Error Down to 22 nm Design Rule in SRAMs

Figure 6.26 shows cumulative flux used for simulations of SEE by electrons, protons, muons and neutrons. The expression of cumulative flux is more useful than the differential expression. One can immediately see, for example that the total flux of

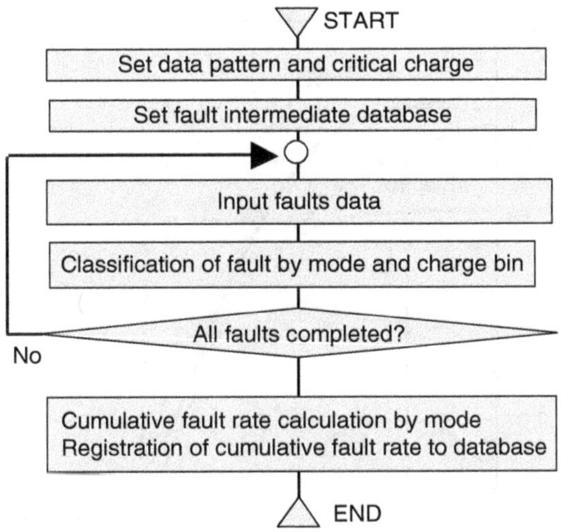

Figure 6.26 Classification flow of faults to obtain a statistics of cumulative single event fault rate

neutrons above 10 MeV is $13\text{n/cm}^2\text{/h}$, which is the JESD89A recommended value for neutron-induced SEU calculation from test results in spallation neutron facilities.

A scaling roadmap is assumed for simulation of SRAM SEU: Lateral two-dimensional scaling is assumed to reduce the area by a factor of 2 by each generation (six generations: 130=>90=>65=>45=>32=>22 nm). Meanwhile, density (Mbit/chip) is assumed to increase by a factor of 2 with each generation. Depth profile is assumed to be constant due to lack of roadmap information and also because of difficulty in making a shallow profile. As parasitic capacitance is basically in proportion to device area, critical charge is also assumed to decrease by a factor of 2 in each generation. Although reduction in the supply voltage V_{cc} is preferable for reducing power consumption, it is actually being limited in order to ensure enough margins from the upper bound of V_{th} variation [15] and, therefore, assumed to be constant.

Figure 6.27 shows typical results on SEU in 130 to 22 nm SRAMs and total FBM (Failed Bit Map) for 22 nm SRAM after 60 000 nuclear spallation reactions at the origin.

It is seen from the figure that:

(i) SEU in SRAMs decreases mildly from 130 down to 22 nm. The decrease is also quite consistent with the recent experimental data [16].
(ii) Although SEU/Mbit decreases beyond 130 nm, SEU/device increases by a factor of as much as 6–7 due to an intense increase in density.
(iii) MCU ratio increase exponentially as scaling proceeds. The result is that the MCU ratio drastically increases as scaling proceeds giving a MNT (Multi-Node

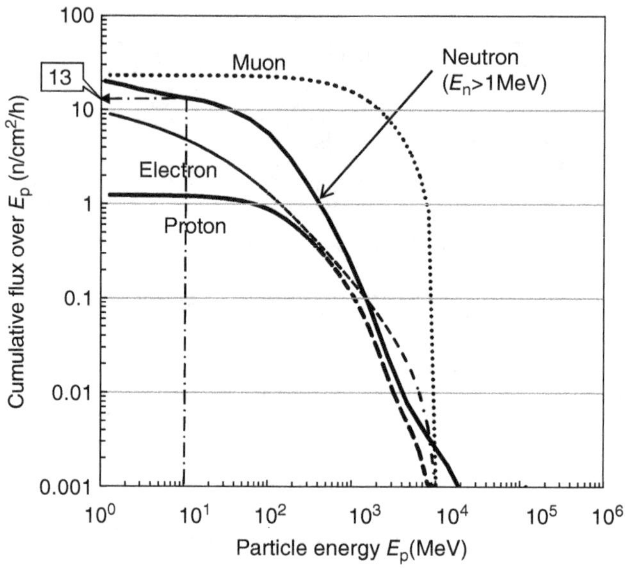

Figure 6.27 Cumulative flux of secondary cosmic rays at NYC sea level

Transient) in which multiple logical nodes of sequential or combinational logic device are corrupted and increase as well. This may cause a serious impact on the reliability design of logic devices since MNTs would make error detection impossible. This would make the redundancy SER mitigation techniques extremely vulnerable to MNT.

(iv) It is seen from the FBM that the area densely affected in 22 nm SRAM covers about as large as 1000 × 1000 bit (1 Mbits) area.

6.13 Evaluation of Effects of Heavy Elements in Semiconductor Devices by Nuclear Spallation Reaction

In CORIMS, it is assumed that semiconductor devices are made of Si only for simulation of nuclear spallation reaction but charge deposition takes place only in actual Si components such as well and source-channel drain. A number of elements, however, are used in Silicon-based semiconductors such as W, Cu, Ti, and so on. These elements which are heavier than Si may have more impact on neutron soft-error rate because they produce heavier secondary particles that cause dense charge deposition in Si. In particular, since WSi_2 plug contacts almost directly with the Si diffusion layer, secondary particles from W may have a significant impact.

In order to evaluate this point, the multi-element neutron soft-error simulator SEALER was developed and applied with virtual multi-element components as shown in Figure 6.28. Databases for calculation of total nuclear reaction section and

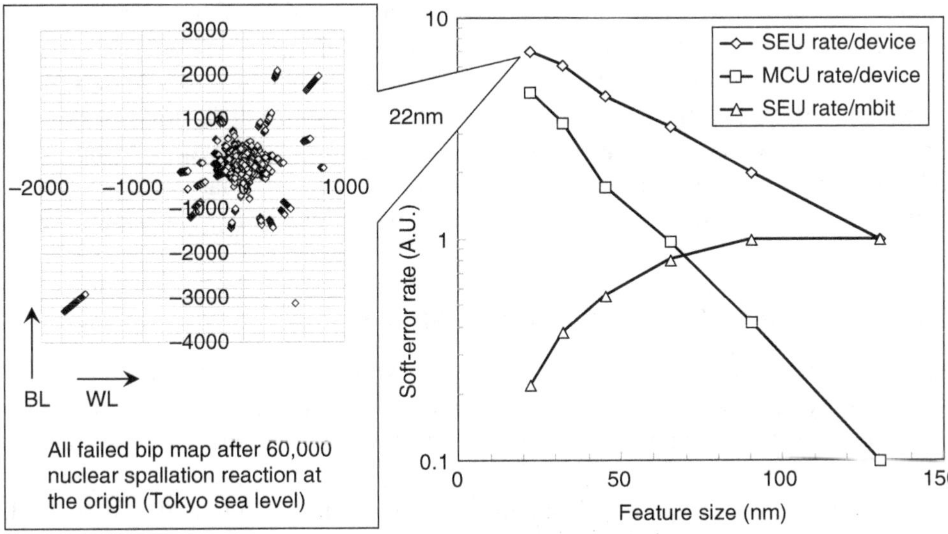

Figure 6.28 An example of simulation results for the effects of scaling by terrestrial high-energy neutron

LETs are introduced in detail in [1]. A component where nuclear spallation reaction takes place is chosen based on the sum of total nuclear cross section of the component. An element with which nuclear spallation reaction takes place is chosen based on the sum of total nuclear reaction cross section of the element in the component.

Energy spectra of secondary particles at the boundary of the Si p-well are evaluated for 100 MeV neutrons as shown in Figure 6.29 in the element base. It is seen that the energy and frequency of heavy secondary particles such as Ta and W at the well

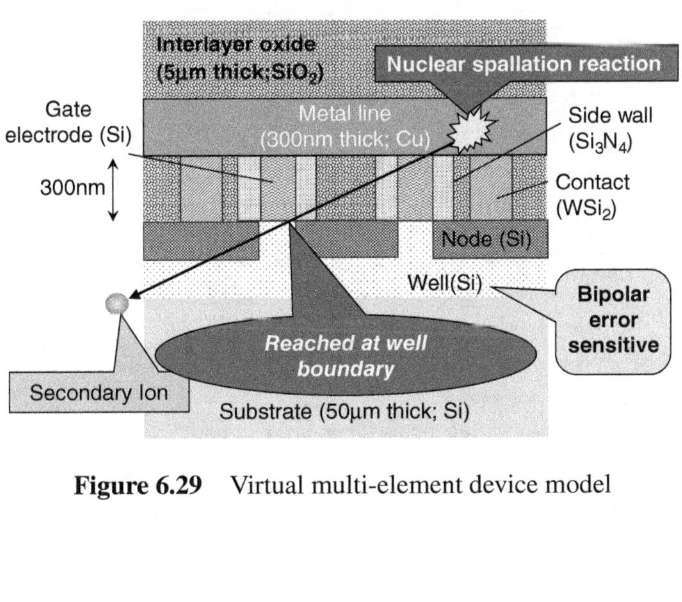

Figure 6.29 Virtual multi-element device model

boundary are quite low compared to other secondary particles from lighter elements such as Si, O and Cu. Even maximum energies of W and Ta are below 1 MeV, because they lose their energy by dense charge deposition before reaching the Si well, or even in WSi_2 plug itself [8]. In [8], Ibe *et al.* did not include nuclear fission reaction of W in their model. Clemens *et al.* also carry out similar simulation with a 65 nm SRAM model with a nuclear fission model of tungsten and conclude that W does not have a significant impact on SEU or MCU with a reasonable range of Q_{crit} (1-2fC) [17].

6.14 Upper Bound Fault Simulation Model

The particle penetrates from outside the device with random position and direction determined by the MC method. Charges produced in the p-well region in each SRAM cell hit by a charged particle are calculated and tabulated, so that multiple faults are tabulated for each single particle. Care must be paid at this point. The statistics in the following are not 'event-base'. All faults in an event are tabulated as independent faults. Light particles with long ranges like protons, muons and electrons would give much larger fault numbers than the actual number of single events. The total flow chart of upper bound fault simulation is summarised in Figure 6.30, starting with the intermediate fault database.

The cumulative number of faults N_{f}^{s} for the s-th charge bin is calculated from

$$N_{\text{f}}^{\text{s}} = \sum_{\text{i=s}}^{M} n_{\text{i}} \tag{6.72}$$

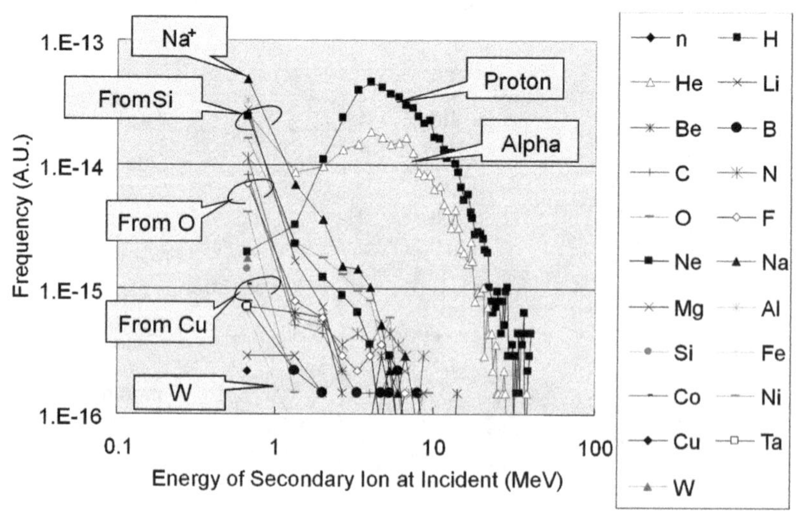

Figure 6.30 Energy spectrum of each element at the well boundary

where,

n_i: the number of faults in a i-th charge bin
M: the number of charge bins.

The cumulative upper bound SER_{UB}^s beyond Q_s (s-th charge bin) is calculated from

$$SER_{UB}^s = \frac{N_f^s}{T_{irrad}} \times 10^9 \text{ (FIT/bit)} \tag{6.73}$$

where, T_{irrad}: actual duration of particle irradiation (h).

As for charged particles with direct ionisation like muons, electrons and protons without spallation, T_{irrad} can be calculated from,

$$T_{irrad} = \frac{N_p}{\phi_{cum} \times S_c} \text{ (h)} \tag{6.74}$$

where,

N_p: total number of particles injected in the simulation
ϕ_{cum}: cumulative flux from the minimum energy (n/cm^2/h)
S_c: area of a SRAM cell (cm^2).

As for particles with nuclear reaction, T_{irrad} can be calculated from,

$$T_{irrad} = \frac{n_p^j}{\phi_j \times N_{Si} \times \sigma_j \times \Delta E_j} \text{ (h)}, \tag{6.75}$$

where,

n_p^j: number of particles injected in the j-th energy bin for simulation
ϕ_j: field differential flux for the j-th energy bin (n/cm^2/h/MeV)
N_{Si}: number of Si nuclei in a SRAM cell
σ_j: cross section of the relevant nuclear reaction (cm^2)
ΔE_j: width of the j-th energy bin (MeV).

The number j can be chosen appropriately.

6.15 Upper Bound Fault Simulation Results

6.15.1 Electrons

Figure 6.31 shows simulation results for the cumulative $SEFR_{UB}$ (Upper Bound Single Event Fault Rate) by electrons for cell, bipolar, bit, ALPEN and total modes. Cumulative SER_{UB} means total number of faults produced when the critical charge

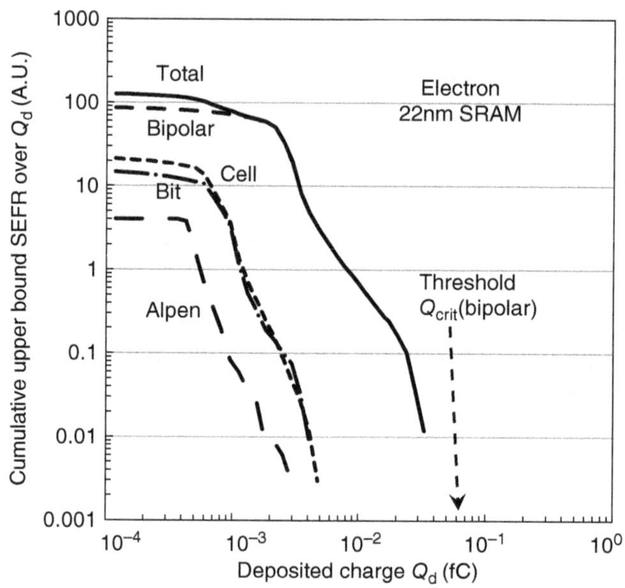

Figure 6.31 Simulation result of cumulative upper bound fault rate in a 22 nm SRAM by terrestrial electrons at NYC sea level

Q_{crit} of SRAM equals the deposited charge Q_d on the x-axis. It can be seen that contribution of the bipolar mode is much larger than other modes, because the volume of sensitive zone is much larger than any other mode. In reality, bipolar mode by electron is less likely to take place than other particle's case since deposited charge has to be very low to trigger bipolar action.

It is also seen from the figure that if the Q_{crit} is larger than 0.04 fC, any fault is not produced at all for all fault modes including the bipolar mode and eventually failure is extremely unlikely to occur. If the Q_{crit} is larger than 0.005 fC, the cell-mode fault is not produced at all. We refer to this amount of charge as threshold Q_{crit} (mode). Therefore, threshold Q_{crit} (bipolar) = 0.04 fC and threshold Q_{crit} (cell) = 0.005 fC for electron.

The impacts of ALPEN and bit mode faults are negligibly small as seen in other particle results.

6.15.2 *Muons*

Figure 6.32 shows the simulation results for the cumulative $SEFR_{UB}$ by muon for cell, bipolar, bit, ALPEN and total modes.

Threshold Q_{crit} (bipolar) = 0.2 fC. In the case of bipolar mode, the threshold is rather high compared to electros, but the charge density along with muon is below a few

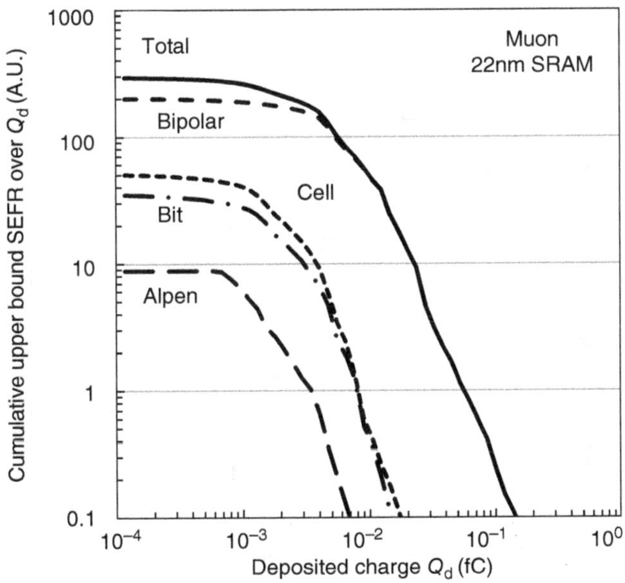

Figure 6.32 Simulation result of cumulative upper bound fault rate in a 22 nm SRAM by terrestrial muons at NYC sea level

femtocoulombs per micrometre and much smaller than secondary ions from nuclear spallation reaction with neutron/proton as seen in Chapter 3 so that bipolar mode by muons rarely takes place. Threshold Q_{crit} (cell) = 0.02 fC. This means that muon does not cause a cell mode fault at all.

6.15.3 Direct Ionisation by Proton

Figure 6.33 shows cumulative $SEFR_{UB}$ by direct ionisation of proton. Threshold Q_{crit} (bipolar) = 0.5 fC. This means that bipolar action of 22 nm SRAM due to direct ionisation by proton is a possible threat. Threshold Q_{crit} (cell) = 0.1 fC. This means that cell-mode fault may be a possible threat for 22 nm SRAMs.

6.15.4 Proton Spallation

Figure 6.34 shows cumulative $SEFR_{UB}$ by spallation reaction of protons. Threshold Q_{crit} (bipolar) = 300 fC and threshold Q_{crit} (cell) = 40 fC although the $SEFR_{UB}$ is very low compared to direct ionisation by electrons, muons and protons. This implies that more accurate and realistic upper bound estimation is necessary to take the next steps – ignore or mitigate.

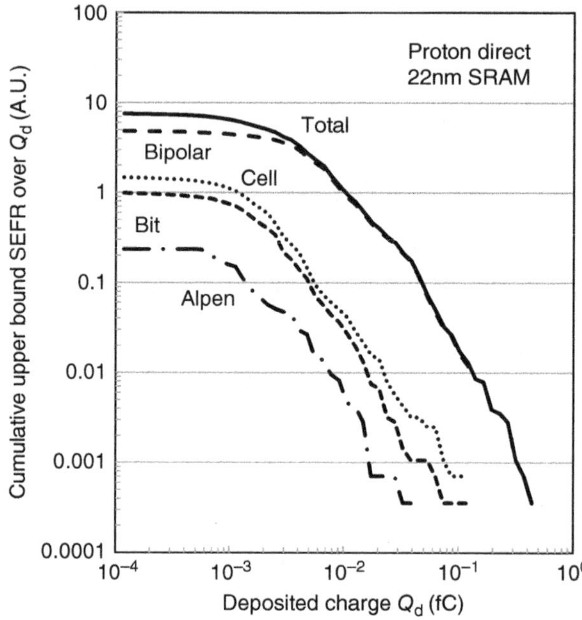

Figure 6.33 Simulation result of cumulative upper bound fault rate in a 22 nm SRAM by direct ionisation of terrestrial protons at NYC sea level

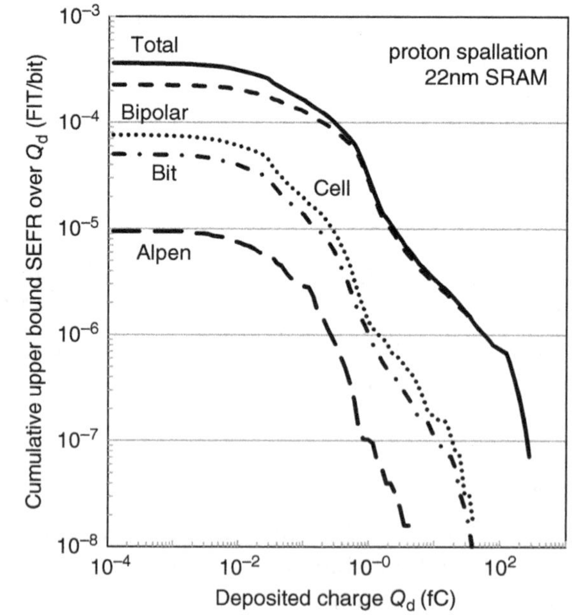

Figure 6.34 Simulation result of cumulative upper bound fault rate in a 22 nm SRAM by spallation reaction of terrestrial protons at NYC sea level

6.15.5 Low-Energy Neutron

Figure 6.35 shows the simulation results for the cumulative $SEFR_{\mathrm{UB}}$ by low-energy (<1 MeV) neutron capture reaction of contaminated ^{10}B in the 5 μm thick upper layer with a concentration of 10^{20} isotopes/cm^3 above the MOSFET transistor layer. Compared with high-energy (>1 MeV) neutron results shown in Figure 6.36, the cumulative $SEFR_{\mathrm{UB}}$ is much lower than that of a high-energy neutron. Threshold Q_{crit} (bipolar) = 0.4 fC and threshold Q_{crit} (cell) = 0.04 fC.

6.15.6 High-Energy Neutron Spallation

Figure 6.36 shows simulation results for the cumulative $SEFR_{\mathrm{UB}}$ by high-energy (>1 MeV) neutron. Threshold Q_{crit} (bipolar) = 100 fC and threshold Q_{crit} (cell) = 20 fC. The $SEFR_{\mathrm{UB}}$ is much higher than that of proton spallation. This implies that high-energy neutrons are still a major contributor in 22 nm SRAMs to SEE at the ground.

6.15.7 Comparison of Secondary Cosmic Rays

Figure 6.37 compares total cumulative $SEFR_{\mathrm{UB}}$ for all radiations. The following can be pointed out:

(i) Cumulative $SEFR_{\mathrm{UB}}$ by proton spallation are much lower than that by neutron spallation mainly due to the difference in flux, but the spectrum shape is almost consistent with that of neutrons except for higher deposited charge (more than a few femtocoulombs). This may be the slight difference in the secondary particles since total charge in the reaction system is slightly different (in one atomic unit).

(ii) Protons, low-energy neutrons and high-energy neutrons can be major contributors to faults since the deposited charge is larger than other light particles.

(iii) If the critical charge is reduced to below 0.1 fC, electrons, muons and direct ionisation by protons can be a big threat since their SEFRs are potentially very high. Bipolar mode, however, is not likely from these radiation sources since the deposited charge is very low.

6.16 Upper Bound Simulation Method for SOC (System On Chip)

An upper bound fault rate calculation can be applied to any kind of device. This means the method can be applied to an entire chip and a board. Figure 6.38 shows the flow of upper bound simulation of SOC (System On Chip). First, the data of types of devices and their numbers in the chip must be obtained. Second, the layout of each device should be obtained from, for example GDS-II files. Depth profile information should be obtained or reasonably assumed. Based on the layout and depth profiles, a

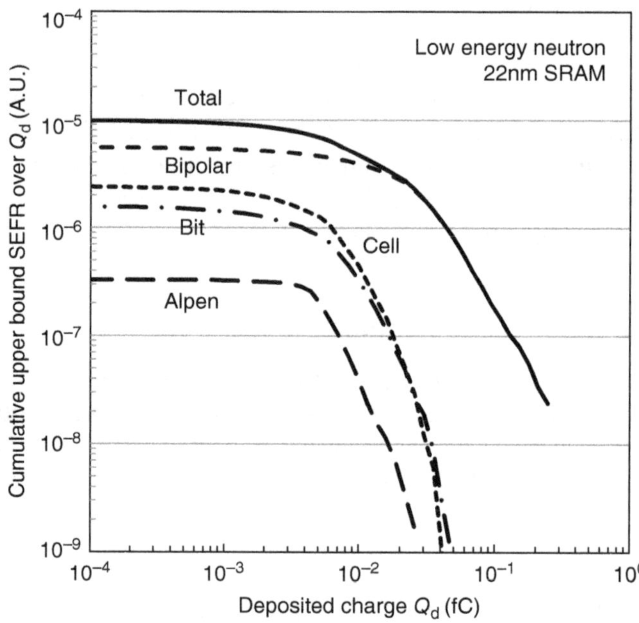

Figure 6.35 Simulation result of cumulative upper bound fault rate in a 22 nm SRAM by neutron capture reaction of ^{10}B with low-energy neutrons at NYC sea level

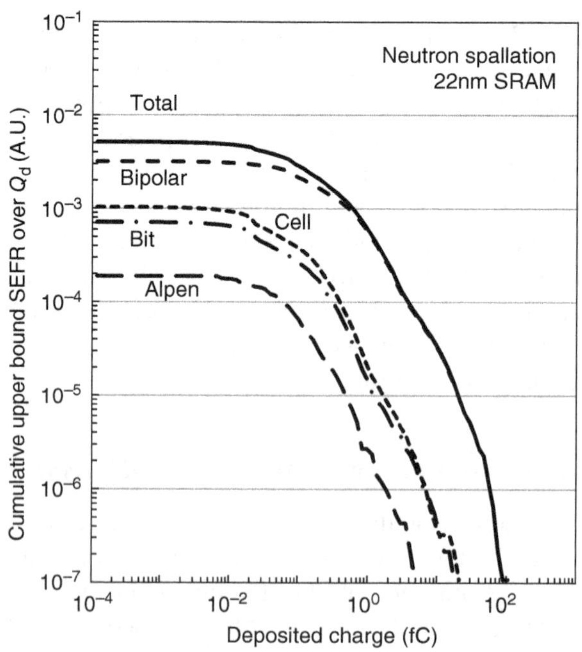

Figure 6.36 Simulation result of cumulative upper bound fault rate in a 22 nm SRAM by neutron spallation reaction with high-energy neutrons at NYC sea level

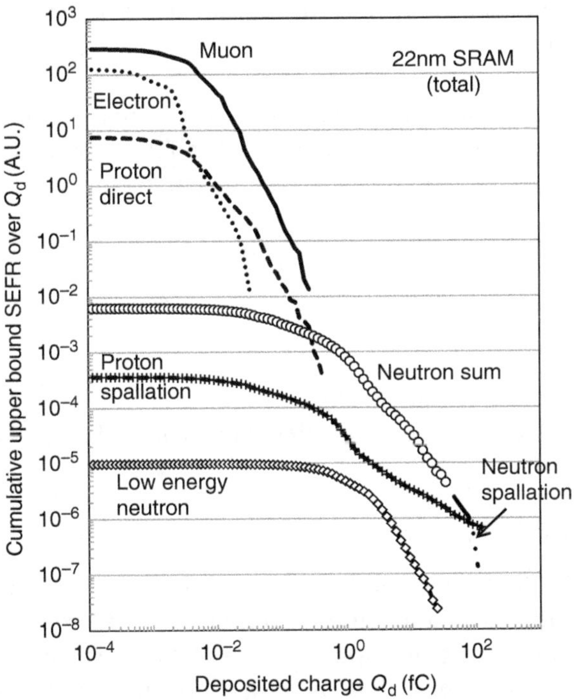

Figure 6.37 Comparison of simulation result of cumulative upper bound fault rate in a 22 nm SRAM by all secondary cosmic rays at NYC sea level

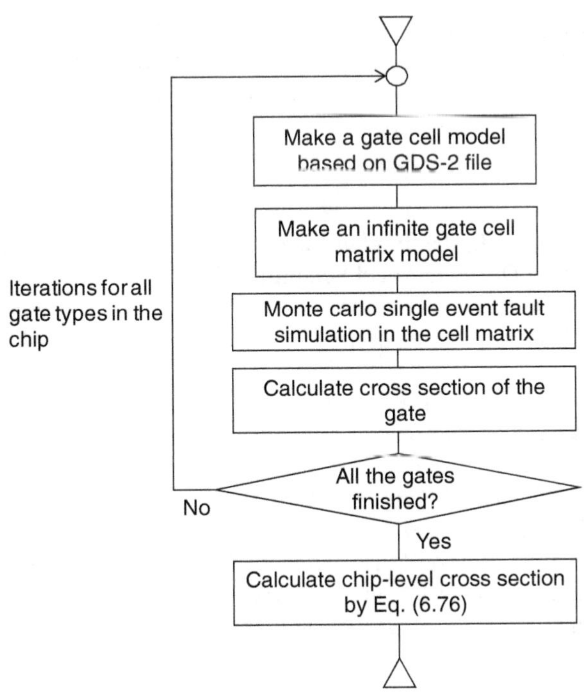

Figure 6.38 Simulation flow of cumulative upper bound fault rate in SOC (system on chip)

three-dimensional model can be constructed for each device by using the automated tool. Then, a virtual infinite matrix of each device is made and MC fault simulation is carried out.

From the simulation result, an upper bound fault cross section of each device is obtained. After carrying out these procedures for all types of devices, an upper bound fault cross section σ_{chip} of an entire SOC can be obtained by,

$$\sigma_{\text{chip}} = \sum_{\text{i}} N_{\text{i}} \times \sigma_{\text{i}}, \tag{6.76}$$

where,

N_{i}: the number of i-th device in the chip
σ_{i}: upper bound fault cross section of i-th device obtained by the simulation.

6.17 Summary of Chapter 6

The first major modelling of relativistic binary collision, INC, evaporation, charge collection/deposition, nuclear capture reaction by ^{10}B are shown together with automated device modelling and ion tracking techniques. Some typical results of soft-error simulation in 22 nm SRAMs under terrestrial high-energy neutrons and upper bound fault simulation results of 22 nm SRAMs are also shown under terrestrial electron/muon/proton/low-energy (<10 MeV) neutron/high-energy neutrons with four fault modes (cell/bit/ALPEN/Bipolar). The impacts of all terrestrial radiation sources are evaluated in terms of cumulative upper bound SEFR and threshold Q_{crit} (fault mode). The impacts of ALPEN and bit mode are shown to be negligible. It is important to evaluate the impacts of cell and bipolar modes carefully.

References

[1] Nakamura, T., Baba, M., Ibe, E. *et al.* (2008) *Terrestrial Neutron-Induced Soft-Errors in Advanced Memory Devices*, World Scientific, Hackensack, NJ.

[2] Kanekawa, N., Ibe, E., Suga, T. and Uematsu, Y. (2010) *Dependability in Electronic Systems - Mitigation of Hardware Failures, Soft Errors, and Electro-Magnetic Disturbances*, Springer.

[3] Ibe, E., Ishibashi, K., Osada, K. *et al.* (2011) *Low Power and Reliable SRAM Memory Cell and Array Design*, Springer.

[4] Ibe, E., Taniguchi, H., Yahagi, Y. *et al.* (2010) Impact of scaling on neutron-induced soft error in SRAMs from a 250 nm to a 22 nm design rule. *IEEE Transactions of Electron Devices*, **57** (7), 1527–1538.

[5] Bertini, H.W., Culkowski, A.H., Hermann, O.W. *et al.* (1978) High energy(E≦100 GeV) intranuclear cascade model for nucleons and pions incident on nuclei and comparisons with experimental data. *Physical Review C*, **17** (4), 1382–1394.

[6] Tang, H.H.K., Srinivasan, G.R. and Azziz, N. (1990) Cascade statistical model for nucleon-induced reactions on light nuclei in the energy range 50-MeV-1GeV. *Physical Review C*, **42** (4), 1598–1622.

[7] Hu, C. (1982) Alpha-particle-induced field and enhanced collection of carriers. *IEEE Electron Device Letters*, **EDL-3** (2), 31–34.

[8] Ibe, E., Yahagi, Y., Yamaguchi, H. and Kameyama, H. (2005) SEALER: novel Monte-Carlo simulator for single event effects of composite-materials semiconductor devices. 2005 Radiation and its Effects on Components and Systems, 19–23 September, Palais des Congres, Cap d'Agde, France (E-4).

[9] Wen, S., Pai, S.Y., Wong, R. *et al.* (2010) B10 findings and correlation to thermal neutron soft error rate sensitivity for SRAMs in the sub-micron technology. IEEE International Integrated Reliability Workshop, Stanford Sierra, California, 17–21 October, pp. 31–33.

[10] Wen, S., Wong, R., Romain, M. and Tam, N. (2010) Thermal neutron soft error rate for SRAMs in the90nm-45nm technology range. 2010 IEEE International Reliability Physics Symposium, Anaheim, CA, 2–6 May (SE5.1), pp. 1036–1039.

[11] Sreepad, H.R. Fermi Gas Model, http://www.gcm.ac.in/downloads/elearning/Fermi %20gas%20model(2c).pdf (accessed 7 April 2013).

[12] Furihata, S. (2004) Parameters used in GEM. Development of a generalized evaporation model and study of residual nuclei productions. PhD Thesis, Tohoku University, pp. 18–20.

[13] Perley, C.M. and Perley, F.G. (1976) *Compilation of phenomenological optical-model parameters*, 1954–1975. **17**(1).

[14] Takeda, E., Hisamoto, Dai and Toyobe, T. (1988) A new soft-error phenomenon in VLSIs. The alpha-particle-induced source/drain penetration (ALPEN) effect. Annual Proceedings. IEEE International Reliability Physics Symposium, 26, pp. 109–112.

[15] Sugii, N., Tsuchiya, R., Ishigaki, T. *et al.* (2008) Comprehensive study on Vth variability in silicon on thin BOX (SOTB) CMOS with small random-dopant fluctuation: finding a way to further reduce variation. IEEE International Devices Meeting, San Francisco, CA, 15–17 December, pp. 249–253.

[16] Dixit, A., Heald, R. and Wood, A. (2009) Trends from ten years of soft error experimentation. 2009 IEEE Workshop on Silicon Errors in Logic-System Effects, Stanford University, California, 24, 25 March.

[17] Clemens, M.A., Sierawski, B.D., Warren, K.M. *et al.* (2011) The effects of neutron energy and high-Z materials on single event upsets and multiple cell upsets. *IEEE Transactions on Nuclear Science*, **58** (6), 2591–2598.

7

Prediction, Detection and Classification Techniques of Faults, Errors and Failures

7.1 Overview of Failures in the Field

Table 7.1 summarises recent concerns in various fields of industries related to SEEs (Single Event Effects) with applications, possible root causes and observable failure symptoms [1–22].

In avionics, TMR (Triple Modular Redundancy) is applied to critical components to avoid FTFLs (Fatal Failures) and faults can be found through error flags from TMRs [1]. IGBTs (Insulated Gate Bipolar Transistors) are used in trains, automotives [2, 19, 20] and avionics for inverters or DC-DC converters. They are subject to destructive SEE mode, SEB (Single Event Burnout) due to electron avalanche by a charged particle penetration in pnp structures. SEEs in servers [3, 4] and routers [5] have been of major concern in networks because they use a tremendous amount of memory, in particular, SRAMs (Static Random Access Memories) that turn out to be very prone to radiations [24, 25]. Power supplies for network systems like DC-DC converters are also vulnerable to terrestrial radiation in the destructive mode [7].

Power consumption is one of the major metrics that determines the mitigation techniques to be applied to large-scale systems like data centres and supercomputers [6, 8–10]. 'Dark silicon' refers to such constraints on performance and reliability due to power consumption [6, 11]. Space redundancy techniques such as TMR and DMR (Double Modular Redundancy) may not be applied to such large data systems in order to avoid large power consumption. An Exa-flop supercomputer that can executes 10^{18} instructions per second or exascale data centre is being exploited and there is serious concern over the trade-off between power consumption and resilience [8–11].

In large scale supercomputing, SDCs (Silent Data Corruptions), by which incorrect results are obtained without any evidence of error after extremely massive and costly calculations [9], is one of the most serious concerns over resilience. DUEs (Detected Unrecoverable Errors) are also a major concern in microprocessors, in particular, for

Terrestrial Radiation Effects in ULSI Devices and Electronic Systems, First Edition. Eishi H. Ibe.

Table 7.1 State-of-the-art failure reports by terrestrial neutrons in various industries

Field	Application	Root cause	Failure symptom	References
Airplane	X by wire	SEU/SEL/SEFI	Reboot	Matthews [1]
Railway	GTO[a] IGBT[b]	SEB[c]	Out-of-service	Normand [22]
Network	Server	SEU[d]/MCU[e]/SEL[f]	Data corruption/reboot	Slayman [3], Schindlbeck [4]
	Router	SEU/MCU/SEL	Reboot/address change	Shimbo [5]
	Data centre	SEU/MCU/SEL	Power consumption due to redundancy	Falsafi [6]
	Core network	SEU/MCU	Not Available	http://www.ntt.co.jp/news2013/1303e/130321a.html
	Power supply (DC-DC converter)	SEB/SEL	Out-of-service	Rivetta [7]
Super computer		SDC[g]	Unrecognisable wrong calculation/power consumption by redundancy	Geist [8] and Daly [9]
Automobile	Brake by wire	SEU/MCU/SEL	Non-stop/sudden stop	Skarin [12] and Baumeister [13]
	Steering	SEU/MCU/SEL	Stick/unexpected rotation	Vaskova [16]
	Engine control	SEU/MCU/SEL	Sudden acceleration/No operation	Nakata [23]
	CAN[h]/LIN[i]	SEU/MCU/SEL	Communication error	Lopez-Ongil [15] and Vaskova [16]
	Pedestrian detection by using GPU[j]	SEU/MCU/SEL	Missing pedestrians	Rech [18]
	IGBT	SEB	Out-of-service	Shoji [19] and Nishida [20]
PDA	Smart phone	SEU/MCU	Freeze/mail address corruption	Chen [21]
	Tablet	SEU/MCU	Freeze/mail address corruption	
	Desktop PC	SEU/MCU	Freeze/mail address corruption	

[a] Gate turn-off thyristor.
[b] Insulated gate bipolar transistor.
[c] Single event burnout.
[d] Single event upset.
[e] Multi-cell upset.
[f] Single event laytchup.
[g] Silent data corruption.
[h] Controller area network.
[i] Local interconnect network.
[j] Graphic processing unit.

the systems that are not allowed to stop. It is said that the rate of DUEs is higher than that of SDC in cache memory [26]. Capture and recovery of SDCs and DUEs impacts is a top-priority issue in *real time*, safety critical systems.

Time redundancy techniques such as checkpointing and rollback (retry) are regarded as the most likely countermeasure against SEE for extra-large systems from the viewpoint of suppressing power consumption [9], but concerns are addressed over such time redundancy techniques because error rate may exceed checkpointing frequency [8].

Concerns over automotive reliability are spreading rapidly and widely, for brake by wire [12, 13], engine control [14], power steering [16], air bags, communication protocol in cars like CAN (Controller Area Network) [15], FlexRay and LIN (Local Interconnect Network) [16]. GPUs (Graphic Processing Units) or GPGPUs (General Purpose GPUs) are widely used in many circumstances including nodes for supercomputers thanks to their ability parallel compute. Extension of GPUs to mission critical or even to real time systems, such as cars, causes concern over accidents due to SEEs in GPUs. Rech *et al.* carried out the neutron beam testing at ISIS for 40 nm GeForce GTX480 and showed high susceptibility of GPUs to terrestrial neutrons [17]. Pedestrian detection at night is one of the most important safety requirements ('Five stars' in Euro NCAP (European New Car Assessment Programme) rating) where in near future GPUs will be used [18].

Even portable digital applications, like smart phones and tablet PCs, have been tested under neutron irradiation [21]. Obviously these days, no electronic equipment is free from threat due to SEE through terrestrial radiation.

There are a number of fault/error/failure modes that reflect their root cause mechanisms that depend on the applications. Prediction and estimation techniques of faulty condition rate and mode have a primary importance in designing devices, circuits and electronic systems. Techniques for detecting faulty conditions must be established so that the nature of the faults/errors/failures can be established based on their underlying physical mechanisms and the prediction and estimation techniques validated. Cost-effective resilient techniques may be assigned according to the nature of each failure and application. The following sections introduce some examples of prediction/estimation and the detection and classification techniques of fault/error/failures.

7.2 Prediction and Estimation of Faulty Conditions due to SEE

Prediction/estimation techniques of the faulty conditions induced by SEEs are tabulated in the second column of Table 7.2 in various stack layers of the silicon industry from semiconductor substrate to application levels.

7.2.1 *Substrate/Well/Device Level*

At substrate and device level TCAD (Technology Computer-Aided Design) simulation with a three-dimensional physical model is applied to analyse a transient pulse

Table 7.2 Summary of fault/error/failure prediction/estimation and in-situ detection methods

Stack layer	Prediction/estimation	In-situ detection
Application	Simulation-based fault injection Emulator-based fault injection	Anormal software behaviour (e.g. SWAT) SBST (software0-based self test)
OS[a]	Emulator-based fault injection Log-based FFDA[b] (grouping symptoms)	Detection mechanism in the kernel
Board	Full/partial board irradiation	Watch-dog timer
Chip/processor	Simulation-based fault injection Emulator-based fault injection Statistical approach to quantify masking effects Irradiation test	BIST (built-in self test) DMR (double module redundancy) TMR (triple module redundancy) Replication On-chip monitor CRC[c] Lock-step operation Additional components to microprocessor

[a]Operating system.
[b]Field failure data analysis.
[c]Cyclic redundancy code.

produced by a charged particle in a cell. This TCAD approach is becoming erroneous as device scaling proceeds, because a secondary particle may affect more than two cells. Sometimes, mixed-mode TCAD simulation where some parts of the device are evaluated with a simplified equivalent circuit model. Three-dimensional TCAD model for more than two cells is, however, required for that kind of situation [27–29]. The model should cover a wide enough area (cells) to give fixed and stable boundary conditions [27, 28]. The three-dimensional TCAD model is also used to get a mathematical expression of the transient pulse form that can be used for the input of a circuit simulator like SPICE by which the circuit level response to SEE can be evaluated or the Q_{crit} used for Monte-Carlo soft-error simulator can be obtained [30, 31]. Transient noises are often modelled by the following double-exponential equation in the circuit models [30],

$$I(t) = I_0[\exp(-\alpha t) - \exp(-\beta t)], \tag{7.1}$$

where,

$I(t)$: transient current
I_0: transient current constant
α,β: decay constants.

Methods to measure SER^{M} are established as in [32, 33]. Assuming Poisson distribution for memory error rates, memory SER (soft-error rate), SER^{M} is given using the following equation with a statistical margin in a field or accelerator test results [33]:

$$SER^{M} = \frac{\chi^{2}_{(\alpha/2):k}}{2NT_{r}} \times 10^{9} \quad \text{(FIT)} \tag{7.2}$$

where,

$\chi^{2}_{(\alpha/2):k}$: chi-square value given according to parameters α and k
α: ratio of population outside the Poisson distribution, depending on the confidence level (CL). If $CL = 95\%$, 95% of the population of SER in memory devices is below SER^{M}, and $\alpha = 0.1$ since $CL = 1 - \alpha/2(\%)$
k: degree of freedom $= 2(N_{event} + 1)$
N_{event}: number of events found at the time T_{r} (hours)
N: number of devices tested
FIT: Failure In Time.

Methods to measure SER in logic gates SER^{G} have not been fully established, but logic gate-chain methods are gaining popularity and seem attractive. In the gate-chain methods, logic gates like inverters [34–38], NAND gates [37] or NOR gates [34] are used. The gates are typically connected in series and all gates are irradiated simultaneously and uniformly in a DUT (Device Under Test) as illustrated in Figure 7.1. SET (Single Event Transient) pulses generated in the DUT propagate to the detector where pulse count and width are measured. In the detector, an SET propagates through a series of delays (commonly buffers with delay time τ). FFs (Flip-Flops)

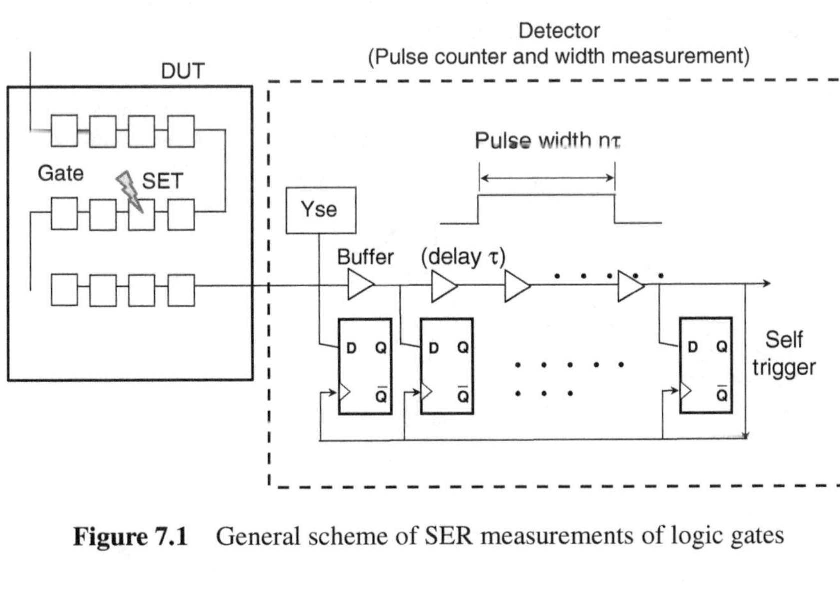

Figure 7.1 General scheme of SER measurements of logic gates

are connected in parallel with the delay line and a certain location of the edge of the output pulse of the delay is used as a clock trigger of FFs in the upstream of the delay. SET pulse width is measured from the product of the number of FFs with data '1' in the series of delays and the delay time τ.

PIPB (Propagation Induced Pulse Broadening), however, through a large number (hundreds to thousands) of logic gates, in particular inverters, make pulse width wider than actual SET pulses and has recently been recognised as an intrinsic problem to be solved in this method [39, 40]. The PIPB mechanism is not fully understood, but is thought to be relevant to the FBEs (Floating Body Effects). The number of gates in a chain is, therefore, limited in many cases [41, 42].

A Monte-Carlo soft-error simulator like CORIMS (COsmic Radiation IMpact Simulator) [43] is used for SEU (Single Event Upset that includes SBU (Single Bit Upset) and MCU (Multi-Cell Upset)) simulation with an incident particle model, nuclear reaction models, charge deposition/collection models and relevant databases as explained in Chapter 6. Raw SERs (SERs without masking effects as explained later) of memory and logic devices can be obtained by using such a MONTE-CARLO soft-error simulator.

7.2.2 *Circuit Level*

The circuit level simulator like SPICE (Simulation Program with Integrated Circuit Emphasis) can be utilised to see the circuit response to an SEE or to obtain Q_{crit} of a unit circuit, for example an SRAM [44].

SER mechanisms of sequential (memory elements such as FFs and latches) logic gates can be threefold: (i) direct incidence in the sequential gates, (ii) transient noise from the global control line, such as clock or set/reset channel and (iii) indirect incidence to make a SET in the combinational part as indicated in Figure 7.2. As for

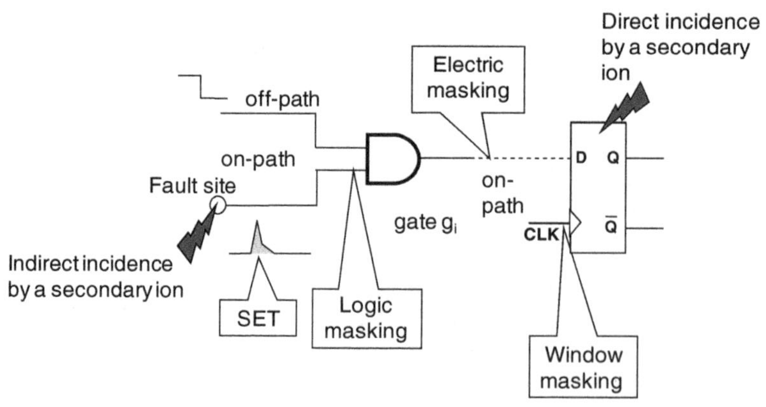

Figure 7.2 Typical masking mechanisms of single event transient in a logic circuit

indirect incidence (iii), SET noise may not be latched to the FFs due to three masking mechanisms such as window (timing), logic, electric masking. In logic masking, fault is cancelled out in AND gate if the other input is '0', for example. In electric masking, the pulse height of the fault decays down below the threshold level of a gate during propagation in a circuit. Timing window masking take place if the input gate of a FF is closed when the fault arrives.

Here, we define the intrinsic SET rate, SET_{int} as the rate of SET generated in the combinational logic circuit between two FFs that have enough pulse height and width to cause SEU in the FF in the immediate downstream. A certain ratio of SET_{int} is captured and change to SEUs in the FF in the downstream in proportion to the operating frequency as indicated in Figures 5.14 and 5.15. Therefore, SEU rate in FFs can be expressed by,

$$\begin{aligned} SER_{FF} &= SER_{FF}^{Direct} + SER_{FF}^{Captiured} \\ &= SER_{FF}^{Direct} + TVF \times SET_{\text{int}}, \end{aligned} \tag{7.3}$$

where,

SER_{FF}: total SER in the FF
SER_{FF}^{Direct}: SER in the FF by direct hit from a charged particle
$SER_{FF}^{Captiured}$: SER in the FF by capturing intrinsic SET in the combinational circuit
TVF: timing vulnerability factor that is proportional to operating frequency [44].

It should be noted that:

1. $SER_{FF}^{Captiured}$ is dependent on the level of SET_{int} as indicated in Figure 5.15 and does not exceed SET_{int}. When SET_{int} is negligibly low, compared to $SER_{FF}^{Captiured}$, the dependency of SER_{FF} on the operating frequency should not be visible as previously indicated in Figure 5.15b.
2. SET_{int} should vary depending on the circuitry in the combinational circuits so that TVF is the only local metric and cannot be a global metric as AVF (Architectural Vulnerability Factor)/ DF (Derating Factor)/ MF (Masking Factor) above.

Masking effects can be evaluated by using the circuit simulation techniques as explained.

SET pulses are injected at a selected timing and location in the circuit to evaluate the fatality of the fault or masking effects during propagation.

Such simulations generally take a very long execution time even for rather simple circuits in benchmarks like ISCAS (International Symposium on Circuits and Systems)85 [46] and SPEC CINT2000 [47], while the fault coverage is not so high.

The simulation results also vary depending on the application working on the benchmark circuits. In the most simple case, it can be a simple read/write operation, typically the Marching test [48] (see Appendix A.2) or a more complicated and realistic application.

7.2.3 Chip/Processor Level

Quantification in the chip-level SER may start with total raw (average or average with statistical margin) FLR_{raw} (raw failure rate) of the system as expressed in Equation (7.4),

$$FLR_{raw} = \sum_j (SET_j^G \times N_j^G) + \sum_i (SER_i^M \times N_i^M) \tag{7.4}$$

where,

SET_j^G: SET/fault rate of a j-th gate that has enough pulse height and width to cause SEU in a FF in the downstream when the SET pulse is captured
N_j^G: number of j-th gates
SER_i^M: upperbound SER of an i-th memory
N_i^M: number of i-th memories.

Usually, a tremendous number of gate types are implemented in ULSI (Ultra Large Scale Integration) chip so that obtaining all SET_j^G by irradiation experiments is obviously impossible. Instead, SEE fault simulation tools like CORIMS validated in the field can be used to evaluate SET_j^G as explained in Chapter 6.

In actual electronic circuits/systems as illustrated in Figure 7.2, a large portion of SETs created in combinational circuits disappear by using the masking mechanisms. There are three kinds of metrics that express the degree of attenuation of original SETs as follows:

1. *AVF* [49, 50]

$$AVF = \frac{N_{failure}}{N_{fault}} \tag{7.5}$$

   where,

   N_{fault}: the number of faults generated or injected in the system/component irrespective of the pulse height and width

   $N_{failure}$: the number of failures caused by the total faults.

2. *DF* [51, 52]

$$DF = \frac{N_{fault}}{N_{failure}} = \frac{1}{AVF} \tag{7.6}$$

   Kellington *et al.* evaluate *DF* of IBM Power 6 processor by proton beam irradiation. They divide *DF* into *MD* (Machine Derating) and *AD* (Application Derating) and gave $AD = 6.7$ for SDC when bzip2 is used for the benchmark and $MD = 500$

for unexpected errors. The majority of the literature seems to use *DF* as defined in Equation (7.6), but there is some exception where *DF* is used in the same manner as *AVF* [53]. Standardisation may be required.

3. *MF* [54]

   Although the definition of *MF* does not seem to be given in the literature, it can be naturally expressed from its meaning by,

$$MF = \frac{N_{fault} - N_{failure}}{N_{fault}} = 1 - AVF \tag{7.7}$$

   When ten failures take place in a system as a consequence of 100 faults, *AVF* is 0.1 while DF is 10 and MF is 0.9. This is a very confusing situation and *AVF* in Equation (7.5) is used in this book for simplicity. *AVF* may partially include *TVF* as previously mentioned.Therefore,

$$FLR_{real} = AVF \times FLR_{raw}, \tag{7.8}$$

   where, FLR_{real}: Actual failure rate of electronic system.

A processor most frequently accesses register files so that errors generated in the register files should propagate easily, in particular, in embedded systems. Fazeli *et al.* reported that *AVF* of register files in the embedded device was 0.15 and AVF of the read counter was 0.01 [55]. error correction code (ECC) or TMR, however, is not applied to protect register files basically with a priority of saving power consumption and achieving better performance. To find out which vulnerable register files are frequently accessed, an emulator-based fault injection technique can also be applied [56].

Ando *et al.* carried out the neutron irradiation test at the RCNP (Research Center for Nuclear Physics) of a SPARC64 microprocessor with some workloads; they obtained *AVF* of 0.5 for L2 cache without ECC and 0.064 for latches [57] and processors [57–59]. In the case by Kellington *et al.* above, *AVF* of a IBM Power 6 microprocessor with some workloads is 0.15 with the bzip2 workload and 0.002 for unexpected errors [51].

Emulator-based fault injection [29] with a hardware-based model, which is generally implemented in FPGAs (Field Programmable Gate Arrays), is commonly utilised for estimation of *AVF*. Transient pulses are injected into circuit nodes randomly while the circuit is operating a certain workload on some of the benchmark circuits. Compared to software-based fault injection, emulator-based fault injection is much more efficient and has better fault coverage.

Farazmand *et al.* evaluated *AVF* of the GPGPU by fault injection, and reported that *AVF* of the register and memory are 0.15 and 0.03 with some workloads, implying that *AVF* of registers and cache in CPU (Central Processing Unit) are, in general, 0.15 and 0.25, higher than those of the GPGPU [49].

There are some proposals for other types of masking mechanisms:

Alexandrescu *et al.* propose a chip-level masking mechanism, which is an *opportunity window* that relates to a life time of data in memories [60].

Frequency of access to memory elements can also be a masking mechanism [55]. As an extreme example, AVF = 0 if the memories are not accessed at all.

Costenaro *et al.* propose the hierarchical SET analysis system in which nuclear spallation models, device models and RTL (Register Transfer Level) circuit models are implemented [61]. They demonstrated fault injection results for the logical DF in relatively complicated circuits (blocks) like adders, multipliers, encoders and decoders. They further estimated a SET rate of ALU (Arithmetic-Logic Unit) by assembling the results for blocks.

Instead of using a circuit simulation approach for masking metrics, stochastic or probabilistic approaches are being proposed as being much faster simulation techniques [62]. Chen *et al.* applied such techniques to evaluate the re-convergence effect, by which a transient pulse is divided into more than two pulses and overlapped in the downstream to produced a narrower width pulse [63].

As for chip-level SER, since its variation originates from specific circuits and applications, and it is not a random process, such a statistical theory seems to be very difficult to establish although some proposals are made [64].

7.2.4 Board Level

Irradiation tests can be applied to any stack layer to quantify and classify faults, errors and failures.

For a more integrated/complicated system like a board, a partial/full board irradiation test is useful to find the vulnerable part of the board [5, 65]. An overall PCB (printed circuit board) under certain operation of a workload is irradiated to see the overall failure rate and to obtain a cross section of the failure with its mode. The failure mode may depend on the application programs. Such irradiation tests are made on routers [5, 66] and servers.

When a certain mode of failures is fatal and *FLR* is relatively high, a partial irradiation test may be carried out to identify the part that causes the failure mode and may be target part to be hardened.

An adjustable beam size irradiation facility with collimators like QMN (Quasi-Monoenergetic Neutron) [67] or ANITA (Atmospheric-like Neutrons from thIck TArget) [68] in The Svedberg Laboratory (TSL) is useful for this kind of test.

7.2.5 Operating System Level

Log-based FFDA (Field Failure Data Analysis) is applied to verify dependability in the OS (Operating System). Pecchia *et al.* propose a method to filter out unnecessary or duplicate information from the FFDA data in a multimode computer system by grouping the data in order to make the FFDA more effective [69]. In a typical grouping method, two faults found within a certain period of time are regarded as having originated from one event (tuple). If the two faults do not originate from one event, the situation is called 'collision'. Inaccurate grouping affects the evaluation of OS dependability.

Fault injection is another effective method to evaluate dependability of OS. Partitioning of OS is a technique used to prevent propagation of a fault from one task to another. Barbosa *et al.* showed the benchmarking of three types of partitioned OSs by fault injection and found the most vulnerable function in OS exception routines [70].

Skarin *et al.* evaluate the failures in a brake control system and find the most fatal brake failures like (i) brake does not work at necessary timing, (ii) unexpected sudden stop and (iii) lockage of tyres are caused by errors in the stack pointer by using an emulator-based fault injection system.

7.2.6 Application Level

Cook and Zilles evaluate the masking mechanisms due to software (instruction level masking, by which correct results are obtained after operation of instructions with wrong data), and propose six masking modes as follows:

1. *Value-comparison*: In the case where a result of comparison is compressed into 1 bit expression, many data bits with the wrong values are eliminated.
2. *Sub-word operation*: In the case where part of the bits in a word are used to execute an operation, the correct results are obtained when the wrong values are not included in the set of bits.
3. *Logical operations*: The same as logical masking.
4. *Overflow/precision*: A bit with the wrong value is eliminated by some operation like shifting bits.
5. *Lucky load*: Correct values are obtained occasionally from data in the wrong address.
6. *Dynamically dead*: After some instructions are executed using the wrong data, the results are overwritten by the subsequent instructions [71].

Kost *et al.* proposed *PVF* (Program Vulnerability Factor)asa rather independent factor of programs run on the system [72] and also proposed the following by software-based detection methods: (i) executable assertion, (ii) use of independent program to detect failures in the running program, (iii) control flow check and (iv) execution of the redundant program.

Rivers *et al.* propose *AVF* evaluation techniques based on RTL simulation of M1 (the first metal layer) architecture [73]. The concept of *residency* (ratio of component operating time to total operating time) is also proposed as part of *AVF*s.

Fault injection techniques are also applied to identify the seriousness of a failure.

Accuracy of simulation-based fault injection or emulator-based fault injection depends on many factors like timing information in the netlist used in the fault injection and the actual workload [74], so that care must be taken when checking.

Cho *et al.* also pointed out the inaccuracy of the emulator-based fault injection test with a variety of injection modes like:

1. injection to a register file;
2. injection to a register file while data is written;
3. injection into a program variable and
4. injection into a program variable while data is written [75].

There is a large variation between less than 1% up to 16% in 'mismatch rate' (SDC) depending on the mode and benchmark in SPEC INT2008.

7.3 In-situ Detection of Faulty Conditions due to SEE

In-situ detection techniques of faulty conditions due to SEE are tabulated in the third column of Table 7.3 for various stack layers from semiconductor substrate to application levels.

7.3.1 Substrate/Well Level

7.3.1.1 Indirect Fault Detection Method

As explained in Section 1.2, all faulty conditions due to terrestrial radiation originate from fault/transient/noise in a well or substrate of a semiconductor device. Because the fault, however, may appear randomly in the layout of a device and propagate very rapidly in the device, it is very difficult to capture the fault. Therefore, given that it is not necessary to detect all the faults, the only possible way to capture a fault is to probe the physical disturbances due to the fault. A common method is to capture a symptom

Table 7.3 Summary of fault/error/failure prediction/estimation and in-situ detection methods

Stack layer	Prediction/estimation	In-situ detection
Circuit	Circuit simulation Logic masking simulation Irradiation test	Double/triple phase-shift sampling and comparison RAZOR SEM/STEM
Device	SEE Monte-Carlo simulation TCAD simulation Irradiation test	ECC, parity
Substrate/well	TCAD simulation	SAW detector BICS (built-in current sensor) BIPS (built-in pulse sensor)

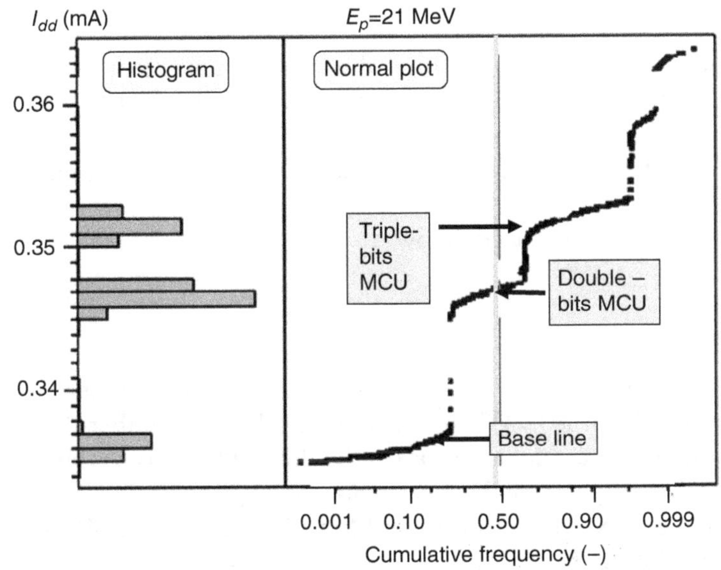

Figure 7.3 Example of stepwise I_{dd} increase produced by quasi-monoenergetic ($E_p = 21$ MeV) neutron irradiation

of error by measuring the current or voltage change in a limited portion of a device [76–80]. Figure 7.3 shows such an example of histogram and cumulative frequency of stepwise noise current peaks due to faults appearing in the I_{dd} current [78]. Each step is believed to be double, triple and quadra MCUs which appear in SRAM devices.

BICS (Built-In Current Sensor) is used to detect fault in the substrate by measuring the potential change due to the SET current [79]. Krimer *et al.* proposed the current detection techniques in which the current is measured using only limited clock timing when the signal is expected to be stable against latchup [81]. Locating the fault is much more difficult, although it is suggested that the matrix structure of the sensors can be used [82].

Upasani *et al.* describe a new fault identification technique which exploits SAW (Surface Acoustic Wave) detectors, by which acoustic waves created in a device by charged particles are captured and MCUs can be identified along with their locations by solving a TDOAs (Time Difference of arrivals) matrix among 5×5 SAW devices [26].

7.3.1.2 Determination of Fatality of a Fault to an Electronic System

In principle, neither faults nor errors matter unless they cause failure in the final output.

In order to identify such a fatal fault or error, a focused laser/ion beam of around 1 μm diameter formed from optical or electrostatic lenses, is utilised to identify the faulty portion and mechanism in a device [83–86].

7.3.2 Device Level

Errors in memories and sequential logic devices are rather easily detected. Errors in memories can be detected by topological methods – parity and ECC (Error Correction Code).

Errors in sequential logic are commonly detected by time/space domain redundancies.

Some examples are introduced below.

7.3.2.1 Parity

Parity has one of two values, either 1 or 0. When a binary number has odd '1's, the parity is 1, otherwise it is 0. When one digit of a binary number flips 0 to 1 or vice versa, the parity changes and an error can be detected. Parity is the simplest way of detecting error in computers and telecommunication. Nonetheless, it is of no use if the even digits in a binary number of a word flip like MBU (Multi-Bit Upset), because parity does not alter in this case.

7.3.2.2 ECC

ECC is used for error detection and recovery in memories by combining additional code bits with data bits. The Hamming code [87] (see Appendix A.2) is the most commonly used in computer applications. Hamming code is SECDED (Single Error Correction and Double Error Detection) [88]. Since ECC has an area and performance overhead, it is not favoured for use in circuits that require speed such as registers and caches.

7.3.3 Circuit Level

Time-domain detection techniques such as double phase-shift and triple phase-shift sampling and comparison can be applied for fault detection in a circuit level [89]. RAZOR FF is principally designed for timing error detection, where a signal from a logical stage is given to the main FF and shadow FF with a delay as shown in Figure 7.4 [90–92]. If the signal from the logical stage L1 meets the clock timing of the main FF, both the main FF and shadow latch have the same output. If some delay occurs in the logical stage and the shadow latch has a capture correct signal, the outputs from the main FF and shadow latch must be different and a timing error is detected. If the delay of the clock for the shadow latch is long enough compared to the SET pulse width and short enough compared to the normal signal width, a SET can also be detected. There are a number of extended ideas of RAZOR. SEM (Soft Error

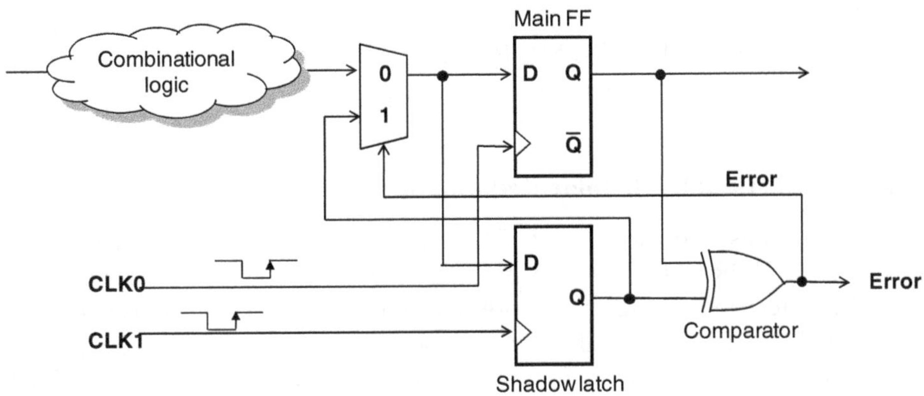

Figure 7.4 Fault detection technique by double sampling in RAZOR. (Ernst *et al.* [91] ©2004 IEEE)

Mitigation /STEM (Soft and Timing Error Mitigation) [93, 94] is also a time domain detection technique.

7.3.4 Chip/Processor Level

7.3.4.1 On-Chip Monitor

On-chip monitors are embedded on a peripheral part of a chip, to monitor voltage and/or current change in this area only [95–98]. Some symptoms of fault/error/failure can be detected by using on-chip monitors as well.

7.3.4.2 CRC (Cyclic Redundancy Code)

In cyclic redundancy code (CRC), a binary bitstream can be regarded as a large integer and CRC code is given by the remainder of the integer divided by a certain prime integer [98, 99]. If an error takes place in the bitstream during propagation the CRC code is changed and any errors may be detected.

7.3.4.3 BIST (Built-In Self Test)

BIST (Built-In Self Test) is commonly applied for testing a logic circuit in boot time (before starting operation). Utilisation of BIST at runtime (under regular operation) is another option as described in [100, 101]. Scan-chains together with a LFSR (Linear Feedback Shift Register) and a MISR (Multiple Input Signature Register) are used as in-situ classification of errors in FFs. LFSR is used to generate a random data pattern and the scan chain is fed in, and MISR is used to compress the scan-chain output to generate a signature from which errors are identified.

7.3.4.4 DMR (Double Modular Redundancy)

Most circuits consist of sequential (FFs, latches) and combinational (AND, OR, NAND, NOR, XOR, ALU, and so on) logic devices [102, 103]. Sequential logic devices keep temporary logical conditions until the next clock pulse reaches the devices. In DMR, two identical circuits (modules) operate simultaneously with the same instructions as shown in Figure 7.5. Two temporal outputs of the identical sequential logic devices in the modules are compared, and an error message is created when the two outputs are different. Fingerprints or signatures that are generated through the encryption of the redundant data may be used for the comparison [104].

Lock-step operation of dual processors can be used for error detection in the chip level [11, 105].

7.3.4.5 TMR (Triple Modular Redundancy)

TMR has three identical modules and operates them simultaneously [105, 106]. A voter is implemented after a set of three sequential logic devices as illustrated in Figure 7.6. If any two module outputs out of three are different, an error message is created, but execution is continued based on the two-out-of-three majority outputs. If the operation is cyclic, not once-through, faulty conditions in the failed module must be recovered before the next cycle.

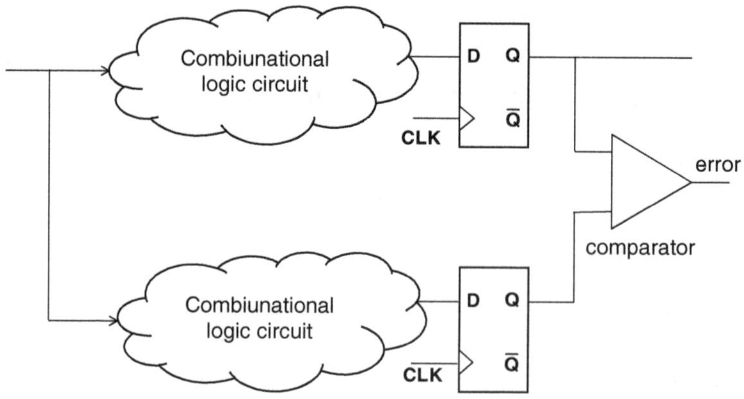

Figure 7.5 Simplified scheme of dual (double) modular redundancy (DMR)

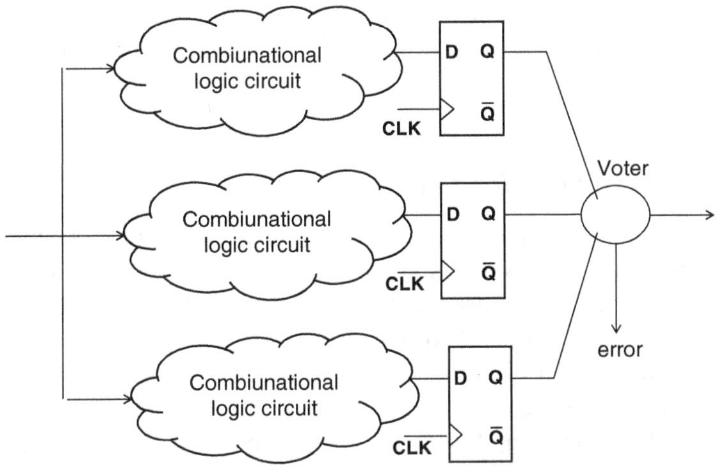

Figure 7.6 Simplified scheme of triple modular redundancy (TMR)

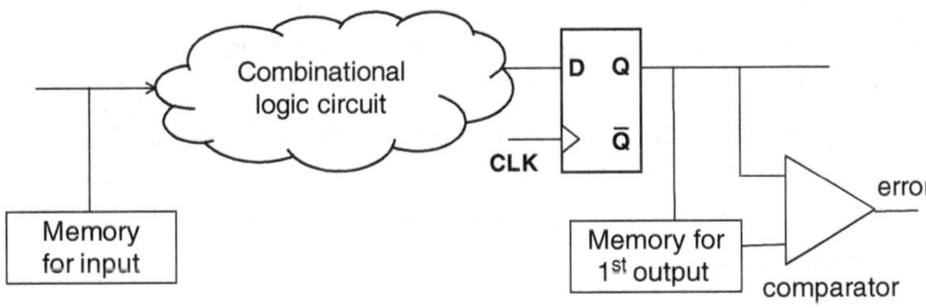

Figure 7.7 Example of implementation in logic circuit for replication

7.3.4.6 Replication

In the replication method, under the same initial conditions, a predetermined logic part is operated once and temporary outputs are retained as memory elements as illustrated in Figure 7.7 [107]. Then, the same operation is made and the outputs are compared with the former outputs. If two outputs are not identical, an error message is generated and the operation is restarted from the previous operation point.

7.3.4.7 Additional Components

Du *et al.* propose that debug interfaces should be used to implement recent processors to detect control flow errors and they demonstrated how the mini MIPS processor would give excellent fault coverage [108].

Carvalho *et al.* also propose using an additional module for sending a diagnosis signal to a 65 nm SoC PowerPC chip through those pins available during a certain period of time [109].

7.3.5 Board/OS/Application Level

A failure is the easiest to detect. It causes system failure, reboot, abnormal operations, and so on. Once it takes place, however, the impacts can be very serious. It may cause long-term outage of a system, recall of products which is termed RMA (Return Material Authorisation) [98], and so on.

Failures can be detected as malfunctions in software or applications like BIST before resulting in hardware failures.

In SWAT [23, 110], abnormal software symptoms such as a specific type of deviation in the control flow and memory address are monitored throughout the entire operation. More specifically, SWAT detects SDCs that may cause fatal trap, hang, application abort, kernel panic (the state when a fatal error takes place in the kernel causing complete shutdown of the OS) and address violation. They suggest that faults do not result in this type of deviation ('data only') are very difficult to detect.

SBST (Software-Based Self Test) is proposed for self-test of the processor by using a test program implemented in the flash memory in the processor chip and can activate the hardware device that cannot be activated by software alone [48, 111, 112].

Brown and Lin use software to evaluate the limitation of two soft-error detection mechanisms. Two of which are:

1. Errors can be detected by space/time redundancy if instructions do not depend on control flow.
2. Errors can be detected by checking the branch instruction decode (ID) before and after the branching instruction according to the control flow.

   They point out six limitations in the second mechanism even when the error is a single bit error:

   i. Instruction for error detection and recovery is skipped occasionally in the branch.
   ii. Intra-block branch: Change in branch ID force to return to the same instruction block.
   iii. Branching takes place without branching instruction.
   iv. Error in PC (Program Counter): Instruction flow is destroyed completely.
   v. In computed 'goto', the correct branch ID is not identified beforehand.
   vi. 'Return' does not always lead to the correct 'return' location.

They propose that branch ID be written twice to the stack register and check them before and after branching to solve the limitations except for (iv) [113].

7.4 Classification of Faulty Conditions

7.4.1 Classification of Faults

SET is a noise produced in a well or substrate due to a charged particle hit. Even though the SET pulse width is generally smaller than a few nano seconds, it may cover more than two clock pulses. Pulse height and width are one of the key parameters that determine the masking effects in a logic circuit [114, 115]. Classification of faults is mainly made by binning faults according to the pulse height and width. Such measurement is only possible in the accelerator experiments with special equipment and high frequency of events. As scaling proceeds, MNT (Multi-Node Transient) it is becoming common to cause MCU in memory elements including FFs [116] and corruption of space redundancy techniques like TMR [117] and DICE (Dual Interlocked storage CEll) in combinational logic circuits.

7.4.2 Classification of Errors in Time Domain

Figure 7.8 shows the simplified sequential test algorithm proposed by Ibe *et al.* to classify the nature of SEU in memory. First, when an error in memory is detected, the algorithms are applied to identify whether it is only transient or stable in the *Phase I* classification. If it is not a transient, the timing of occurrence, addresses of errors are recorded. Then, another algorithm is applied to identify whether it is soft error or SEU in the *Phase II* classification. If the error is not soft error, further classification algorithms are applied to identify whether it is SEFI (Single Event Functional Interrupt) or SEL (Single Event Latchup) or permanent (hard) error in *Phase III* classification.

Figure 7.9 illustrates the details of the Phase I classification. One normal *write/read* cycle lasts several seconds for all bits in a DUT. Two or more errors in the same

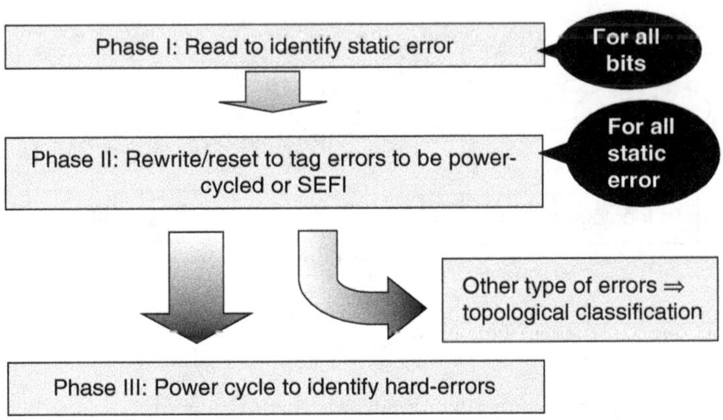

Figure 7.8 Overall procedures for classification of errors in memory elements in time domain

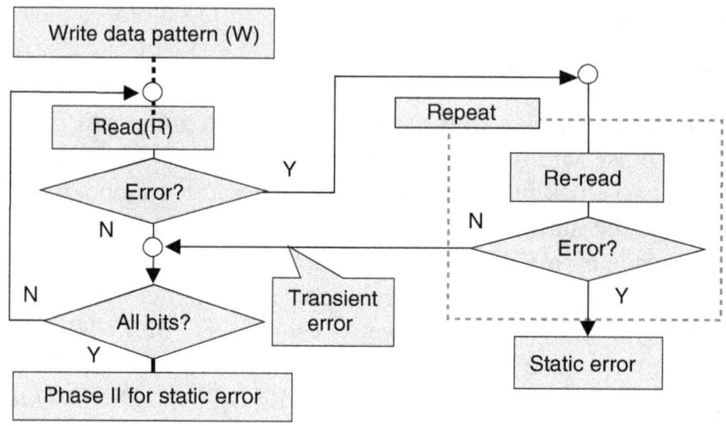

Figure 7.9 Procedure Phase-I: classification of transient/static error in time domain

sampling interval are basically regarded as temporary MCU unless otherwise identified as separate events by the topological classification techniques explained later. Once the data in a certain bit is in error, the error classification algorithm is applied. If the data is recovered by re-reading, the error is regarded as a transient. If the error is not a transient error, it is regarded as 'static error' and complement data is written to the bit in the *Phase II* classification as shown in Figure 7.10. If the bit is re-writable by the complement data, then the error is regarded as soft error. If it is not soft error but can be corrected by resetting FFs to default values, it is regarded as an SEFI. After all the bits are checked and if there are any error bits which are not re-writable or cannot

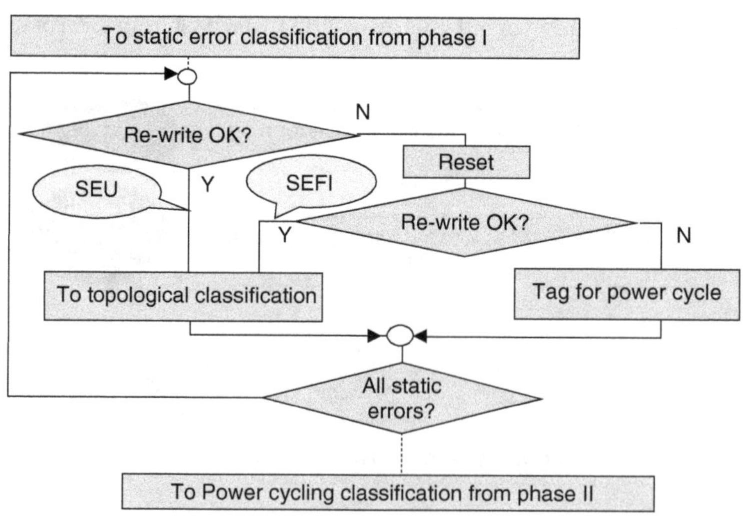

Figure 7.10 Procedure Phase-II: classification of SEU/SEFI and other errors in time domain

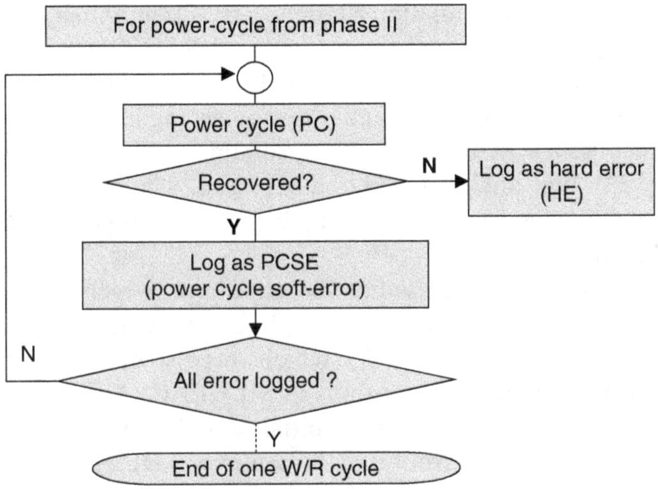

Figure 7.11 Procedure Phase-III: classification of SEL and permanent/hard errors in time domain

be corrected by resetting the FFs, the DUT power is turned off and then turned on (power cycle) to see if the bit is re-writable in the Phase III classification as shown in Figure 7.11. If the bit is re-writable after a power cycle, the errors are categorised as PCSE (Power Cycle Soft Error). Non-destructive SEL is a PCSE mode. If the errors cannot be corrected even by the power cycle, then those errors may be classified as permanent and HE (Hard Error). I_{DD} current and device temperature may be measured with a certain time interval independently to see if the PCSE is SEL or not. To identify other types of HE, such as SEB and SEGR (Single Event Gate Rupture) microscopic measurements by optical or scanning electron microscopy may be applied.

7.4.3 MCU Classification Techniques of Memories in Topological Space Domain

Historically, MCUs are understood as taking place when two or more storage nodes are hit by one charged particle in a device. As device scaling down proceeds, novel MCU modes are being reported as 'charge sharing' among memory storage nodes in the vicinity [119] or bipolar effects in the p-well [24].

Ibe *et al.* have proposed MCBI (Multi-Coupled Bipolar Interaction) for one of the bipolar MCU mechanisms that is regarded as a parasitic thyristor effect triggered by a SESB (Single Event SnapBack) in the p-well and causes MCU multiplicity of more than 10 bits [24]. It is also reported that the MCU physical address pattern differs depending on written data patterns typically between the groups 'solid' (All '1' or All '0') and Checkerboard (CHB or its complement CHBc).

Data corruption in RHBD (Radiation Hardened-By-Design) FFs such as DICE [120] is now being recognised as a real threat due to the MNT mechanisms caused by the charge-sharing or potential elevation in wells during bipolar events. Some novel RHBD FF designs are proposed to counter this emerging threat [121, 122].

Errors due to glitches in the global control line such as the clock [28, 122]/ SET/RESET [123] line are also being recognised. Understanding the nature of MCUs/MNTs using classification techniques is very important for establishing mitigation techniques.

A space-domain topological classification algorithm, which automatically identifies and classifies MCUs within a single sampling time window, is implemented in a specially designed program MUCEAC(), which was also developed by Ibe *et al.*.

Figure 7.12 outlines the basic algorithms in MUCEAC. Among errors in the same time interval, one error is selected, and the initial AOI (Area OF Interest) is determined in the memory matrix with the error bit in the centre of AOI. The size (width along Bit Line (BL) and width along Word Line (WL)) of AOI is pre-determined. Other error bits found in the AOI are generally regarded as being included in the same MCU. Then the AOI is expanded around the new error bits and the search of other error bits is continued until no more error bits are found in the expanded AOI. The error is defined as SBU when only one error bit is included in the initial AOI. If two or more errors are included in the expanded AOI, they are regarded as being included in the single MCU (one event).

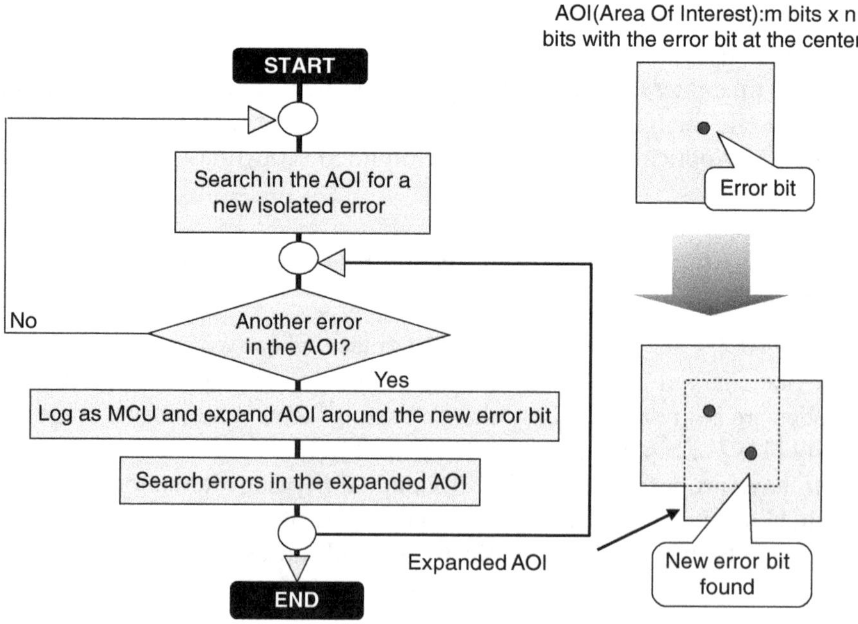

Figure 7.12 Procedure for MCU topological (in space domain) classification

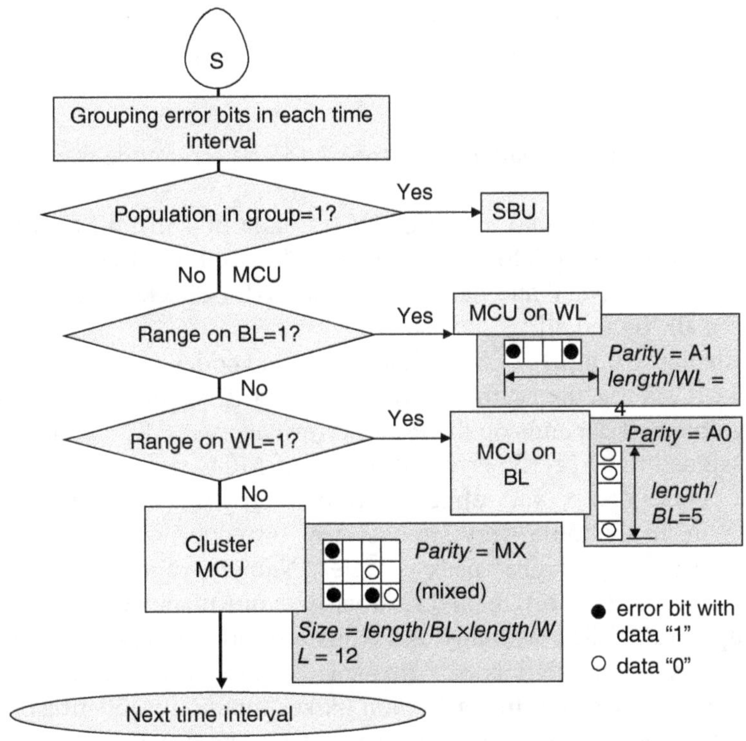

Figure 7.13 Procedure for MCU type classification: MCU on WL, MCU on BL and cluster

Once errors are identified as being members of an MCU, the following MCU classification rules are further applied in MUCEAC as shown in Figure 7.13.

1. The MCU pattern is classified into three basic categories like a single line along BL (Bit Line) (category 'b'), a single line along WL (Word Line) (category 'w') and cluster (an MCU that has two or more bits along with both BL and WL directions; category 'c'). In category 'b', errors align along with only one BL. In category 'w', errors align along with only one WL. In category 'c', errors are contained in two or more bit and WLs.
2. To express in more detail the nature of MCU, MCU code that can be almost uniquely expressed and that is relevant to a physical address pattern in an MCU is given as shown in Figure 7.14:

$$\text{MCU code has a format of } C\_N_1\_N_2\_N_3\_N_4\_P,$$

where,

C: category (b/w/c)
N_1: MCU size ($=N_3$xN_4)
N_2: Bit multiplicity in an MCU
N_3: Width in the BL direction (bits)

N_4: Width in the WL direction (bits)
P: Parity (A1 initial data in MCU bits are all '1', A0 initial data are all '0', MX initial data are a mixture of '0' and '1'.

In the case of the error bit pattern in Figure 7.14, the error code is c_12_4_3_4_any (A0/A1/MX).

MUCEAC has been applied to the QMN test data of 130 nm SRAMs with triple well structure measured at TSL (The Svedberg Laboratory) [67]. Peak energies E_p of 21, 47, 96 and 176 MeV and data patterns of CHB (CheckerBoard), CHBc (CHB complement), all '0' and all '1' are used.

Table 7.4 depicts examples of MCU categories and codes for the size of 10–20 bits out of 2564 MCUs together with the error bit patterns. Whether or not the table can be instantly obtained depends on the PC's performance. A MCU code can be almost uniquely assigned to a specific error bit pattern if MCU size is not too large. MCU categories or codes can be very effective for obtaining hints to identify the underlying mechanism. In the table, very strange error bit patterns in group A (CHB data pattern) can be seen for higher peak energies. Namely, two adjacent error bits are aligned along with a BL (horizontal direction) with one blank WL – leap frog pattern. As for group B (all0/all1), as many as 12 bits are failed along with one BL. These error bit patterns are clarified as a consequence of bipolar action in the p-well [25]. Once a heavy secondary ion like a Mg ion passes though pn-junctions in the p-well in a triple-well structure, electrons may escape from the p-well according to the funnelling effect and holes remain in the p-well which results in increased potential in the p-well. This causes the parasitic transistor (SESB) to switch on and triggers the

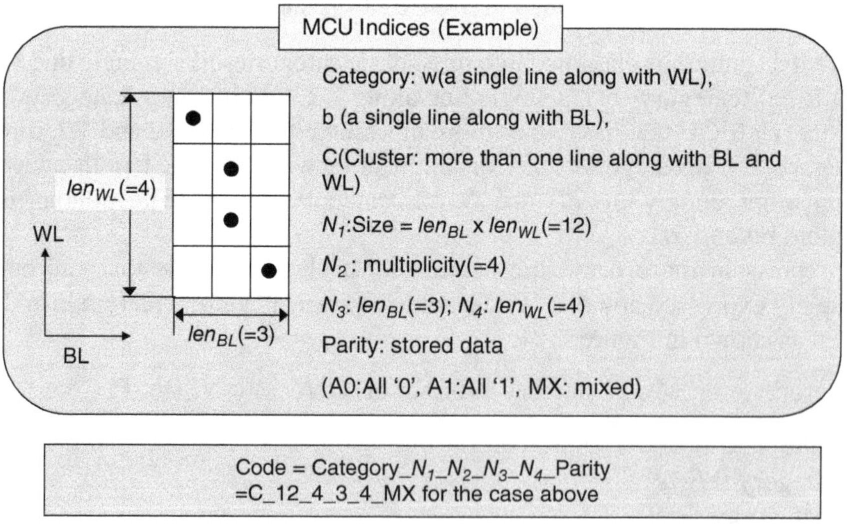

Figure 7.14 Definition of MCU code

Table 7.4 MCU codes, frequency and error bit patterns in 130 nm SRAM matrix observed in neutron irradiation experiments at TSL

Size	Basic category	Code	Data Pattern: CHB E_p=21	CHB 47	CHB 96	CHB 176	CHBc 21	CHBc 47	CHBc 96	CHBc 176	All0 21	All0 47	All0 96	All0 176	All1 21	All1 47	All1 96	All1 176	Typical observed pattern (● : failed bit;data"1" ○ : failed bit;"data"0") Group A: CHB/CHBc	Group B: ALL0/ALL1
10	Cluster	C_10_4_5_2_6			2	1			3										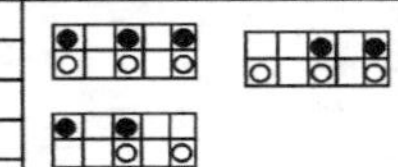	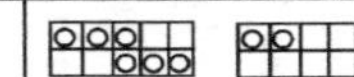
	Cluster	C_10_6_5_2_4				5			1	1				1						
	Cluster	C_10_5_5_2_5				4				1										
	Cluster	C_10_2_5_2_8		1																
	Cluster	C_10_3_5_2_7											1							
12	BL	b_12_12_12_1_0									1									
	Cluster	C_12_4_6_2_8				1													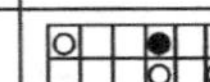	
14	Cluster	C_14_7_7_2_7			2														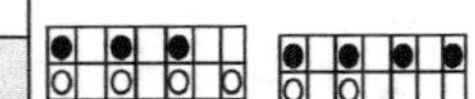	
	Cluster	C_14_4_7_2_10																1		
	Cluster	C_14_6_7_2_8		1		1														
15	Cluster	C_15_7_5_3_8				1														
18	Cluster	C_18_8_6_3_10				1													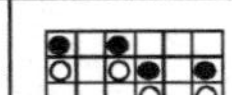	
20	Cluster	C_20_3_5_4_17				1				1									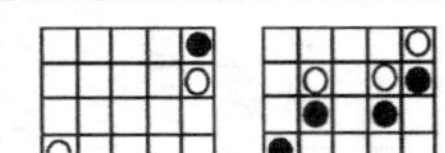	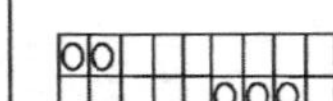
	Cluster	C_20_6_10_2_14											1							
	Cluster	C_20_7_5_4_13				1														

parasitic thyristor in the pnpn structure in the triple well which results in a continuous current flow as shown in Figure 5.3. This overall mechanism is known as MCBI.

Nakauchi *et al.* also observed npn parasitic bipolar MCU events in 65 nm SRAMs under a high-energy neutron irradiation test at RCNP [124]. MCUs spread only across the adjacent 2 bits in a WL direction, while they spread across a maximum range of 15 bits in the BL direction and they show that events can be suppressed by a back bias of −2 V.

Figure 7.15 illustrates how the error bit patterns in Table 7.4 appear in the case of MCBI. A set of 4 bit SRAM cells are shown for CHB and all '1' data pattern. In the case of SRAMs, the difference of data '1' and '0' corresponds to 'high' node location, left or right. As can be seen in Figure 7.15a for the data pattern of CHB, a pair of high nodes appears every other WL. When an MCBI takes place in a certain area of the p-well, two adjacent SRAM bits can fail in every other WL. Meanwhile, as in Figure 7.15b for the data pattern of all solids, one line of SRAM bits along with one BL can fail simultaneously using the MCBI mechanism.

Space domain classification techniques are useful for distinguishing SEL and MCBI. While in MCBI, most failed bits are contained in the two adjacent BLs as explained so far, failed bits in SEL can spread over a wide range of BLs due to the 'Domino' effects [125].

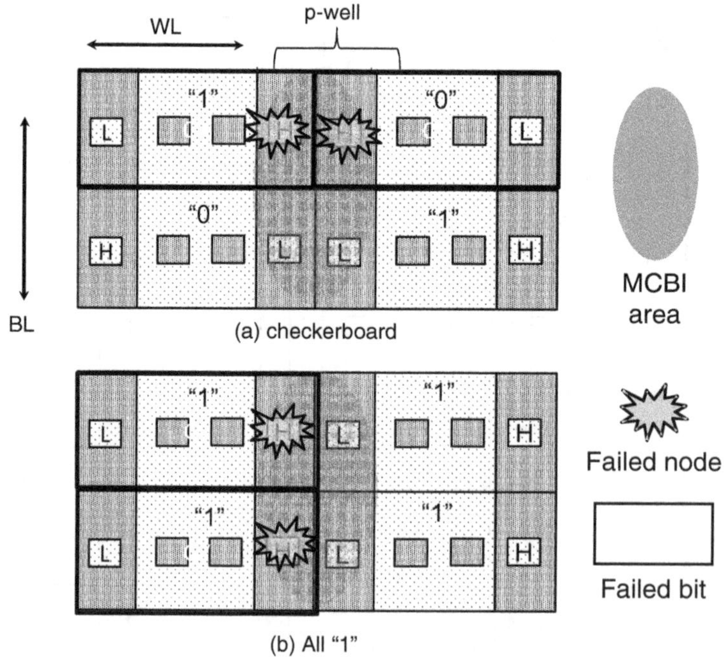

Figure 7.15 Mechanism of MCU abnormal bit patterns due to MCBI. (a) Checkerboard and (b) all '1'

7.4.4 Classification of Errors in Sequential Logic Devices

As for sequential logic like FFs, error can be detected and evaluated by utilising FF scan chain in static mode, in which known data are written in each FF before irradiation and FFs in scan chain with flipped data can be checked after irradiation. But under the actual operation mode of chips, this method cannot be applied because there is no way of knowing what correct state is in each FF. Taking this setback into consideration, statistical approaches are pursued to estimate SER in sequential logic devices by Yoshimura *et al.* [126].

Yamamoto *et al.* made a matrix of 180, 150 and 130 nm FFs like memory matrix and measured direct SEUs in FFs by using RCNP [116]. They showed SER of 130 nm FFs is comparable (about 1/2) to that of 130 nm SRAMs.

7.4.5 Classification of Failures: Chip/Board Level Partial/Full Irradiation Test

A part or over all chip of PCB is irradiated while operating the benchmark application and failure modes and vulnerable parts are identified [5, 52–55, 120–122]. The most typical failure mode is to reboot the entire system. By measuring the FTF (fluence to failure) Φ_i, cross section σ_{fail} can be calculated through MFTF (Mean Fluence To Failure) as proposed by Shimbo *et al.* [5, 127].

$$MFTF = \frac{1}{n}\sum_{1}^{n} \Phi_i. \tag{7.9}$$

Then,

$$\sigma_{fail} = \frac{1}{MFTF}, \tag{7.10}$$

where, n: total number of cycles for the relevant chip or board.

This approach can be very effective when it comes to evaluating all the actual failure modes including the consequences of MNT.

7.5 Faulty Modes in Each Hierarchy

7.5.1 Fault Modes

Table 7.5 summarises various known modes of faults, which covers permanent modes such as traps in oxide that cause TID (Total Ionising Dose) effects. An SET is a noise produced in a well due to a charged particle hit. Even though SET pulse width is generally smaller than a few nanoseconds, it may cover more than two clock pulses. Pulse height and width are one of the key parameters that determine the masking effects in a logic circuit [115]. The classification of faults is mainly made by binning faults according to the pulse height and width. Such measurement is only possible

Table 7.5 Fault modes including single event effects, total ionising dose effects and displacement effects

Class	Definition	Name	Characteristics	Source	Affected area	In-situ detection method	In-situ recover/mitigation method
Transient/ noise	Transient in electric potential and/or current in a chip	SET[a]	Single transient due to charge collected to the diffusion layer in the chip. Pulse width is below a few nano second, and can long more than two clock pulses	Well/substrate	Random but limited to single well	Time and/or space redundancy such as DMR[b], TMR[c]	Time and/or space redundancy
		MNT[d]	Simultaneous SETs in more than two diffusion layers. Mainly, MNTs take place in a single well due to charge sharing or bipolar action. Space redundancy techniques such as DICE[e], TMR may not work against MNTs	Well/ substrate	Random but limited to single well	Monitoring the well potential and/or current	None
		RILC[f]	When a charged particle passes through tunnel oxide of a floating gate memory (flash memory), a leakage path is formed in the tunnel oxide and the potential in the floating gate may shifts resulting in change of Vth to cause soft-error	Tunnel oxide of a floating gate memory	Random but limited to tunnel oxide	ECC/parity	ECC

		Cross-talk	Noise propagation between close wires via parasitic capacitance	Wire	Random but limited to wire(s)	Time and/or space redundancy	Time and/or space redundancy
		Disturbance in power supply		Power supply line	Unlimited area	Monitor in power supply	None
		EMI[g]	Electromagnetic noise including burst noise	Anywhere	Unlimited area	Electro-magnetic probe	None
Trap in oxide	Holes trapped in oxide. They may cause leakage current and may disappear in time	TID[h]	Parasitic level by the trap may cause Vth shift, or potential due to holes in oxide may cause leakage path formation in adjacent semiconductor portion	Oxide	Random in a small area	Vth measurement	Annealing may work
Defect	Lattice defects	–	Damage or interstitials in crystal, which may deteriorate device functionality. They may cause stack at '0/1' error and can be permanent error	Anywhere/ tunnel oxide	Random in a small area	None	None

[a]Single event transient.
[b]Double modular redundancy.
[c]Triple modular redundancy.
[d]Multi-node transient.
[e]Double interconnect cell.
[f]Radiation induced leakage current.
[g]Electro-magnetic interference.
[h]Total isonisation dose effects.
Re-display of Table 1.1.

in accelerator experiments using special equipment and with a high frequency of events. As scaling proceeds, MNT is more likely to cause MCU in memory elements including FFs [116] and exacerbation of space redundancy techniques like TMR [117] and DICE in combinational logic circuits.

Noises can propagate to different wires by cross-talk through parasitic capacitance between wires [128]. Disturbances in power supply, and EMI (Electro-Magnetic Interference) cause noises across a relatively wide area compared to ionising radiation effects [129].

When a charged particle passes through oxide, electron-hole pairs are produced along with the particle path. Charges remaining along with the track may trigger bipolar action of the parasitic circuit or cause TID effects by remained holes in the oxide [130] or migrated protons to the boundary of the oxide (dangling bond) and semiconductor materials. A direct hit of neutron or charged particles to lattice atoms may cause production of permanent defects in the lattice, which cannot cause single events.

In RILC (Radiation Induced Leakage Current) of floating gate memories, some oxide traps or parasitic levels at Si/SiO2 boundaries may be created when a charged particle passes through a tunnel oxide of a floating gate memory. This can be the origin (fault) of an SEU of a floating gate memory.

7.5.2 Error Modes

Error is defined as data flip in memory elements such as dynamic random access memory (DRAM), SRAM, flash memory, latches and FFs. As explained in Section 5.2.2 for latches and FFs, capture of SETs produced in the combinational logic circuit in the upstream of the cell as well as the direct hit of charged particles can cause errors. SETs in the global clock line and global control line can cause errors in latches and FFs [131, 132].

Table 7.6 summarises various error modes. SEU is an event by which data in memory elements is flipped (soft errors). The number of soft errors in an SEU can be single – SBU or more than two – MCU. There has been a great deal of research into MCUs and in particular their ratios to the total SEU have been studied in detail [25, 133]. Though MBUs can be avoided by a combination of ECC and the interleaving technique [24, 25], MCUs that can be corrected by error detection and correction (EDAC)/ECC can still be problematic in high performance devices such as CAMs (Contents Addressable Memories) [134] or registers used in network processors and routers. In the case of system design, it is, therefore, very important to evaluate MCUs as well as SERs of the device in the design phase.

When the event is due to bipolar action in a well, it is called bipolar events or more specifically, MCBI [25]. If data in more than 2 bits in the same word are flipped, the SEU is known as MBU and may cause hang or system-down. By making the distance between bits in the same word large enough to prevent MBU (interleaving or scrubbing), each error can be converted to a single bit error in a word and can be recovered by the Hamming ECC.

Table 7.6 SEE modes with detection, recovery techniques

Class	Definition	Mode name	Characteristics	In-situ detection	In-situ recover method
SEU[a], soft-error	Data change in memory elements such as SRAMs, flip-fops by a single particle hit (event)	SBU[b]	Single bit error for one event	Parity, ECC[c]	ECC
		MCU[d]	More than 2 bits fail in one event. Data in multiple FFs[e] may be flipped by SET in clock line or SET/RESET lines	Current/potential monitor Address mapping	Interleave + ECC
		MBU[f]	MCU in the same word. They cannot be corrected by simple ECC like Hamming code	Upper grade ECC	Upper grade ECC
		MCBI[g]	More than 2 bits fail locally due to potential disturbance in well by bipolar action	Current/potential monitor	Interleave + ECC
Pseudo hard-error, PCSE[h]	Error that cannot be re-written they can mostly be activated by power cycling	FBE[i]	Main error mode in SOI[j]. Body-tie may suppress this mode	Parity, ECC	Power cycle
		SEL[k]	Re-writing does not work. Current continue to flow by parasitic thyristor effect. Power cycle can be applied to activate the chip	Current/potential monitor	
		SEFI[l]	All in one definition of functional anormalities in logic circuits. Power cycle or resetting FFs can activate the chip. SEFI in decoder in peripheral circuit of memory may be caused by wrong address produced in the decoder	FF parity/ECC	
		Firm Error	Error in configuration memory in SRAM based FPGA[m]	CRC[n]	Partial reconfiguration
		RILC[o]	Soft error in flash memory EEPROM (Electrically Erasable Programmable Read-Only Memory)	Potential shift in floating gate	Isolation of faulty cell

(*continued overleaf*)

Table 7.6 (*continued*)

Class	Definition	Mode name	Characteristics	In-situ detection	In-situ recover method
Hard error/ permanent error	Destructive and permanent error	SEGR[p]	Destruction of gate oxide in power devices mainly due to heavy ions. Flash memory can be failed by this mode as scaling extremely proceeds	Abnormalities in parts	Loading stand-by system
		SEB[q]	Destructive mode in power MOSFET such as IGBT[r]. SEB may take place in IGBTs for trains and automobiles	Abnormalities in parts	Loading stand-by system

[a]Single event upset.
[b]Single bit upset.
[c]Error correction code.
[d]Multicell upset.
[e]Flip flop.
[f]Multi-bit upset.
[g]Multi-coupled bipolar interaction.
[h]Power cycle soft error.
[i]Floating body effect.
[j]Silicon on insulator.
[k]Single event latchup.
[l]Sigle event functional interrupt.
[m]Field programmable gate array.
[n]Cyclic redundancy check.
[o]Radiation induced leak current.
[p]Single event gate rupture.
[q]Single event burnout.
[r]Insulated gate bipolar transistor.

The terms pseudo-hard error and/or Power Cycle Soft Error are assigned to more complicated or specific error modes. FBE is a specific error mode in SOI (Silicon On Insulator) devices and is caused by bipolar action in the silicon body [135]. FBE can be suppressed by body-tie (via the ground), but it causes area penalty. SEL used to be classified as an HE that destroys the device through heat produced by a large continuous current triggered by parasitic thyristor action. As scaling proceeds and resistance in the substrate becomes high, however, such a mode is no longer destructive due to low continuous current. Parasitic thyristor current can be blocked by power cycling. SEL, therefore, is called PCSE as well [32].

The definition of SEFI is somewhat ambiguous. SEFI is said to be an all-in-one definition of functional anomalies in logic circuits [136, 137]. The 'k-cell error' in DRAM [138] can be classified into SEFI and may be caused by errors in peripheral circuits around the DRAM matrix. More specifically, errors in the decoder result in the wrong address to read/write.

Power cycling or resetting FFs can resume the chip. The term 'Firm error' is assigned to data flips in SRAM-based FPGA configuration memory in which a program that describes circuit operation is stored/system, error in the memory should be fatal. Since the configuration memory determines the entire operation of a component, partial or full reconfiguration is required [139–142]. RILC means charge stored in the floating gate in a flash memory flows out through a permanent conduction path produced by the charged particle passing through the tunnel oxide. In RILC, V_{th} shifts are observed to have a certain variation, presumably depending on the hit location [143–145]. This kind of error is detected by measuring V_{th} of all the memory cells.

Major HEs have two modes in power devices like power MOSFET (Metal Oxide Semiconductor Field Effect Transistor), GTO (Gate Turn-Off Thyristor), IGBT. SEGR means destruction of gate oxide by a large leakage current produced along with the charged particle path to cause destruction of the gate oxide [146].

SEB means destruction of Si body in power devices by an electron avalanche triggered by a charged particle penetrated to the pnp structure [147, 148].

7.5.3 Failure Modes

A failure is defined as an observable faulty state in an electronic system, which requires actions for recovery. Faults and errors may be masked sometimes without any countermeasures, but failures are not. To establish a solution for failures, the root cause or physical mechanisms must be identified. Classification of failures is often applied to identify the root cause or root parts/chips in a system board. Such classification ideas vary from application to application as summarised in Table 7.7. The following are examples:

1. Baranowski *et al.* define the faulty conditions as *hazards* that cause malfunctions, wrong results or poor system performance including abortion (no service) [149]. Thirteen hazards are defined mainly related to read/write/enable instructions.

Table 7.7 List of classified instances of failures

Name ():temporal	Major application	Outline of method	Number of modes	Typical symptom or failure modes	Remark	References
(Hazard metrics)	Core-based system	Hazard takes place in the core when the core causes errors in the system or cause poor performance	13	wrong_data_read/write wrong decode wrong address invalid_enable read_on_write/write_on_read spurious_enable/read/write) deadlock no_service	–	Baranowski [149]
SWAT	Commodity systems	Failures are detected by symptoms in software	5	Fatal trap in application or OS Application abort Hang of application or OS Heavy load on OS SDC (silent data corruption)	Other modes: Hyperviser crash, Hardware stall, firmware check, abnormal exit, segmentation fault, core dump, dynamic linker error, and so on.	Li [150] and Pellegrini [23]
(Instruction level derating classification)	Super computing	Fault injection, detection of failure by DMR and comparison with golden results	10	Fail-control (program counter miss) Incorrect store value/address Misaligned load/store address Incorrect register inputs to syscall Memory protection exception due to invalid address	Obtaining golden results will be very difficult in Exascale super-computing	Cook [71]
(Software symptom detection)	Automobile	Failures are detected by symptoms in software	6	Fatal trap Hang Application abort kernel panic Illegal address	OS failure mode	Ramachandran [110]
(Impact based classification)	Matrix calculation in super computer	Failures are defined by impact on system	5	Marginal. error < 10% Problematic. Completed with error > 10% SDC Hang. Long calculation time Abort	–	Bronevetsky [151]

(Criticality for safety)	Bralk-by-wire	Fault injection into emulator	2	Critical (tire lock, no brake) Hang (no command within 1 s)	15% failures are critical	Skarin [12]
(Availability based classification)	Network	Field data	4	Low recovery time (0.001–0.01–3 s/year) Medium recovery time (0.01–3 s/year) High recovery time (3–100 s/year) Unrecoverable (> 100 s/year)	Metric DPM(Soft error FIT) × (recovery time in second) is also proposed	Wen [98]
(Re-configuration based classification)	FPGA (field programmable gate array)	DMR based detection and re-configuration	2	Recoverable by rollback Hard error	Hard-error region is isolated and configuration data are re-mapped	Noji [141]
(Communication trouble)	CAN (controller area network in automobile)	Fault injection	4	Timing mismatch SDC Inactive receiver Latent receipt	CAN is used for critical communication among (about a hundred of) micro control unit (MCU)s in a car	Lopes Ongil [15]
	LIN (local interconnect network)	Fault injection	4	Detected (fault detected and mis-behaviour)) Failure (data corrupted without fault detection) Silent (no failure without fault detection) Latent (no failure but error remained in memory)	LIN is used for minor communication among MCUs (power windows, seat control)	Vaskova [16]

2. Failures are detected in software malfunctions in SWAT [109, 150]. Four failure modes are proposed: (i) Fatal trap in the application or OS detected by the hardware; (ii) Stop of application detected by OS; (iii) Hang of application or OS detected by the hardware hang detector; (iv) High load in OS detected by the hardware performance counter. More failure modes are defined in [23] related to SWAT.
3. Cook *et al.* define 10 faulty conditions in supercomputing by using DMR for faulty condition detection and classifying the state by comparison of golden results [71]. Obtaining golden results will be much more difficult due to SDCs in future, and this may be a major setback of this method.
4. In OS, panic in the kernel, which controls resources and transactions between the software and hardware, is a fatal mode and causes the OS to crash and overall termination of the system [108].
5. Bronevetsky *et al.* [151] give five failure modes of supercomputers for matrix calculation: (i) Marginal. Calculation is converged within 10% error; (ii) SDC; (iii) Calculation is converged but error is more than 10%; (iv) Hang. Unusually long calculation time; (v) Abort. System down.
6. Skarin *et al.* [12] give four response modes for fault injection tests in VOLVO^{TM} brake-by-wire systems: (i) Normal operation; (ii) Fault is detected and recovered by the hardware; (iii) Fault is not detected and wrong commands are generated; Fifteen percent of wrong commands turn out to be critical (Tyres lock, 41%, braking did not work, 51%); (iv) Hang: No commands are generated within 1 second.
7. Wen *et al.* [152] propose a new target metric *DPM* (Defects Per Million) for network routers. *DPM* is determined as

$$DPM = (\text{Soft Error FIT}) \times (\text{Error Recovery Time}). \qquad (7.11)$$

   DPM value has three ranks. (i) 0.0001 *DPM*, (ii) 0.1 *DPM* and (iii) 10 *DPM*.

   Error recovery time is also divided into four ranks. Namely, (i) Low: 0.001–0.01 s/year, (ii) Medium: 0.01–3 s/year, (iii) High: 3–100 s/year and (iv) unrecoverable: more than 100 s/year.
8. In systems where FPGAs are used, a large portion of the chip is taken up with configuration memory by which system operation is defined. The error modes in such systems are given therefore, mainly for configuration memory recovery behaviour. Noji *et al.* [145] define three types of failure modes after detecting failure by DMR: (i) Soft-error in FPGAs other than configuration memory. Soft error can be recovered by rollback and re-computation; (ii) Firm error (Error in the configuration memory). Firm error can be recovered by retention reconfiguration (re-loading program) and rollback; (iii) Other failures are classified into HE and such a region becomes isolated and reconfiguration is done using the rest area.
9. In a communication system, errors can be detected by using detection mechanisms such as parity, ECC, CRC and can be recovered by re-sending messages. Failures take place only when faults are not detected. Lopes Ongil *et al.* define the root

cause of such cases in an automotive CAN system by applying a fault injection [15] that is used for critical networks: (i) Timing miss-match in receiving messages; (ii) Wrong data is received but the fault is not identified (SDC); (iii) Receiver is not active for correct messages; (iv) Latent receipt of message. Error may remain in memory.

Vaskova *et al.* define four symptoms in an automotive LIN system [16] that is used for minor communication in the power window, seat control, and so on: (i) Detected: Faults are detected with malfunction of the system; (ii) Failure: Data are corrupted without fault detection; (iii) Silent: No data corruption without fault detection. (Note that the meaning of the term 'Silent' is different from 'Silent' in SDC in this case); (iv) Latent: Data is corrupted but system function is intact.

Nevertheless, it is seen that there are so many failure modes/symptoms proposed that a more versatile definition of failures is needed.

Table 7.8 shows a possible proposal from the author about a versatile classification of failures based on the fatality of the failure with two key factors such as latency in operation and duration for recovery. If the system gives wrong but recoverable results with marginal latency, we would call this type of failure MNFL (MargiNal FaiLure). When SDCs take place in a large-scale supercomputer, simulation may give the wrong results without any latency. This type of failure may be caused by SDC and is called SLFL (SiLent FaiLure). If the SDCs suffer convergence of matrix calculation or frequent rollback due to error detection, significant time loss may take place in the computer/server system. Some errors may be retained or generated by timing error due to the latency. We call it LTFL (LaTency FaiLure). If the system reboots automatically or requires short-range outage to recover by using the power cycle, for example, we can call the failure LHFL (Light Halt Failure). If the system requires long-range outage, we call the failure HHFL (Heavy Halt FaiLure). If the system is un-recoverable, we call the failure FTFL.

The amount of recovery time required obviously depends on the system. In real-time or safety-critical systems like air planes and cars, *milli-second recovery* may be a challenging theme that ensures safe operation even under critical failures.

Definition of fatality may vary from industry to industry. Nevertheless, the author believes the scheme in Table 7.8 can be applied to any industrial field, and common methodologies for hardening electronic systems based on LABIR (inter-LAyer Built-In Reliability)/Cross Layer Reliability concept may need to be explored further.

7.6 Summary of Chapter 7

In order to identify the root cause of failures and determine the most appropriate prediction/estimation techniques to evaluate the impact of SEE, detection/monitoring of fault/error and classification of fault/error/failures is very important. In this chapter,

Table 7.8 A proposal for failure mode classification

Class	Definition	Mode name	Characteristics	In-situ detection	In-situ recover method
Recoverable failure	Recoverable by a simple procedure with a slight latency	MNFL[a]	Failure caused by SEU in memory elements	Parity, ECC	Re-write, reset of memory elements
None-latency failure	Silent data corruption in data or address that cause wrong simulation results by super computer	SLFL[b]	Soft-error takes a major role. SDC[c] causes un-recognisable mis-calculation or mis-operation	None	(If fault level detection works) Checkpointing + Rollback
Latency failure	Performance of the electronic system is lowered due to over frequent rollback, for example	LTFL[d]	In case time redundancy techniques are applied, this mode is commonly take place. Typically, rollback after fault detection by using double module redundancy technique	DMR[e] error flag, the number of retries	Reboot
Light halt failure	Electronic system can be recovered by short-time operation	LHFL[f]	MBU, MNT, SEL can be the cause	ECC, current/ potential monitor	Reboot/power cycling
Heavy halt failure	Electronic system can be recovered by long-time operation. Some logs and data may be lost	HHFL[g]	Error in the configuration memory in FPGA, SEL can cause this mode	CRC check	Partial reconfigura-tion/power cycling
Unrecoverable (fatal) failure	Destruction of power supply and/or power device. Power supply or overall electronic system maybe exchanged	FTFL[h]	Destruction of IGBT, DC-DC converters due to SEB	System down	None

[a]Marginal failure.
[b]Silent failure.
[c]Silent data corruption.
[d]Latency failure.
[e]Double module redundancy.
[f]Light halt failure.
[g]Heavy halt failure.
[h]Fatal failure.

such techniques including recent challenges which have been viewed and summarised for each stack layer in semiconductor manufacturing. Some recommendations are made by the author.

References

[1] Matthews, D.C. and Dion, M.J. (2009) NSE impact on commercial avionics. 2009 IEEE International Reliability Physics Symposium, Montreal, QC, April 26–30, pp. 181–193.

[2] Asai, H., Sugimoto, K., Nashiyama, I. *et al.* (2011) Terrestrial neutron-induced single-event burnout in SiC power diodes. The Conference on Radiation Effects on Components and Systems, Sevilla, Spain, September 19–23 (PC-3).

[3] Slayman, C. (2005) Cache and memory error detection, correction, and reduction techniques for terrestrial servers and workstations. *IEEE Transactions on Device and Materials Reliability*, **5** (3), 397–404.

[4] Schindlbeck, G. and Slayman, C. (2007) Neutron-induced logic soft errors in DRAM technology and their impact on reliable server memory. IEEE Workshop on Silicon Errors in Logic – System Effects 3, Austin Texas, April 3, 4.

[5] Shimbo, K., Toba, T., Nishii, K. *et al.* (2011) Quantification and mitigation techniques of soft-error rates in routers validated in accelerated neutron irradiation test and field test. 2011 IEEE Workshop on Silicon Errors in Logic – System Effects, Champaign, IL, March 29–30, pp. 11–15.

[6] Falsafi, B. (2011) Reliability in the dark silicon Era. 17th IEEE International On-Line Testing Symposium, Athens, Greece, July 13–15.

[7] Rivetta, C., Allongue, B., Berger, G. *et al.* (2001) Single event burnout in DC-DC converters for the LHC experiments. 6th European Conference on Radiation and Its Effects on Components and Systems, September 10–14, pp. 315–322.

[8] Geist, A. (2012) Exascale monster in the closet. 2012 IEEE Workshop on Silicon Errors in Logic – System Effects, Champaign-Urbana, IL, March 27–28 (5.1).

[9] Daly, J.T. (2013) Emerging challenges in high performance computing: resilience and the science of embracing failure. The 9th Workshop on Silicon Errors in Logic – System Effects, Palo Alto, CA, March 26, 27 (Keynote III).

[10] Loncaric, J. (2011) DOE's exascale initiative and resilience. 2011 IEEE Workshop on Silicon Errors in Logic – System Effects, Champaign, IL, March 29–30.

[11] Esmaeilzadeh, H., Blem, E., Amant, R.St. *et al.* Dark Silicon and the End of Multicore Scaling, ftp://ftp.cs.utexas.edu/pub/dburger/papers/ISCA11.pdf (accessed 12 May 2014).

[12] Skarin, D. and Sanfridson, J. (2007) Impact of soft errors in a brake-by-wire system. IEEE Workshop on Silicon Errors in Logic – System Effects 3, Austin Texas, April 3, 4.

[13] Baumeister, D. and Anderson, S.G.H. (2012) Evaluation of chip-level irradiation effects in a 32 – bit safety microcontroller for automotive braking applications. 2012 IEEE Workshop on Silicon Errors in Logic – System Effects, Champaign-Urbana, IL, March 27–28 (2.2).

[14] Nakata, Y., Ito, Y., Sugure, Y. *et al.* (2011) Model-based fault injection for failure effect analysis – evaluation of dependable SRAM for vehicle control units. 5th Workshop on Dependable and Secure Nanocomputing, Hong Kong, China, July 27.

[15] Lopez-Ongil, C., Portela-Garcia, M., Garcia-Valderas, M. *et al.* (2012) SEU sensitivity of robust communication protocols. IEEE International On-Line Testing Symposium, Sitges, Spain, June 27–29, 2012 (9.4), pp. 188–193.

[16] Vaskova, A., Portela-Garcia, M., Garcia-Valderas, M. *et al.* (2013) Hardening of serial communication protocols for potentially critical systems in automotive applications: LIN bus. 19th IEEE International On-Line Testing Symposium, Chania, Crete, July 8–10 (1.3), pp. 13–18.

[17] Rech, P., Aguiar, C., Ferreira, R. *et al.* (2012) Neutrons radiation test of graphic processing units. IEEE International On-Line Testing Symposium, Sitges, Spain, June 27–29, 2012 (3.3).

[18] Rech, P. and Carro, L. (2013) Experimental evaluation of neutron-induced effects in graphic processing units. The 9th Workshop on Silicon Errors in Logic – System Effects, Palo Alto, CA, March 26, 27 (5.3).

[19] Shoji, T., Nishida, S., Ohnishi, T. *et al.* (2010) Neutron induced single-event burnout of IGBT. The 2010 International Power Electronics Conference, Sapporo, Hokkaido, June 21–24, pp. 142–148.

[20] Nishida, S., Shoji, T., Ohnishi, T. *et al.* (2010) Cosmic ray ruggedness of IGBTs for hybrid vehicles. The 22nd International Symposium on Power Semiconductor Devices & ICs, Hiroshima, Japan, June 6–10, 129–132.

[21] Chen, Y. (2013) Cosmic ray effects on cellphone and laptop applications. The 9th Workshop on Silicon Errors in Logic – System Effects, Palo Alto, CA, March 26, 27 (5.4).

[22] Normand, E., Wert, J.L., Oberg, D.L. *et al.* (1997) Neutron-induced single event burnout in high voltage electronics. *IEEE Transactions of Nuclear Science*, **44**, 2358–2368.

[23] Pellegrini, A., Smolinski, R., Chen, L. *et al.* (2011) Crash test'ing SWAT: accurate, gate – level evaluation of symptom – based resiliency solutions. 2011 IEEE Workshop on Silicon Errors in Logic – System Effects, Champaign, IL, March 29–30.

[24] Ibe, E. (2001) Current and future trend on cosmic-ray-neutron induced single event upset at the ground down to 0.1-micron-device. The Svedberg Laboratory Workshop on Applied Physics, Uppsala, Sweden, May 3 (1).

[25] Ibe, E., Chung, S., Wen, S. *et al.* (2006) Spreading diversity in multi-cell neutron-induced upsets with device scaling. The 2006 IEEE Custom Integrated Circuits Conference, San Jose, CA, September 10–13, 2006, pp. 437–444.

[26] Upasani, G., Vera, X. and Gonzalez, A. (2013) Achieving zero DUE for L1 data caches by adapting acoustic wave detectors for error detection. 19th IEEE International On-Line Testing Symposium, Chania, Crete, July 8–10 (5.2).

[27] Yamaguchi, H., Ibe, E., Yahagi, Y. and Kameyama, H. (2006) Novel mechanism of neutron-induced multi-cell error in CMOS devices tracked down from 3D device simulation. International Conference on Simulation of Semiconductor Processes and Devices, Monterey, CA, September 6–8, 2006, pp. 184–187.

[28] Yamaguchi, H., Ibe, E., Yahagi, Y. and Kameyama, H. (2005) 3D device simulation for neutron-induced latch-up in CMOS devices. International Conference on Solid State Devices and Materials, Kobe, Japan, September 13–15, P3 (1), pp. 578–579.

[29] Roche, P. (2010) Industrial impacts of SER on today's consumer electronic arena. 16th IEEE International On-Line Testing Symposium, Corfu Island in Greece, July 5–7 (Keynote 1).

[30] Wang, F. and Xie, Y. (2006) An accurate and efficient model of electrical masking effect for soft errors in combinational logic. The Second Workshop on System Effects of Logic Soft Errors, Urbana-Champain, IL, April 11–12, 2006.

[31] Seifert, N., Zhu, X. and Massengill, L.W. (2002) Impact of scaling on soft-error rates in commercial microprocessors. *IEEE Transactions on Nuclear Science*, **49** (6), 3100–3106.

[32] JEDEC JESD89A. (2006) *Measurement and Reporting of Alpha Particle and Terrestrial Cosmic Ray Induced Soft Errors in Semiconductor Devices (JESD89A)*. JEDEC Standard JESD89A, JESD, pp. 1–94.

[33] JEITA (2005) JEITA SER Testing Guideline. EIAJ (EIAJ EDR-4705), pp. 1–62.

[34] Makino, T., Kobayashi, D., Hirose, K. *et al.* (2009) Soft-error rate in a logic LSI estimated from SET pulse-width measurements. *IEEE Transactions on Nuclear Science*, **56** (6), 3180–3184.

[35] Ahlbin, J.R., Bhuva, B.L., Gadlage, M.J. *et al.* (2009) Single event transient pulse quenching in 130 nm CMOS logic. *IEEE Transactions on Nuclear Science*, **56** (6), 3050–3056.

[36] Baze, M., Wert, J., Clement, J. *et al.* (2006) Propagating SET characterization technique for digital CMOS libraries. *IEEE Transactions on Nuclear Science*, **53** (6), 3472–3478.

[37] Cannon, E.H. and Cabanas-Holmen, M. (2009) Heavy ion and high energy proton-induced single event transients in 90 nm inverter, NAND and NOR gates. *IEEE Transactions on Nuclear Science*, **56** (6), 3511–3518.

[38] Narasimham, B., Ramachandran, V., Bhuva, B.L. *et al.* (2005) On-chip characterization of single event transient pulse widths. 2005 IEEE Nuclear and Space Radiation Effects Conference, Seattle, Washington, July 11–15, 2005 (A4).

[39] Ferlet Cavrois, V., Pouget, V., McMorrow, D. *et al.* (2008) Investigation of the Propagation Induced Pulse Broadening (PIPB) effect on single event transients in SOI and bulk inverter chains. *IEEE Transactions on Nuclear Science*, **55** (6), 2842–2853.

[40] Gadlage, M.J., Ahlbin, J.R., Narasimham, B. *et al.* (2010) Scaling trends in SET pulse widths in Sub-100 nm bulk CMOS processes. *IEEE Transactions on Nuclear Science*, **57** (6), 3336–3341.

[41] Nakamura, H., Tanaka, K., Uemura, T. *et al.* (2010) Measurement of neutron-induced single event transient pulse width narrower than 100ps. 2010 IEEE International Reliability Physics Symposium, Anaheim, CA, May 2–6, pp. 694–697.

[42] Furuta, J., Yamamoto, R., Kobayash, K. and Onodera, H. (2012) Evaluation of parasitic bipolar effects on neutron-induced SET rates for logic gates. IEEE International Reliability Physics Symposium, Anaheim, CA, April 15–19 (SE-5).

[43] Nakamura, T., Baba, M., Ibe, E. *et al.* (2008) *Terrestrial Neutron-Induced Soft-Errors in Advanced Memory Devices*, WorldScientific, Hackensack, NJ, pp. 219–252.

[44] Letaw, J.R. and Normand, E. (1991) Guidelines for predicting single-event upsets in neutron enviroments. *IEEE Transactions on Nuclear Science*, **38** (6), 1500–1506.

[45] Bramnik, A., Sherban, A. and Seifert, N. (2013) Timing vulnerability factors of sequential elements in modern microprocessors. *19th IEEE International On-Line Testing Symposium*, Chania, Crete, July 8–10 (3.3).

[46] http://www.pld.ttu.ee/~maksim/benchmarks/iscas85/verilog (accessed 9 October 2013).
[47] SPEC http://www.spec.org/cpu2000/CINT2000/ (accessed October 9 2013).
[48] Theodorou, G., Kranitis, N., Paschalis, A. and Gizopoulos, D. (2010) A software-based self-test methodology for in-system testing of processor cache tag arrays. 16th IEEE International On-Line Testing Symposium, Corfu Island in Greece, July 5–7 (7.4), pp. 159–164.
[49] Farazmand, N., Ubal, R. and Kaeli, D. (2012) Statistical fault injection-based AVF analysis of a GPU architecure. 2012 IEEE Workshop on Silicon Errors in Logic – System Effects, Champaign-Urbana, IL, March 27–28 (2.1).
[50] Mitra, S., Seifert, N., Zhang, M. *et al.* (2005) Robust system design with built-in soft-error resilience. *Computer*, **38**, 43–52.
[51] Kellington, J. and McBeth, R. (2007) IBM POWER6 processor soft error tolerance analysis using proton irradiation. IEEE Workshop on Silicon Errors in Logic – System Effects 3, Austin, TX, April 3, 4.
[52] Polian, I., Reddy, S.M. and Becker, B. (2008) Scalable calculation of logical masking effects for selective hardening against soft errors. IEEE Workshop on Silicon Errors in Logic – System Effects, University of Texas at Austin, March, 26, 27.
[53] Wissel, L., Pheasant, S., Loughhran, R. *et al.* (2002) Managing soft errors in ASICs. IEEE 2002 Custom Integrated Circuits Conference, pp. 85–88.
[54] Shivakumar, P., Kistler, M.S., Keckler, W. *et al.* (2002) Modeling the effect of technology trends on the soft error rate of combinational logic. International Conference on Dependable Systems and Networks, pp. 389–398.
[55] Fazeli, M., Namazi, A. and Miremadi, S.G. (2009) An energy efficient circuit level technique to protect register file from MBUs and SETs in embedded processors. The 39th Annual IEEE/IFIP International Conference on Dependable Systems and Networks, Estoril, Lisbon, Portugal, June 29–July 2, p. 195.
[56] Aguirre, M.A., Tombs, J.N., Munoz, F. *et al.* (2006) Selective protections using a SEU emulator. Case study: the leon2. Workshop on Radiation Effects on Components and Systems, Athens, Greece, September 27–29, 2006 (PB-2).
[57] Ando, H. and Hatanaka, S. (2007) Accelerated testing of a 90nm SPARC64 V microprocessor for neutron SER. IEEE Workshop on Silicon Errors in Logic – System Effects 3, Austin, TX, April 3, 4.
[58] Uezono, T., Yoneki, S., Toba, T. *et al.* (2013) Evaluation of neutron-induced soft error effects on CPU in automotive microcontrollers. 2013 Convention on Radiation Effects on Components and Systems, Oxford, UK, September 23–27, p. F-5.
[59] Rao, S., Hong, T., Sanda, P. *et al.* (2008) Examining workload dependence of soft error rates. IEEE Workshop on Silicon Errors in Logic – System Effects, University of Texas at Austin, March, 26, 27.
[60] Alexandrescu, D. (2011) A comprehensive soft error analysis methodology for SoCs/ASICs memory instances. 17th IEEE International On-Line Testing Symposium, Athens, Greece, July 13–15 (10.1), pp. 175–176.
[61] Costenaro, E., Evans, A., Alexandrescu, D. *et al.* (2013) Towards a hierarchical and scalable approach for modeling the effects of SETs. The 9th Workshop on Silicon Errors in Logic – System Effects, Palo Alto, CA, March 26, 27 (4.5).

[62] Kumar, R. (2011) Stochastic computing: embracing errors in architecture and design of processors and applications. 5th Workshop on Dependable and Secure Nanocomputing, Hong Kong, China, July 27.

[63] Chen, S. and Liu, B. (2009) The re-convergence's effect to SER estimation in combinational circuits. *IEEE Transactions on Nuclear Science*, **56** (6), 3122–3129.

[64] Takata, T. and Matsunaga, Y. (2011) A robust algorithm for pessimistic analysis of logic masking effects in combinational circuits. 2011 IEEE Workshop on Silicon Errors in Logic – System Effects, Champaign, IL, March 29–30.

[65] Ibe, E., Kameyama, H., Yahagi, Y. and Yamaguchi, H. (2005) Single event effects as a reliability issue of IT infrastructure. 3rd International Conference on Information Technology and Applications, Sydney, Australia, July 3–7, 2005, pp. 555–560.

[66] Silburt, A.L., Evans, A., Burghelea, A. *et al.* (2008) Specification and verification of soft errorperformance in reliable internet core routers. *IEEE Transactions on Nuclear Science*, **55** (4), 2389–2398.

[67] Prokofiev, A.V., Bystrom, O., Ekstrom, C. *et al.* (2005) A new neutron beam facility for SEE testing. 2005 Radiation and Its Effects on Components and Systems, September 19–23, Palais des Congres, Cap d'Agde, France (W-14).

[68] Prokofiev, A.V., Blomgren, J., Majerle, M. *et al.* (2009) Characterization of the ANITA neutron source for accelerated SEE testing at the svedberg laboratory. 2009 IEEE Radiation Effects Data Workshop, Quebec City, Canada, July 20–24, pp. 166–173.

[69] Pecchia, A., Cotroneo, D., Kalbarczyk, Z. and Iyer, R.K. (2011) Improving log-based field failure data analysis of multi-node computing systems. 2011 International Conference on Dependable Systems and Networks, Hong Kong, China, June 28–30, pp. 97–108.

[70] Barbosa, R., Karlsson, J., Yu, Q. and Mao, X. (2011) Toward dependability benchmarking of partitioning operating systems. 2011 International Conference on Dependable Systems and Networks, Hong Kong, China, June 28–30.

[71] Cook, J. and Zilles, C. (2008) Characterizing instruction-level error derating. IEEE Workshop on Silicon Errors in Logic – System Effects, University of Texas at Austin, March, 26, 27.

[72] Kost, R.L., Connors, D.A. and Pasricha, S. (2009) Characterizing the use of program vulnerability factors for studying transient fault tolerance in modern architectures. The 39th Annual IEEE/IFIP International Conference on Dependable Systems and Networks, Estoril, Lisbon, Portugal, June 29–July 2, pp. B9–B14

[73] Rivers, J.A., Bose, P., Kudva, P. *et al.* (2008) Phaser: Phased methodology for modeling the system-level effects of soft errors. *IBM Journal of Research and Development*, **52** (3), 293–306.

[74] Miskov-Zivanov, N. and Marculescu, D. (2010) Modeling and analysis of SER in combinatorial circuits. IEEE Workshop on System Efects of Logic Soft Errors, Stanford University, March 23, 24.

[75] Cho, H., Mirkhani, S., Cher, C.-Y. *et al.* (2013) Understanding inaccuracies of soft error injection techniques. The 9th Workshop on Silicon Errors in Logic – System Effects, Palo Alto, CA, March 26, 27 (13.1).

[76] Siskos, S. (2010) A new built-in current sensor for low supply voltage analog and mixed-signal circuits testing. International On-Line Testing Symposium 2010, Corfu Island, Greece, July 5–7.

[77] Neto, E.H., Kastensmidt, F.L. and Wirth, G.I. (2008) A built-in current sensor for high speed soft errors detection robust to process and temperature variations. Proceedings of the 20st annual Symposium on Integrated Circuits and System Design, September 2007, pp. 190–195.

[78] Ibe, E., Shimbo, K., Toba, T. *et al.* (2011) LABIR: inter – LAyer built – in reliability for electronic components and systems. 2011 IEEE Workshop on Silicon Errors in Logic – System Effects, Champaign, IL, March 29–30.

[79] Wang, T., Zhang, Z., Chen, L. *et al.* (2010) A novel bulk built-in current sensor for single-event transient detection. IEEE Workshop on System Effects of Logic Soft Errors, Stanford University, March 23, 24.

[80] Furuta, J., Yamamoto, R., Kobayashi, K. and Onodera, H. (2011) Correlations between well potential and SEUs measured by well-potential perturbation detectors in 65nm. IEEE Asian Solid-State Circuits Conference, pp. 209–212.

[81] Krimer, E., Crop, J., Erez, M. and Chiang, P. (2013) Replication-free Single-Event Upset (SEU) detection for eliminating silent data corruption in CMOS logic. The 9th Workshop on Silicon Errors in Logic – System Effects, Palo Alto, CA, March 26, 27 (4.2).

[82] Bota, S.A., Torrens, G., Alorda, B. *et al.* (2010) Cross-BIC architecture for single and multiple SEU detection enhancements in SRAM memories. 16th IEEE International On-Line Testing Symposium, Corfu Island, Greece, July 5–7 (7.1), pp. 141–146.

[83] Hirao, T., Laird, J.S., Onoda, S., Shibata, T. *et al.* (2004) Charge collected in Si MOS capacitors and SOI devices p+n diodes due to heavy ion irradiation. The 6th International Workshop on Radiation Effects on Semiconductor Devices for Space Application, Tsukuba, October 6–8, 2004, pp. 105–109.

[84] Dodd, P.E., Shaneyfelt, M.R., Flores, R.S. *et al.* (2011) Single-event upsets and distributions in radiation-hardened CMOS flip-flop logic chains. *IEEE Transactions on Nuclear Science*, **58** (6), 2695–2701.

[85] Takai, M., Abo, S., Iwamatsu, T., Maegawa, S. *et al.* (2006) Evaluation of Soft errors in SOI-SRAM using nuclear probe. The 7th International Workshop on Radiation Effects on Semiconductor Devices for Space Application, Takasaki, Japan, October 16–18, pp. 79–84.

[86] Weulersse, C., Bezerra, F., Miller, F. *et al.* (2007) Probing SET sensitive volumes in linear devices using focused laser beam at different wavelengths. 9th European Conference Radiation and Its Effects on Components and Systems, Deauville, France, September 10–14 (A-1).

[87] Baker, J.M., Lum, G.K., Robinette, L. *et al.* (2005) Analysis of single event effects in CMOS devices using heavy ion microbeam and ion electron emission microscope techniques. 2005 IEEE Nuclear and Space Radiation Effects Conference, Seattle, Washington, July 11–15, 2005 (F–2).

[88] Mazumder, P. (1992) An on-chip ECC circuit for correcting soft errors in DRAM's with trench capacitors. *IEEE Journal of Solid-State Circuits*, **27** (11), 1623–1633.

[89] Vera, X., Abella, J., Carretero, J. *et al.* (2009) Online error detection and correction of erratic bits in register files. 15th IEEE International On-Line Testing Symposium, Sesimbra-Lisbon, Portugal, June 24–26 (4.3).

[90] Reddy, K.K., Amrutur, B. and Parekhji, R. (2008) False error study of on-line soft error detection mechanisms. International On-Line Testing Symposium 2008, Greece, July 6–9, 2008 (3.1), pp. 53–58.
[91] Ernst, D., Das, S., Lee, S. *et al.* (2004) Razor: circuit-level correction of timing errors for low-power operation. *IEEE Micro*, **24** (6), 10–20.
[92] Blaauw, D., Kalaiselvan, S., Lai, K. *et al.* (2008) Razor II: in situ error detection and correction for PVT and SER tolerance. 2008 IEEE International Solid-State Circuit Conference, San Francisco, CA, February 4–6.
[93] Avirneni, N.D.P., Subramanian, V. and Somani, A.K. (2009) Low overhead soft error mitigation techniques for high-performance and aggressive systems. The 39th Annual IEEE/IFIP International Conference on Dependable Systems and NetWorks, Estoril, Lisbon, Portugal, June 29–July 2, pp. 185–194
[94] Julai, N., VYakovlev, A. and Bystrov, A. (2012) Error detection and correction of single event upset tolerant latch. IEEE International On-Line Testing Symposium, Sitges, Spain, June 27–29, 2012 (1.1).
[95] Noguchi, K. and Nagata, M. (2007) An on-chip multichannel waveform monitor for diagnosis of systems-on-a-chip integration. *IEEE Transactions of on Very Large Scale Integration (VLSI) Systems*, **15** (10), 1101–1110.
[96] Azais, F., Larguier, L., Bertrand, Y. and Renovell, M. (2008) On the detection of SSN-induced logic errors through on-chip monitoring. International On-Line Testing Symposium 2008, Greece, July 6–9, 2008 (10.2), pp. 233–238.
[97] Yoshikawa, K., Hashida T. and Nagata, M. (2011) An on-chip waveform capturer for diagnosing off-chip power delivery. International Conference on IC Design and Technology, Kaohsiung, Taiwan, May 2–4, 2011.
[98] Wen, S., Silburt, A. and Wong, R. (2008) IC component SEU impact analysis. IEEE Workshop on Silicon Errors in Logic – System Effects, University of Texas at Austin, March, 26, 27.
[99] Alderighi, M., Angelo, S.D., Mancini, N. *et al.* (2005) SEU sensitivity of virtex configuration logic. 2005 IEEE Nuclear and Space Radiation Effects Conference, Seattle, Washington, July 11–15, 2005 (PF-2).
[100] Sanyal, A., Alam, S. and Kundu, S. (2008) A built-in self-test scheme for soft error rate characterization. International On-Line Testing Symposium 2008, Greece, July 6–9, 2008 (3.3), p. 65.
[101] Prejean, S. (2010) Neutron soft error rate testing of AMD microprocessors. IEEE Workshop on System Efects of Logic Soft Errors, Stanford University, March 23, 24.
[102] Yao, J., Watanabe, R., Yoshimura, K. *et al.* (2011) An efficient and reliable 1.5-way processor by fusion of space and time redundancies. 5th Workshop on Dependable and Secure Nanocomputing, Hong Kong, China, July 27.
[103] Grosso, M., Reorda, M.S., Portela-Garcia, M. *et al.* (2010) An on-line fault detection technique based on embedded debug features. 16th IEEE International On-Line Testing Symposium, Corfu Island, Greece, July 5–7, pp. 167–172.
[104] Gold, B., Smolens, J.C., Falsafi, B. and Hoe, J.C. (2006) The granularity of soft-error containment in shared-memory multiprocessors. The Second Workshop on System Effects of Logic Soft Errors, Urbana-Champain, IL, April 11–12, 2006.

[105] Wood, A., Jardine, R. and Bartlett, W. (2006) Data integrity in HP nonstop servers. The Second Workshop on System Effects of Logic Soft Errors, Urbana-Champain, IL, April 11–12, 2006.

[106] Rezgui, S., Carmichael, C., Moore, J. *et al.* (2004) Dynamic testing of the xilinx virtex-II family Input Output Blocks (IOBs). 2004 Nuclear and Space Radiation Effects Conference, Atlanta, Georgia, July 20–24 (E-3).

[107] Aggarwal, N., Jouppi, N., Ranganathan, P. *et al.* (2008) Reducing overhead for soft error coverage in high availability systems. IEEE Workshop on Silicon Errors in Logic – System Effects, University of Texas at Austin, March, 26, 27.

[108] Du, B., Reorda, M.S., Sterpone, L. *et al.* (2013) Exploiting the debug interface to support on-line test of control flow errors. 19th IEEE International On-Line Testing Symposium, Chania, Crete, July 8–10 (5.3), pp. 98–103.

[109] Carvalho, M., Bernardi, P., Sanchez, E. *et al.* (2013) Increasing fault coverage during functional test in the operational phase. 19th IEEE International On-Line Testing Symposium, Chania, Crete, July 8–10 (3.1).

[110] Ramachandran, P., Hari, S.K.S., Adve, S. and Naeimi, H. (2011) Understanding when symptom detectors work by studying data – only application values. 2011 IEEE Workshop on Silicon Errors in Logic – System Effects, Champaign, IL, March 29.

[111] Ballan, O., Bernardi, P., Yazdani, B. and Sanchez, E. (2013) A software-based self-test strategy for on-line testing of the scan chain circuitries in embedded microprocessors. 19th IEEE International On-Line Testing Symposium, Chania, Crete, July 8–10 (5.2), pp. 79–84.

[112] Carloa, S.D., Gambardellab, G., Indacoc, M. *et al.* (2013) Increasing the robustness of CUDA fermi GPU-based systems. 19th IEEE International On-Line Testing Symposium, Chania, Crete, July 8–10 (6.3).

[113] Brown, A. and Lin, C. (2006) Limitations of software solutiuons for soft-error detection. The Second Workshop on System Effects of Logic Soft Errors, Urbana-Champain, IL, April 11–12, 2006.

[114] Furuta, J., Hamanaka, C., Kobayashi, K. and Onodera, H. (2011) Measurement of neutron-induced SET pulse width using propagation-induced pulse shrinking. 2011 IEEE International Reliability Physics Symposium, Monterey, California, April 12–14 (5B2).

[115] Benedetto, J.M., Eaton, P., Mavis, D. *et al.* (2006) Digital single event transient trends with technology node scaling. *IEEE Transactions on Nuclear Science*, **53** (6), 3462–3465.

[116] Yamamoto, S., Kokuryou, K., Okada, Y. *et al.* (2004) Neutron-induced soft-error in logic device using quasi-monoenergetic neutron beam. 2004 IEEE International Reliability Physics Symposium, Phoenix, AZ, April 25–29, pp. 305–309.

[117] Quinn, H., Morgan, K., Graham, P. *et al.* (2007) Domain crossing events: limitations on single device triple-modular redundancy circuits in xilinx FPGAs. International Nuclear and Space Radiation Effects Conference, Honolulu, HI, July 23–27 (C-5).

[118] Ibe, E., Chung, S., Wen, S. *et al.* (2006) Valid and prompt track-down algorithms for multiple error mechanisms in neutron-induced single event effects of memory devices. Workshop on Radiation and Its Effects on Components and Systems, Sevilia, Spain, September 19–23 (D-2).

[119] Seifert, N. and Zia, V. (2007) Assessing the impact of scaling on the efficacy of spatial redundancy based mitigation schemes for terrestrial applications. IEEE Workshop on Silicon Errors in Logic – System Effects 3, Austin, TX, April 3, 4.

[120] Calin, T., Nicolaidis, M. and Velazco, R. (1996) Upset hardened memory design for submicron CMOS technology. *IEEE Transactions on Nuclear Science*, **43** (6), 2874–2878.

[121] Uemura, T., Tosaka, Y. and Matsuyama, H. (2010) SEILA: soft error immune latch for mitigating multi-node-SEU and local-clock-SET. IEEE International Reliability Physics Symposium 2010, Anaheim, CA, May 2–6 (9), pp. 218–223.

[122] Lee, H.-H.K., Lilja, K., Relangi, P. *et al.* (2010) LEAP: layout design through error-aware placement for soft-error resilient sequential cell design. IEEE International Reliability Physics Symposium 2010, Anaheim, CA, May 2–6 (240).

[123] Cabanas-Holmen, M., Cannon, E., Kleinosowski, A. *et al.* (2009) Clock and reset transients in a 90 nm RHBD single-core tileraprocessor. 2009 IEEE Nuclear and Space Radiation Effects Conference, Quebac, Canada, July 20–24 (PG-3).

[124] Nakauchi, T., Mikami, N., Oyama, A. *et al.* (2008) A novel technique for mitigating neutron-induced multi-cell upset by means of back bias. IEEE International Reliability Physics Symposium, Anaheim, CA, April 15–19 (2 F.2), pp. 187–191.

[125] Voldman, S., Gebreselasie, E., Zierak, M. *et al.* (2005) Latchup in merged triple well structure. 2005 IEEE International Reliability Physics Symposium Proceedings, San Jose, CA, April 17–21, 2005, pp. 129–136.

[126] Yoshimura, M., Akamine, Y. and Matsunaga, Y. (2011) An SER analysis method for sequential circuits. 2011 IEEE Workshop on Silicon Errors in Logic – System Effects, Champaign, IL, March 29–30.

[127] Ibe, E., Shimbo, K., Taniguchi, H. *et al.* (2011) Quantification and mitigation strategies of neutron induced soft-errors in CMOS devices and components-the past and future. 2011 IEEE International Reliability Physics Symposium, Monterey, CA, April 12–14 (3C2).

[128] Jagirdar, A. and Oliveira, R. (2007) Efficient flip-flop designs for SET/SEU mitigation with tolerance to crosstalk induced signal delays. IEEE Workshop on Silicon Errors in Logic – System Effects 3, Austin, TX, April 3, 4.

[129] Kanekawa, N., Ibe, E., Suga, T. and Uematsu, Y. (2010) *Dependability in Electronic Systems-Mitigation of Hardware Failures, Soft Errors, and Electro-Magnetic Disturbances-. Dependability in Electronic Systems*, Springer.

[130] Schrimpf, R., Warren, K.M., Weller, R.A. *et al.* (2008) Reliability and radiation effects on IC technologies. IEEE International Reliability Physics Symposium, Anaheim, CA, April 15–19 (2C.1), pp. 97–106.

[131] Mavis, D.G. and Eaton, P.H. (2007) SEU and SET modeling and mitigation in deep submicron technologies. 2007 IEEE International ReliabilityPhysics Symposium, Phoenix, AZ, April 15–19, 2007 (4B.1).

[132] Radaelli, D., Puchner, H., Chia, P. *et al.* (2005) Investigation of multi-bit upsets in a 150 nm technology SRAM device. *IEEE Transactions on Nuclear Science*, **52** (6), 2433–2437.

[133] Gasiot, G. and Roche, P. (2006) Alpha-induced multiple cell upsets in standard and radiation hardened SRAMs manufactured in a 65 nm CMOS technology. *IEEE Transactions on Nuclear Science*, **53** (6), 3479–3486.

[134] Abbas, S.M., Baeg, S. and Park, S. (2011) Multiple cell upsets tolerant content-addressable memory. 2011 IEEE International Reliability Physics Symposium, Monterey, CA, April 12–14 (SE1).

[135] Makihara, A., Midorikawa, M., Yamaguchi, T. *et al.* (2005) Hardness-by-design approach for 0.15 μm fully depleted CMOS/SOI digital logic devices with enhanced SEU/SET immunity. *IEEE Transactions on Nuclear Science*, **52** (6), 2524–2530.

[136] Irom, F. and Nguyen, D.N. (2007) Single event effect characterization of high density commercial NAND and NOR nonvolatile flash memories (TNS). *IEEE Transactions on Nuclear Science*, **54** (6), 2547–2553.

[137] Graham, P. and Rezgui, S. (2005) Radiation-induced multi-bit upsets in SRAM-based FPGAs. 2005 IEEE Nuclear and Space Radiation Effects Conference, Seattle, WA, July 11–15, 2005 (PF-1).

[138] Schindlbeck, G. and Slayman, C. (2007) Neutron-induced logic soft errors in DRAM technology and their impact on reliable server memory. IEEE Workshop on Silicon Errors in Logic – System Effects 3, Austin, TX, April 3, 4.

[139] Azambuja, J.R., Sousa, F., Rosa, L. and Kastensmidt, F. (2009) Evaluating large grain TMR and selective partial reconfiguration for soft error mitigation in SRAM-based FPGAs. 15th IEEE International On-Line Testing Symposium, Sesimbra-Lisbon, Portugal, June 24–26 (5.3).

[140] Rusu, C., Anghel, L. and Avresky, D. (2010) RILM: reconfigurable inter-layer routing mechanism for 3D multi-layer networks-on-chip. 16th IEEE International On-Line Testing Symposium, Corfu Island, Greece, July 5–7 (6.2), pp. 121–126.

[141] Noji, R., Fujie, S., Yoshikawa, Y. *et al.* (2010) An FPGA-based fail-soft system with adaptive reconfiguration. 16th IEEE International On-Line Testing Symposium, Corfu Island, Greece, July 5–7 (6.3), pp. 127–132.

[142] HowStuffWork http://computer.howstuffworks.com/encryption7.htm (accessed 5 December 2013).

[143] Cellere, G., Paccagnella, A., Visconti, A. *et al.* (2007) Traces of errors due to single ion in floating gate memories. International Conference on IC Design and Technology, Austin, TX, May 18–20.

[144] Cellere, G., Paccagnella, A., Visconti, A. *et al.* (2006) Single event effects in NAND flash memory arrays. *IEEE Transactions on Nuclear Science*, **53** (4), 1813–1818.

[145] Gerardin, S., Bagatin, M., Paccagnella, A. *et al.* (2010) Scaling trends of neutron effects in MLC NAND flash memories. IEEE International Reliability Physics Symposium 2010, Anaheim, CA, May 2–6, pp. 400–406.

[146] Hands, A., Morris, P., Ryden, K. *et al.* (2011) Single event effects in power MOSFETs due to atmospheric and thermal neutrons. *IEEE Transactions on Nuclear Science*, **58** (6), 2687–2694.

[147] Kuboyama, S., Matsuda, S., Kanno, T. and Ishii, T. (1992) Mechanism for single-event burnout of power MOSFETs. *IEEE Transactions on Nuclear Science*, **39** (6), 1698–1703.

[148] Saiz-Adalid, L.-J. (2009) Intermittent faults: analysis of causes and effects, new fault models, and mitigation techniques. The 39th Annual IEEE/IFIP International Conference on Dependable Systems and NetWorks, Estoril, Lisbon, Portugal, June 29–July 2.

[149] Baranowski, R. and Wunderlich, H.-J. (2011) Fail-safety in core-based system design. 17th IEEE International On-Line Testing Symposium, Athens, Greece, July 13–15 (13.3), pp. 278–283.

[150] Li, M., Ramachandran, P., Sahoo, S., Adve, S., Adve, V., Zhou, Y. (2008) Understanding the Propagation of Hard Errors to Software and Implications for Resilient System Design, Proceedings of International Conference on Architectural Support for Programming Languages and Operating Systems (ASPLOS), 2008. (http://rsim.cs.illinois.edu/Pubs/08ASPLOS.pdf)

[151] Bronevetsky, G. and deSupinski, B. (2007) Soft error vulnerability of iterative linear algebra methods. IEEE Workshop on Silicon Errors in Logic – System Effects 3, Austin, TX, April 3, 4.

[152] Wen, S. (2008) Systematical method of quantifying SEU FIT. International On-Line Testing Symposium 2008, Greece, July 6–9, 2008, pp. 109–116.

8

Mitigation Techniques of Failures in Electronic Components and Systems

8.1 Conventional Stack-layer Based Mitigation Techniques, Their Limitations and Improvements

There are a number of conventional mitigation techniques mainly in device/circuit layers [1, 2]. They are mainly against alpha-ray induced soft error in DRAMs (Dynamic Random Access Memories). As neutron-induced soft error becomes imminent, a number of additional mitigation techniques have been proposed in each stack layer basis. Such layer-based mitigation techniques and their limitations are reviewed in this section.

8.1.1 Substrate/Device Level

Table 8.1 summarises traditional and recent prevention, in-situ recovery and off-line recovery techniques in the substrate/device level [3–21].

In the lowest stack layer, substrate and well layer, a fault is created as a consequence of electron/hole pair production and charge deposition in the p/n well or the substrate. When a substantially large amount of charge is collected to the storage node/diffusion layer, conventional mode soft error or charge collection mode soft error takes place. When a substantial density of minor carriers remain in the well region, the parasitic transistor near the region is triggered to be 'ON' to cause a bipolar mode soft error such as SESB (Single Event SnapBack) or MCBI (Multi-Coupled Bipolar Interaction) [15, 21]. One of the most important characteristics of the bipolar mode is that the charged particle does not necessarily pass through the depletion layer below the storage node. Instead, it is often the case that it simply passes through more than one pn junction in the p/n wells like the side wall of a p-well.

Terrestrial Radiation Effects in ULSI Devices and Electronic Systems, First Edition. Eishi H. Ibe.

Table 8.1 List of prevention, in-situ recovery and off-line mitigation techniques against failures in substrate/device level

Layer	Prevention	In-situ recovery	Off-line recovery
Substrate/ well	Confinement of charge collection volume[1] Optimisation of well structure/size[1] Implementation of fault detectors (BICS[2], SAW type detector, on-chip monitor[3])	Zone isolation[9]	Zone isolation[13]
Device	Addition of resistor and/or capacitor (high-*k* gate[4]) Interleaving of memory cells in the same word[5] Gate/transistor sizing[6] Addition of guard electrode/contact[7] Exchange critical devices to hardened devices[8]	ECC (SBU only)[10] Memory page retirement and re-mapping[11] Partial reconfiguration of CRAM[12]	Power cycling[14] Reconfiguration of CRAM in FPGA[15]
References	1. Ibe [3] 2. Wang [4], Neto [5], Bota [6] and Siskos [7] 3. Upsani (2013) 4. Noguchi [8], Azais [9] and Yoshikawa [10] 5. – 6. Ibe [11] 7. Zhou (2004) and Bastos [12] 8. Black [13] and Narasimham [14] 9. Valderas [15] and Shimbo [16]	10. – 11. – 12. Slayman [17] and Schndbeck [18] 13. Pillot (2008) and Nakka (2009)	14. – 15. JESD890A [19] 16. Allen [20]

The following exemplifies the substrate/well layer mitigation techniques that should cover both modes to suppress the faulty conditions above:

(i) Reduction of charge collection volume or p-well volume in triple well structure by making the three-dimensional size such that the total amount of charge deposited in the region was less. It is noteworthy that this is effective only for the charge collection mode of soft error. This has an adverse effect on bipolar mode soft error such as MCBI because holes created in the p-well cannot flow out from the p-well or distances among p-n junctions become shorter so that bipolar action takes place more easily. Shallow p-well (depth < 0.4 μm) is shown to be effective to prevent soft error based on the Monte-Carlo simulation by using CORIMS (COsmic Radiation IMpact Simulator (CORIMS) [3].

(ii) Modification of STI (Shallow Trench Isolation) structure/layout to limit charge flow.

Figure 8.1 shows a typical cross section of a 6T-SRAM (Static Random Access Memory). As the source-drain channel is made in parallel with the bit line and both sides of the channel are sandwiched by the STI structures, the charge collection rarely takes place when the charge particle penetrates the channel in a horizontal (WL – Word Line) direction, and, in addition, the chances of passing through the depletion layer (necessary condition for charge collection by funnelling mechanism) are very scarce. Charge collection by drift diffusion may take place, but the amount of collected charge may be small. This is one of the important reasons that soft error in SRAMs is not spread over more than 3 bits along with the same (single) WL. Deeper STI is also effective for preventing soft error [3].

Techniques (i) and (ii) may cause difficulty in manufacturing processes. Actually, shallow wells are in general challenged by limitations in implantation techniques.

(iii) If some circuits are implemented in part of the substrate or a well has a faulty condition, the zone may be electrically isolated in order to keep the overall function of the system intact as in real-time or safety critical systems. Locating

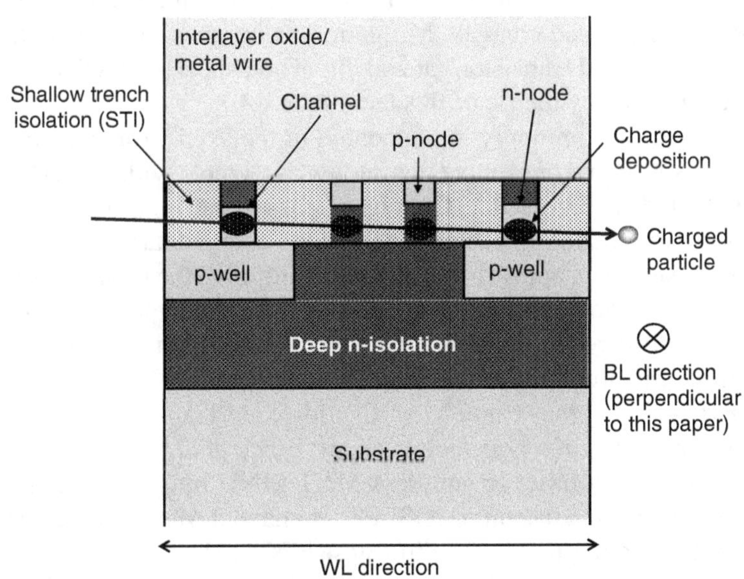

Figure 8.1 Cross section of a SRAM along with word line with a penetrating charged particle. Charge correction to the storage nodes is limited in this situation

faults is primarily the most important technique for this kind of approach and techniques introduced in Section 7.3.1 may be utilised. Isolating such an area is much more difficult and challenging and can be one of the key techniques required in the prompt protection/recovery technologies necessary in real-time systems.

In the device layer, some other techniques are aimed at increasing Q_{crit}:

(iv) Addition of a capacitor to increase Q_{crit}. This is a general device-level approach for DRAMs and some SRAMs. High-k gate is another way to increase capacitance and eventually Q_{crit} [22].

(v) Addition of resistor between two storage nodes in a SRAM to suppress a tendancy to flip.

Together with adding a capacitor, adding a resistor between two storage nodes in a SRAM increases the time constant of SRAMs required to reduce the speed of transient of status, and thus the driving force from the V_{cc} line can effectively prevent flip.

Hirose *et al.* applied resistors and capacitors to fully depleted 200 nm Silicon On Insulators (SOI)s SRAMs with body tie to suppress any floating body effect and optimise them to have threshold LET_{th}, beyond which the SRAMs are sensitive to SEE (Single Event Effect) levels as high as 45 MeV cm^2/mg [23].

(vi) Transistor sizing

In general, wide width of gate channel gives more robust characteristics to SETs (Single Event Transients) by strengthening the driving force of V_{cc} or dispersing deposited charges. Mitigation techniques that utilise these characteristics are called transistor/gate sizing. The aspect ratio of the gate width to gate length is a key metric of this technique.

C-element is commonly implemented in the asynchronous circuits, and has two inputs A, B and one output commonly associated with a SRAM-type keeper as illustrated in Figure 8.2. C-element passes the input data only when two inputs A and B match and keeps the previous output data when two inputs do not match. Bastos *et al.* apply this technique to improve the robustness of C-element to SET [12]. Zhou *et al.* propose optimised design techniques, *geometric programming framework*, including gate sizing for combinational circuits [24].

(vii) Addition of guard contact/electrode

Charge-sharing may cause MNT (Multi-Node Transient)/MNU (Multi-Node Upset) when Q_{crit} of a node/cell is small. Black *et al.*, proposes suppression of charge-sharing in order to suppress MNT/MNU by guard-contact and n-well isolation between adjacent pMOSFET (p-channel Metal Oxide Semiconductor Field Effect Transistor) in the flip-flops (FFs) [13].

Narasimham *et al.* [14] evaluate multiple charge collection in adjacent 130 and 180 nm pMOSFET/pMOSFET (n-channel Metal Oxide Semiconductor Field Effect Transistor) and its suppression by the guard ring and guard drain

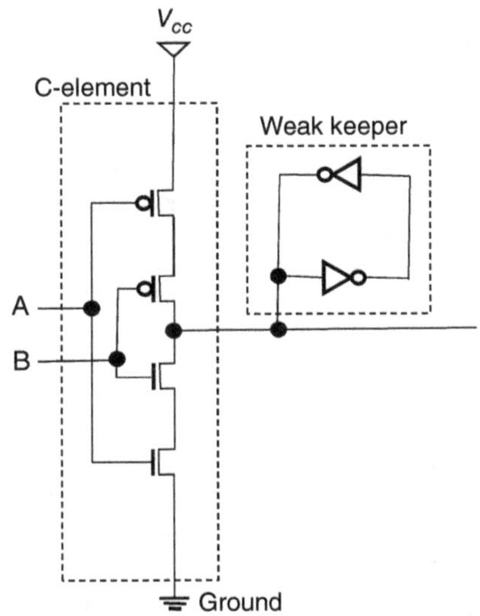

Figure 8.2 Circuit of C-element. Only when inputs A and B match, is the output changed to A(= B) and stored in the weak keeper

by using TCAD (Technology Computer-Aided Design) simulation and heavy ion irradiation. A factor of around 2 suppression in the SEU (Single Event Upset) cross section is observed in the DUTs (devices under test) with the guard rings.

(viii) Exchange critical devices to hardened devices

Shimbo *et al.* screened critical chips (SRAM, CPU (Central Processing Unit) FPGA (Field Programmable Gate Array)) for failure in a network router board and found the SRAM chip is the most vulnerable to 65 MeV neutrons at CYRIC (CYclotron and RadioIsotope Center) [16]. They replaced some SRAMs that do not require speed by DRAMs with ECC (Error Correction Code) and 1/9–1/10 reduction in failure rate is achieved both in accelerator and field tests.

Valderas *et al.* screened critical FFs by FPGA emulator-base fault injection on a PIC18™ microprocessor with some workloads [15]. They found that only 102 (17% of total) FFs for pipeline control, program counter, port register, and so on, are responsible for failures and failure rate could be reduced by 99% by applying partial TMR (Triple Modular Redundancy. See Section 8.1.2.9) schemes to these FFs (Flip-Flops)

8.1.2 Circuit/Chip/Processor Layer

Table 8.2 summarises traditional and recent prevention, in-situ recovery and off-line recovery techniques in the circuit/chip/processor level.

Table 8.2 List of prevention, in-situ recovery and off-line mitigation techniques against failures in circuit/chip/processor level

Layer	Prevention	In-situ recovery	Off-line recovery
Circuit	Enhancement of masking effects in logic parts[1] Space/time redundancy (DICE[2], SEUT[3], BISER[4], SEILA[5], LEAP[6], BCDMR[7], and RAZOR[8])	Space/time redundancy (DICE, SEUT, BISER, SEILA, LEAP, BCDMR and RAZOR) Hardened PLL/DLL[11,12] Double clock lines[13] Partial TMR[14] Multiple clocking[15]	Reset of flip-flops for SEFI[23] Power cycle[24]
Chip/processor	Fault injection to identify critical error/failure mode[9] Selective hardening[10]	Lock-step operation of dual processors[16] Partial reconfiguration[17] Data mirroring and re-writing into memory[18] DMR (double module redundancy)/DCC (Duplication + Comparison + Checkpointing)[19] RCC (Replication + Comparison + Checkpointing-)[20] TMR (triple module redundancy)[21] Utilisation of on-chip debugger for fault detection[22]	Reboot[25] Power cycle[26] Reconfiguration[27]

References	1. Almukhaizim [25], Sierawski [26] and Makihara [27] 2. Calin [31] 3. Hazucha [32] 4. Mitra [33] 5. Uemura [34] 6. Lee [36] 7. Furuta [40] 8. Ernst [41], Ernst [42] and Blaauw *et al.* [43] 9. Roche [44], Skarin and Karlsson [45], and many others 10. Polian [46], Azambuja *et al.* [47] and Valderas *et al.* [15]	16. Wood [28], Subramanian [29] and Baumeister [30] 17. Pillot (2008) and Nakka (2009) 18. – 19. Nicolov [48], Wang [4] and Skarlin (2009) 20. Aggarwal [35] 21. Pratt [37], Inoue [38] and Yang [39] and many others 22. Grosso [49]	23. – 24. JESD89A [19] 25. – 26. JESD89A [19] 27. Allen [20]

8.1.2.1 Interleaving Bits in the Same Word to Suppress MBU

If a MBU (Multi-Bit Upset) takes place in memory elements, detection of error may not be successful and a critical failure such as hang/rebooting may take place in the system.

Combination of ECC and interleaving is the most effective mitigation technique for SEU (including MCU (Multi-Cell Upset) and thus MBU) in SRAMs and DRAMs [15]. It is seen that only 3 or 4 bits interleaving distance (<8 μm) along a WL will be enough to suppress MBU below 0.01 FIT/Mbit (FIT = failure in time). In fact, low power SRAM with lower Q_{crit} and a bit interleaving distance of 2 bits in a WL direction showed no MBUs in the accelerated test [22]. The conclusion is reached that an interleaving distance much shorter (smaller than 10 μm s) than commonly applied in interleave techniques (several tens of μms, comparable to secondary ion range) can be applied to suppress MBU. Maiz *et al.* has shown experimental results at LANSCE (Los Alamos Neutron Science CEnter), which support this conclusion [50]. As memory bits in the same word in the memory matrix are placed at certain intervals typically consistent to the number of word bits (8, 16 and 32 bits) in a single WL, interleaving is naturally applied in a memory matrix.

This technique, however, is not principally effective for CAMs (Content Addressable Memories) which consist of two SRAMs, in general. Abbas *et al.* develop a novel ECC scheme for 512 × 72 bits CAMs to detect and recover m-bit MCU by using m-bit ECC bits and 85% of parity bits are shown to be enough compared to conventional methods that are based on a Hamming distance [51].

But in logic circuits, bits in the same word are placed on a side-by-side basis so that interleaving can be a heavy constraint in the overall design.

In addition, the interleave techniques may not work if Q_{crit} decreases down to below 1 fC [52, 53]. In this Q_{crit} range, secondary alpha particles and protons play a major role, so that SEU ranges well beyond 100 μm from the incident location, have an effect over 1 million bits area.

8.1.2.2 Enhancement of Masking Effects

SER (Soft-Error Rate) can be reduced by enhancing the masking effects in a circuit as described in Figure 7.2. An optimisation technique of masking effects by using SPFD (Sets of Pairs of Functions to be Distinguished) for circuit design (arrangement of gates) is introduced and 20% reduction in SER is demonstrated [25].

Sirawski *et al.* propose optimisation of masking effects to reduce SER by 10–90% with a proper area penalty ranging 5–44% which increases by approximation of the circuit function [26]. Approximation is mainly a result of the addition of 'don'tcare' circuitries to enhance logical masking effects.

Makihara *et al.* propose an SET free inverter in which double pMOSFETs and double nMOSFETs in series are connected to the output logical node as illustrated in

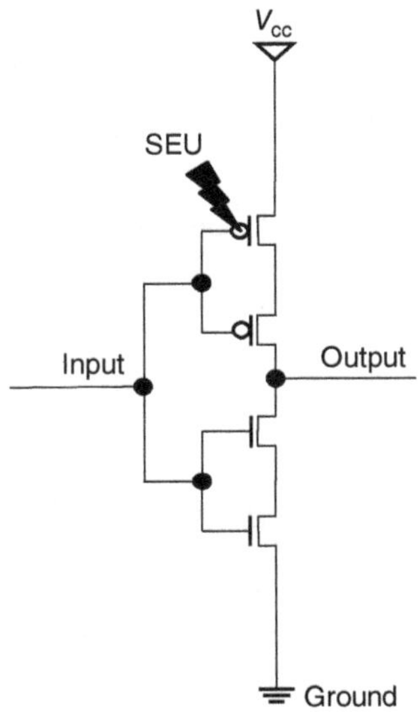

Figure 8.3 Hardened inverter. (Makihara *et al.* [27] ©2005 IEEE)

Figure 8.3 [27]. Single hit by one of pMOSFET or nMOSFET cannot cause a SET with this structure. They demonstrate that latches, NANDs, NORs made with their SET free inverters do not show any SETs up to a LET (linear energy transfer) of 64 MeV/mg/cm^2.

8.1.2.3 Space Redundancy in Circuit Level

There is a number of space/time redundancy techniques proposed to prevent those errors which are mainly related to FFs. The following are such examples.

DICE has two redundant nodes [31] as simply illustrated in Figure 8.4. When one node is about to flip, a feedback loop from the other node suppresses the flip to maintain the initial condition. Seifert *et al.* describe SEUT (Single Event Upset Tolerant) with an extended concept of DICE [54]. These mechanisms, however, are vulnerable to a simultaneous hit like MNT of the two redundant nodes. Figure 8.5 shows such an example in SEUT [32, 55]. SER in SEUT increases to the same level (1.0 in the vertical axis) of FFs without hardening as the device scaling proceeds down to 22 nm. Figure 8.5 also shows how a SET in local clock lines causes simultaneous error in more than two FFs in a circuit. Mitigation techniques as a result of changing the layout of the relevant nodes in DICE to avoid simultaneous hits in a set of redundant nodes are proposed independently in SEILA (Soft Error Immune LAtch) [34], and LEAP (Layout design through Error Aware Placement) DICE [36]. A prevention technique against SETs in local clock lines is also outlined in [34]. Mitra *et al.* describe BISER

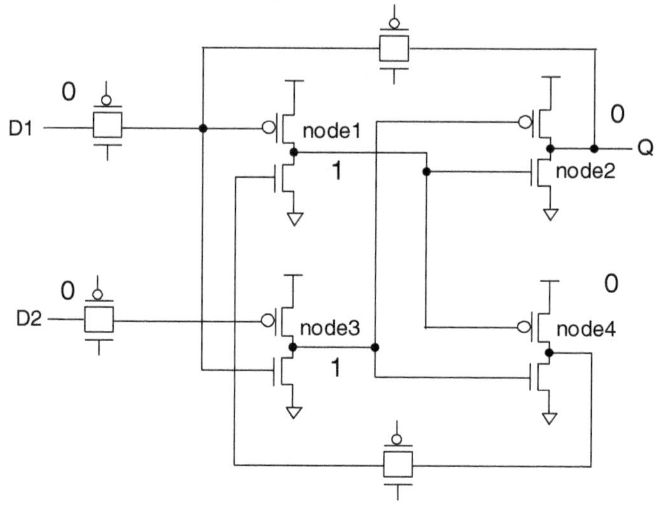

Figure 8.4 DICE circuit

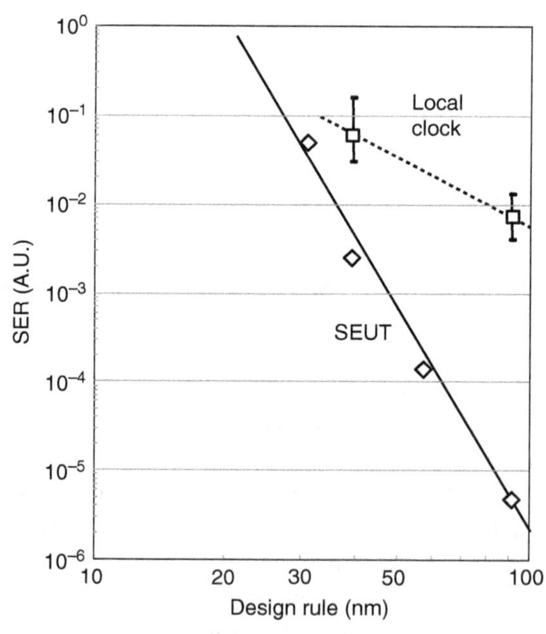

Figure 8.5 Increase in SER in SEUT and SER of FF due to SET in clock line. (Seifert *et al.* [55] ©2007 IEEE)

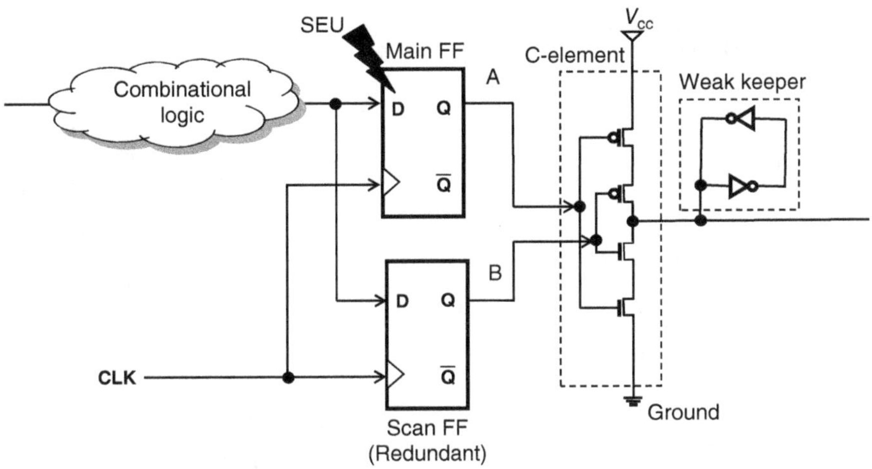

Figure 8.6 Circuit of BISER

(Built-In Soft Error Resilience) that utilises a scan FF as the redundant FF and this is illustrated in Figure 8.6 with a C-element and a weak keeper to prevent propagation of erroneous input to FFs [33]. BISER is, however, said to be vulnerable to a direct hit to the C-element as the clock frequency increases.

Furuta *et al.* suggest that the dual-modular FF, BCDMR (Bistable Cross-coupled Dual Modular Redundancy) could be the possible HBD (Hardened By Design) extension of BISER to avoid such an erroneous operation of the C-element by applying redundant C-elements and cross-coupled weak keepers [40, 56].

Some FFs have SET/RESET inputs. When SET takes place in this kind of global control line, errors in multiple FFs take place similarly as is seen in local clock lines. Figure 8.7 shows how this works in heavy ion irradiation tests [57]. Terrestrial neutrons and protons may cause this type of error in multiple FFs in a circuit since LET (Linear Energy Transfer) of secondary particles from nuclear spallation reaction is in the corresponding hatched range as shown in Figure 8.7.

Clock pulses are produced in DLL (Delay-Locked Loops) or PLL (Phase-Locked Loops). A typical block diagram of PLL is shown in Figure 8.8. A charged particle that strikes DLL or PLL causes clock jitter or race [58, 59]. Clock jitter is a phenomenon where the edge of a clock pulse is shifted and thus the data is not latched to the FF for the correct timing. A SET pulse with enough pulse width may activate a FF gate over a timing in-between regular clock pulses. Clock race is a phenomenon, in which such a 'ghost' clock may send a signal beyond the correct pipeline [59]. A SET pulse may erase a clock pulse in the PLL or DLL circuits. Mailland *et al.* described the radiation-hardened-by-design DLL that implements a radiation hardened voltage-controlled delay line and charge pump, which are the most vulnerable parts in the DLL, to suppress missing pulses completely [59].

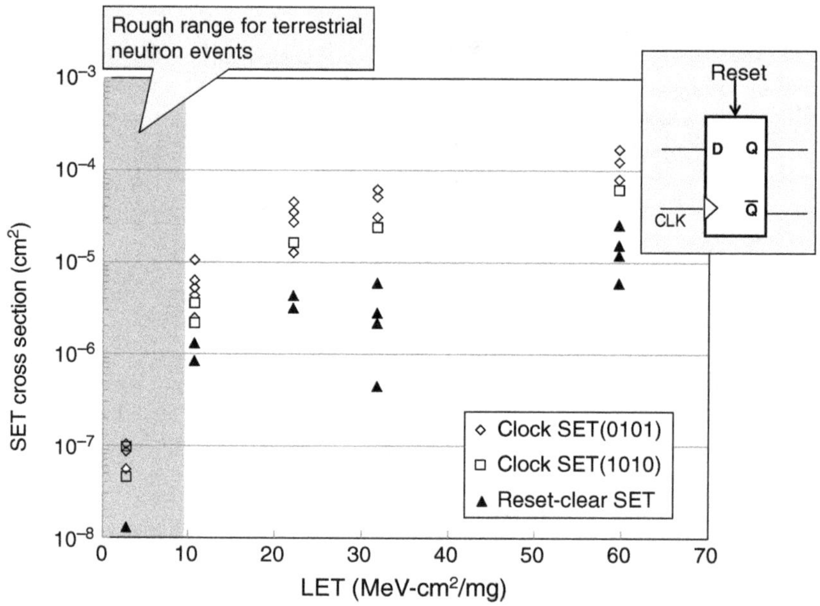

Figure 8.7 SER due to SET in global control lines. (Edited from Cabanas-Holman *et al.* [57] ©2009 IEEE)

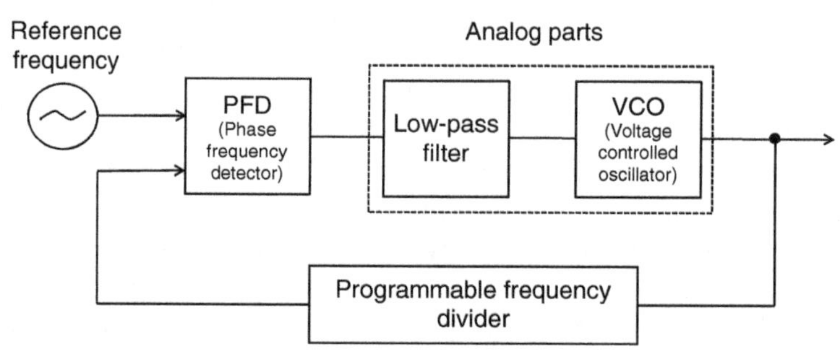

Figure 8.8 Block diagram of digital PLL

Kim *et al.* focus on the difference in disturbance mode between particle hits on digital or analogue parts in a *digital* PLL [60]. When a particle strikes a digital part like a divider or a phase frequency detector, disturbance is delayed from particle hit and persists. When a particle strikes at an analogue part like a charge pump or voltage-controlled oscillator, a disturbance appear immediately and then disappears over a short period of time. Two types of detectors have been developed, each of which are sensitive to either type of disturbance. Once a disturbance is detected in a PLL, a clock pulse source is switched to the other PLL in DMR (Double Modular Redundancy. See Section 8.1.2.8) structure and thus the system can operate safely.

Srinivasan *et al.* apply the selective hardening to a combinational logic implemented on a 4-bit ALU (Arithmetic Logic Unit) data path of a LEON2 SPARC V8 processor™ [61]. They used a SPICE (Simulation Program with Integrated Circuit Emphasis)-type simulation tool to identify vulnerable nodes and those hardened by transistor sizing.

8.1.2.4 Time Redundancy in Circuit Level

There are also time redundancy techniques to prevent SETs in a circuit. Figure 8.9 simply illustrates a typical mechanism of time redundancy technique, double sampling [62]. When a SET pulse with pulse width Δ propagates in a wire, the pulse is distributed to two (double sampling) or more wires. Some of the wires have buffers with delay time τ_1, τ_2, and so on. If the pulse width Δ is short enough compared to the delay time, two or more sampled pulses do not overlap and thus the SET pulse is masked/eliminated by an AND gate.

RAZOR is a technique that utilises two clock lines with different clock timing [43]. RAZOR has a branch with an additional shadow latch in the input line as illustrated in Figure 8.10. The input data is retained with some delay compared to the additional

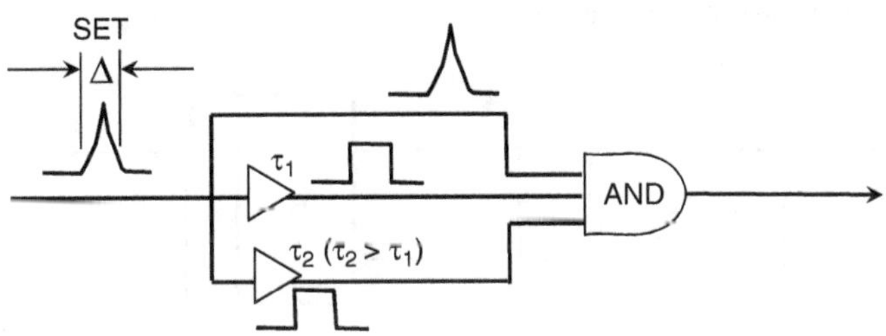

Figure 8.9 Elimination of SET pulse by multiple delay lines

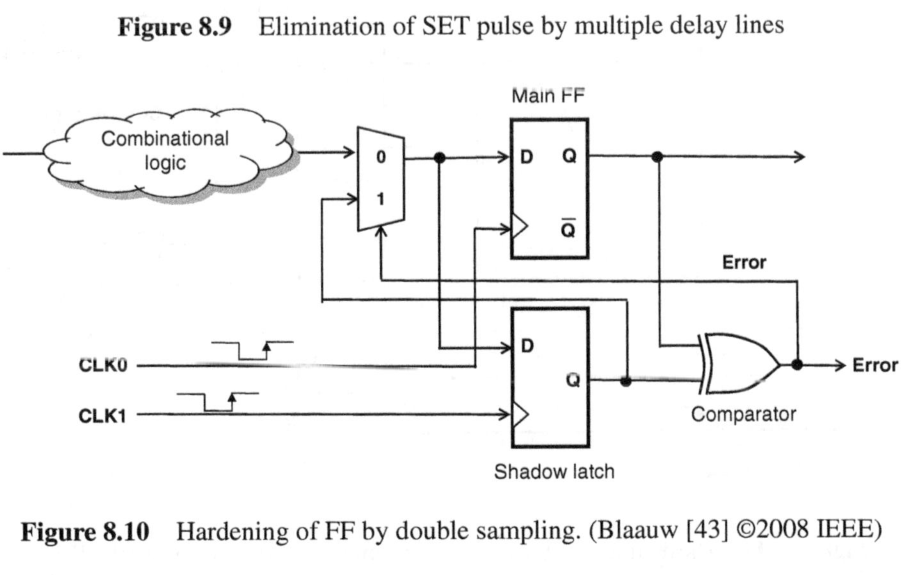

Figure 8.10 Hardening of FF by double sampling. (Blaauw [43] ©2008 IEEE)

latch. If a SET pulse passes through within the delay length, the outputs of the FF and the latch are different and thus the SET can be detected. In this case, delayed latch output is always correct so that the output of the FF can be resumed to feed the latch data to the input of the main FF through the multiplexer. This type of detection and ability to resume technique by shifting and branching the phase of input is activated only when the fault is detected so that speed and area penalties can be minimum compared with other space redundancy techniques. However, this type of technique may become helpless if the SET pulse width is lengthened due to scaling or lowering operational voltage.

Avirnen proposes the SEM (Soft Error Mitigation) cell and STEM (Soft and Timing Error Mitigation) cell, using the approach of multiple clocking of data to protect combinational logic blocks from soft errors [63].

SEM is based on triple sampling with three registers R_1, R_2, R_3 with delays τ_1 and τ_2 (large enough compared to SET pulse width) for R_2 and R_3 respectively as shown in Figure 8.11. In a similar way to RAZOR, a multiplexer connected to the input and the output of R_3 is inserted before the input of register R_1 If the output of R_1 and R_2 is different, an error flag is generated and the operation is put on hold. With enough delay at R_3, the output of R_3 is supposed to always be correct and the feedback to the multiplexer and operation is to reboot.

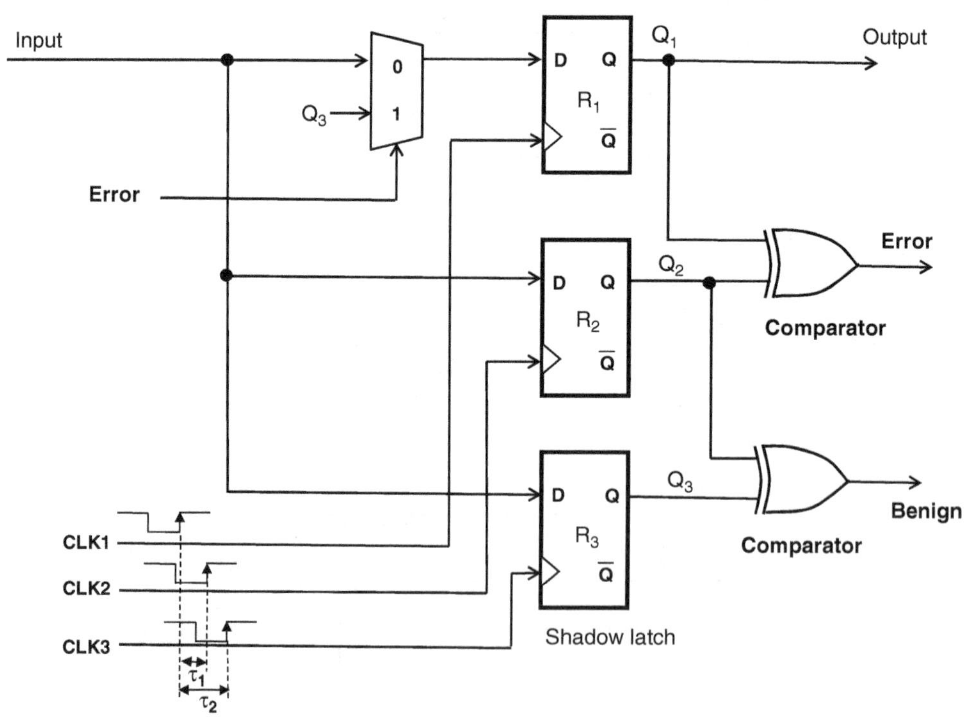

Figure 8.11 Hardening of FF by triple sampling. (Avirnen [63] ©2012 IEEE)

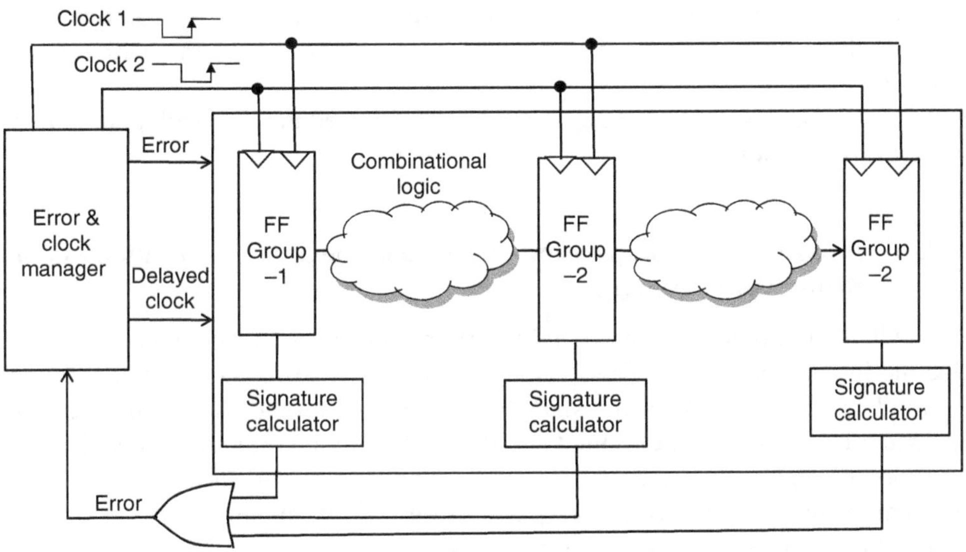

Figure 8.12 Hardening of pipeline by making signature to FF groups. (Prasanth *et al.* [64] ©2011 IEEE)

Prasanth *et al.* propose a detection/recovery technique of having SETs in pipeline logic circuits where a check bit (parity) or a signature encrypted from the data is given to groups of FFs that are not necessarily in the same word as illustrated in Figure 8.12 [64]. They apply double clocking to activate check bit or signature calculator in the upstream of a group of FFs by a clock and in the downstream of the group of FFs by a delayed clock and two check bits or signatures are compared to detect errors in the group of FFs. Intelligent optimisation techniques for grouping are also introduced and demonstrated by using ITC99 benchmark circuits, concluding that the proposed methods have smaller overheads than SEMs.

8.1.2.5 Re-configuration of FPGA System

When a firm error (error in configuration memory) is detected in a FPGA system, the configuration data is restored from a hardened or flash memory and the system is re-started [20]. This method is called re-configuration. Recovery time by re-configuration is increasing to, in some cases, an unacceptable level as the device scaling proceeds, so that partial re-configuration techniques are pursued [65, 66].

Noji *et al.* developed an *adaptive fail-soft system* for the FPGA system in which an error is recovered by three adaptive steps: (i) rollback and re-computation. The mode is soft error if recovered; (ii) Rollback and reconfiguration. The mode is firm error if recovered; (iii) If the system cannot be recovered by steps (i) or (ii), the error is regarded as a hard error and the relevant portion is isolated and reconfiguration is done. By these steps, the system with an error can continue to operate with minimum damage [67].

At the chip/processer level, a number of space/time redundancy techniques are proposed as listed below.

8.1.2.6 Duplication + Comparison + Checkpointing (DCC or Double Modular Redundancy, DMR)

Two redundant modules are used and a certain number of checkpoints are set in the pipeline instruction flow in DCC (Duplication + Comparison + Checkpointing. See Figure 7.5). Necessary data to restart the execution from the checkpoint are stored. The same instructions are executed simultaneously in the redundant modules to the checkpoint (duplication), and results are compared (comparison). If the results are different, error is supposed to take place in one of two modules, and execution is resumed from the previous checkpoint (rollback).

DCC is also vulnerable to MNT and area, and the power penalties are ×2. Some speed penalty exists due to saving checkpoint data and restarting from the checkpoint and delay in comparator.

Nikolov *et al.* point out concerns on performance penalty due to frequent checkpointing (checkpointing overhead: time consumed by data storage for checkpointing) or checkpointing with a long interval (time consumed by long re-execution), which is crucial for systems with a deadline to complete [68]. They propose a statistical optimisation technique for a DMR system to minimise average execution time with required confidence level.

If checkpointing overhead is short enough, frequent checkpointing is preferable for system reliability. Wang *et al.* propose a *VM-μ checkpointing method* where the incremental checkpointing (relevant only to the difference between subsequent checkpoints), copy-on-write to dirty-page (vulnerable page to fault), prediction of dirty-page and in-place recovery within 1 second [4]. They applied the method to Xen VMM™ and demonstrate a 6.3% performance penalty with a 50 ms checkpointing interval.

Skarin *et al.* applied selective DCC to detect errors in the stack pointer in the CPU register together with checking the integrator state of the automotive brake-controller [45]. Verification is made by fault injection to all CPU registers and memory locations in the main memory.

8.1.2.7 Lockstep Operation of Dual Microprocessors

A dual core micro-control unit with one-clock lockstep operations are generally used in automotive systems as a detection and recovery method.

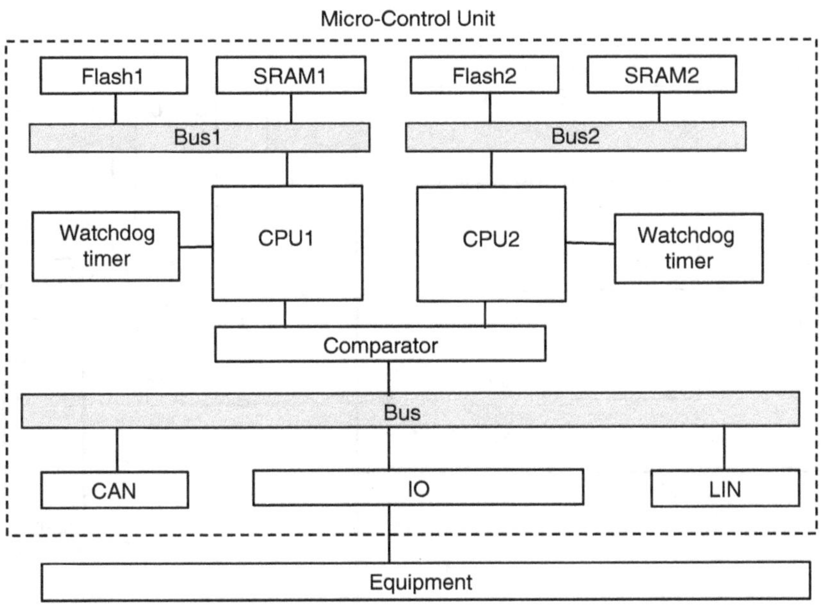

Figure 8.13 Simplified block diagram of lockstep micro control unit

The same instruction is operated in dual microprocessors with a normal delay of one clock interval and the results are compared [30]. A simplified architecture of a lock-step dual microprocessor is illustrated in Figure 8.13. If the results do not match, the instruction is re-executed. This operation is called lockstep operation and is commonly applied to systems where outage is not acceptable.

Wood *et al.* introduced such techniques applied to HP nonstop servers together with confinement techniques of faults [28]. The part where one or more errors are detected is isolated immediately (fail-fast). Scrubbing memory is regularly executed in the background and if an error is detected re-writing is done. If re-writing is failed, the error is regarded as a hard error and the relevant portion of memories is isolated. They also point out that a performance penalty in this lockstep operation may not be accepted for the system where stoppage in a flash is hazardous.

Subramanian *et al.* propose an architecture with a conjoined dual-pipeline processor, which is close to lock-step operation [29]. Figure 8.14 shows the dual pipelines that consist of a leading pipeline and a shadow pipeline, both stages that receive the same input, and execute the same instruction but which save the results in a different register with different clock timing. The shadow pipeline trails the leading pipeline with a certain delay. If the results of both pipelines in the same stage match (no error), execution is continued and the results in the shadow pipeline are stored in the shadow register. The data in the shadow register is regarded as correct because the results in both pipelines match. If results do not match, clocks to registers in both pipelines are stalled and the three-cycle recovery operation is initiated. With the shadow pipeline

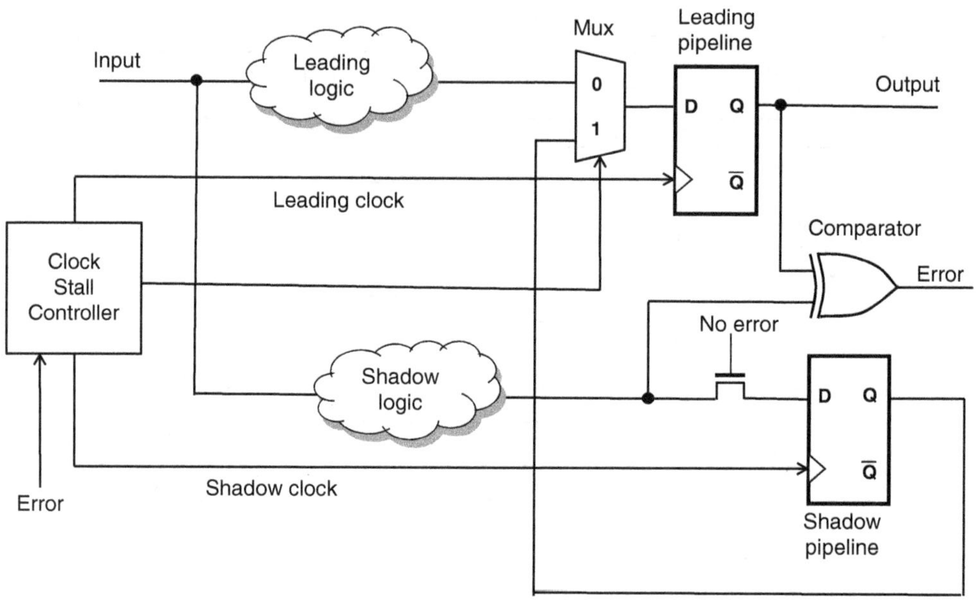

Figure 8.14 Hardening circuit by double pipelines. (Subramanian *et al.* [29] ©2008 IEEE)

stalled, the data in the shadow register is loaded to the corresponding leading register and then execution is re-started during the cycle.

8.1.2.8 Replication + Comparison + Checkpointing (RCC)

The same instruction is executed twice in one module (Replication). Initial execution results are stored in the keeper and compared with the second execution results (see Figure 7.7). If the results are different, instruction is restarted from the previous step (rollback). This method has a ×2 speed penalty.

Aggarwal *et al.* evaluate the methods used to reduce the overhead when detecting logic errors in multi-thread CMP (Chip Multi-Processors). They propose that replication may be applied only to written pages at relevant timing and point out that overhead in replication depends on the granularity [35].

8.1.2.9 TMR (Triple Modular Redundancy)

TMR has three modules and one voter (see Figure 7.6). One instruction is executed in the three modules simultaneously and the voter executes two-out-of-three majority voting. The highest reliability may be achieved by TMR.

SRAM-based FPGAs are known as being vulnerable to soft error. Logic circuits are implemented in FPGAs so that TMR is commonly applied to mitigate errors/failures in SRAM-based FPGA systems [69].

TMR has, however, significant power and area penalties and some speed penalty exists in the voter so that partial/selective hardening or partial TMR is pursued by detecting paths that involve vulnerable nodes to minimise the penalties. Pratt *et al.* use the automatic partial TMR design tool, BLTmr™ tool, to design coarse level TMR [70]. The partial TMR are selectively implemented in the feedback sections that are generally vulnerable to faults, and are validated by fault injections and 65 MeV proton irradiation tests at IUCF (Indiana University Cyclotron Facility).

Azambuja *et al.* applied selective re-configuration and checkpointing techniques to Xilinx™ FPGAs and demonstrated short recovery time [47]. Large grain TMR and scrubbing (regular refreshing of configuration memory) are applied in order to prevent a routing error that may affect the neighbouring two domains.

As device scaling proceeds and operation frequency increases, SET pulse width may exceeds the clock cycle. Inoue *et al.* propose an automatic synthesising technique of TMR for such a multi-cycle transient [38].

Pilotto *et al.* applied large grain TMR with a special voter signalising the faulty module in order to reconfigure only the faulty domains in SRAM-based FPGA designs. They were also checkpointing to minimise downtime during recovery while the other two modules are kept running.

Yang *et al.* develop HHC (Hierarchical Hardware Checkpointing) including on-chip (high-speed) checkpointing and off-chip (conventional, low-speed) checkpointing to reduce MTTR (Mean Time To Repair) that is used for saving data and recovery actions for checkpointing [39]. Most SEE faults can be recovered by high speed on-chip checkpointing and MTTR can be reduced by 94.3% compared to the conventional off-chip checkpointing.

It must be noted that reliability improvement in total is possible only when redundant modules and the voter have a certain level of high reliability. This technique is also helpless when MNT cause errors in two redundant modules simultaneously [71].

8.1.3 Multi-core Processor

Applications, in which multi- or many-core one-chip processors are used, have increased ever rapidly. SoC (System on Chip) [37, 41], NoC (Network on Chip) [72, 73], GPU (Graphic Processing Unit) [74–76] are among such applications. In addition to reliability in each core, resiliency in communication between cores is of great concern. In multi-core processors, a hierarchical core layer structure may be applied such as a root core/master cores/slave cores [73]. Cores in the lower layers periodically send 'alive' signals (watchdog timer) to the root core or the scheduler. If the scheduler does not receive the 'alive' signal from the core in the lower layer, the core is regarded as 'dead' and the mission assigned to the core is re-mapped to another 'alive' core. In this kind of system, a failure in the scheduler may have a fatal impact and redundancy may be applied to the scheduler.

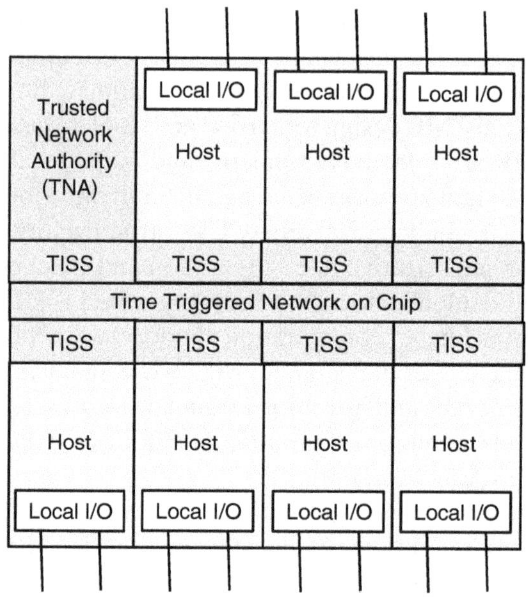

Figure 8.15 Time triggered architecture. (Kopetz *et al.* [77] ©2007 IEEE)

Kopetz *et al.* develop a SoC chip with TTA (Time Triggered Architecture) where each core is connected to a common bus TTNoC (Time-Triggered Network-on-Chip) via TISS (Trusted Interface Subsystem) as illustrated in Figure 8.15. Communication between cores is made through TTNoC and controlled by TNA (Trusted Network Authority) that is also implemented in the chip [77]. Error detection of TNA is made by a TNA watchdog timer and ECC. Recovery of TISS, TTNoC, TNA is made by restart. System error detection and recovery are made by TMR-based architecture and a TTA cluster protected by the guardian.

Gold *et al.* evaluate the confinement of errors in a multiprocessor with shared memory in order to suppress the propagation of errors [78]. There is a tradeoff between small confinement granularity and difficulty in recovery design. Confinement in the core as in the IBM Z series, confinement in a set of core and cache and confinement in a set of core as in the HP nonstop server, confinement in a set of core, cache and shared memory as HP NSAA (Nonstop Advanced Architecture) are evaluated. It is suggested that an error may propagate the entire multiprocessor system if it has a shared memory and communication and timing between cores become increasingly difficult as the range of confinement gets larger.

Vaskova *et al.* point out that error correction techniques such as TMR allow system recovery, but usually do not completely remove latent faults in unprotected memory elements and this can result in accumulated multiple errors [79]. TMR may not work properly with this type of error. This is particularly true for a satellite in the harsh environment of space where high-energy proton flux is as high as

1.6×10^9 particles/cm^2/day. They demonstrate a significant number of latent faults remaining in a global TMR in real-time microprocessor systems by fault injection and suggest that additional protection schemes, such as (i) nMR solution with $n > 3$; (ii) periodic refreshment of critical data; (iii) periodical power-cycle of selected module; (iv) an increase in the window of comparison and (v) periodic scrubbing for reprogrammable devices, are necessary.

8.1.4 Board/OS/Application Level

Table 8.3 summarises traditional and recent mitigation techniques in board/OS (Operating System)/application levels.

Recovery from failures like kernel panic in OSs is generally done simply by rebooting.

Rebooting of the OS may cause a long downtime. Yamakita *et al.* suggest that rebooting can be optimised by phase separation in the rebooting process, known as phase-base rebooting [92]. Occasionally, the system state just after the previous rebooting is the same as the state of the system just after the current rebooting, so the former state is re-used in the phase-base rebooting. The prototype phase-based rebooting scheme is implemented in a virtual machine monitor Xen 3.4.1™ on which Linux 2.6.18 works. The prototype recovers from kernel failures with a significant reduction in downtime.

A RTOS (Real Time Operating System) that works on battery and requires high-speed operation with low voltage in the safety critical real-time embedded system, and thus they are vulnerable to fault. Silva *et al.* propose RTOS-G in which the hardware-based *Guardian* is added to how the behaviour of RTOS is monitored instead of conventional application-level detection techniques [80]. In the real time systems, the detection of faults and recovery must keep to their deadlines. RTOS-G have four functions such as (i) memory management, (ii) device management, (iii) time management by which scheduling is monitored and (iv) interrupt handling by which the execution flow can be changed to keep to the deadlines. A much higher detection coverage and shorter latency by RTOS-G as opposed to the conventional RTOS is demonstrated in the fault injection campaign.

Current supercomputers have a number of processors and inevitably are very prone to soft errors. SDC (Silent Data Corruption) is the biggest problem to be found in scientific supercomputing [93, 94]. Detection of the wrong simulation results caused by SDC is very difficult or even impossible. Kumar proposed stochastic computing, in which, instead of solving matrix calculation,

$$A\vec{x} = B, \tag{8.1}$$

directly, the minimising error δ approach,

$$\delta = |A\vec{x} - B|, \tag{8.2}$$

Table 8.3 List of prevention, in-situ recovery and off-line mitigation techniques against failures in board/OS/application level

Layer	Prevention	In-situ recovery	Off-line recovery
OS	Fault injection to identify critical mode of OS failures[1] Log analysis[2]	SMT (simultaneous multithreading)[9] Optimisation of area and lifetime in memory space[10] Addition of external OS monitor[11]	Reboot[17]
Board	DOUB (design on upper bound)[3] Replacement of critical devices/chips by hardened ones[4] Cross-layer reliability based on common RTL level SER information format[5]	Hardware symptom detection and checkpointing/rollback (LABIR)[12] Conjoined HW/SW detection and recovery[13]	Reboot[18]
Application	Probabilistic/approximate calculation[6] Fault injection to identify vulnerable instructions[7] Log analysis[8]	Software symptom detection and checkpionting-rollback[14] SBST (software-based self test)[15] Paradigm shift in software[16]	Reconfiguration[19] Optimisation of algorithms[20]
References	1. Pellegrini [81] and Morgan (2005) 2. Silva *et al.* [80] 3. Ibe *et al.* [82] 4. Shimbo (2010) and Valderas *et al.* [15] 5. Evans *et al.* [83] 6. Kumar [85] Raghunathan and Roy [86] and Sierawski *et al.* [26] 7. – 8. –	9. Kopetz [77], Gold *et al.* [78] and Bizot *et al.* [73] 10. Alexndrescu [87] 11. Silva *et al.* [80] 12. Ibe *et al.* [53, 82] 13. Subramanian *et al.* [29], Wen [84] and Slayman *et al.* [6] 14. Li and Ramachandran [88], Hari *et al.* [89] and Wang and Patel [90] 15. Theodorou *et al.* [91], Carvalho *et al.* [92] and Ballan *et al.* [93] 16. Gerist (2012)	17. Amanita (2011) 18. – 19. Allen *et al.* [20] 20. –

is recommended to obtain acceptable *good-enough* results even when SDCs are involved in the simulation [85].

Raghunathan and Roy propose approximate computing in many industrial areas to obtain *good enough* results or quality of calculation [86]. At the algorithm/architecture level, strict algorithms are applied to only significant computation while approximate computation is applied to non-significant computation.

Sirawski *et al.* propose SET reduction methods to enhance logical masking in combined logic circuits [26]. They show some examples of how to approximate a function with enhanced logic masking from the original function. Simulation with LGSynth93 benchmark demonstrated substantial SER reduction up to 90% by using the method with some area penalty.

Aggarwal *et al.* propose configurable redundancy in which grain of redundancy is dynamically adapted due to observed errors and location of data storage is also adapted dynamically [35, 96].

At the logic-circuit level, the number of transistors is reduced by logic complexity reduction. The cross-layer approach (see Section 8.2.3) is where approximate computing is applied to various layers of design abstraction [97].

8.1.5 Real-Time Systems: Automotives and Avionics

Real-time systems or safety-critical systems like cars and avionics must be compliant with safety design rules like IEC61508 (issued in 2000) [98] and ISO26262 (issued in 2011) [99]. Table 8.4 compares the safety classes in IEC61508 and ISO26262 with the corresponding failure probability/rate to be realised. IEC61508 has four SIL (Safety Integrity Level) classes 1–4 in two different types of programmable systems depending on the operating modes. SILs, in the low demand mode of IEC61508, is

Table 8.4 Comparison of IEC61508 and ISO26262 with respect to SIL (safety integrity level)

IEC61508			ISO26262	
SIL	Low demand mode (probability of safety function to fail)	High demand mode (average dangerous failure rate) (h^{-1})	ASIL	Random hardware failure target value (h^{-1})
4	10^{-5} to 10^{-4}	10^{-9} to 10^{-8}	–	–
3	10^{-4} to 10^{-3}	10^{-8} to 10^{-7}	D	$<10^{-8}$
			C	$<10^{-7}$
2	10^{-3} to 10^{-2}	10^{-7} to 10^{-6}	B	$<10^{-6}$
1	10^{-2} to 10^{-1}	10^{-6} to 10^{-5}	A	$<10^{-5}$

SIL in the low demand mode is average probability for safety related function to fail. SIL in the high demand mode and ASIL are dangerous failure rate.

the mode in which the safety related function is most likely to fail. SILs in the high demand mode of IEC61508 and ASILs (Automotive Safety Integrity Levels) have a dangerous failure rate per hour. ISO26262 also has four classes of ASIL A-D irrespective of the operation modes and does not have the class equivalent to SIL4. Instead, the SIL3 class is divided into ASIL-C and ASIL-D. The real-time systems are also distributed systems. In particular, recent cars have a number of MCUs (Micro Control Units), where the ASIL-D (10FIT) level safety is mandated for some parts like power steering, brake-systems, air bags and engine control [99]. Although a 10FIT level error rate is not a realistic target without applying any dedicated protection method like lockstep operation, the failure rate can be kept below the 10FIT level by detection of symptom of failures and recovery to a safe condition. On-line self-tests are also one of the key techniques to achieve this goal [92, 93].

SITR (Self-Imposed Temporal Redundancy) consists of a single pipeline functional unit and executes a lockstep-like operation [100]: the same instruction is executed twice with one clock difference and their results are compared at the final stage of the pipeline (the previous output is kept in the latch for comparison). When two outputs are not the same, the instruction is retried or the pipeline is flushed.

BIST (Built-In Self-Test) is commonly applied each time that the automotive engine is started, but MBIST (Memory BIST) are used for in-situ error detection by applying March tests [91].

SBST (Software-Based Self-Test) is a processor-level BIST and any instruction programmed into the flash memory in the SoC is sent to the processor to check whether the execution results are correct or not [92].

Ballan *et al.* [93] applied the SBST algorithm to suppress failures caused by stack-at-1 fault in scan chains in the 32-bit embedded SoC. Scan chains are not used during normal operation or calculation, but may cause a failure in a system if the stack-at-1 fault takes place in the enable bit of the multiplexer, resulting in capture of a wrong data from a FF in the scan chain to the main circuit as exemplified in Figure 8.16.

8.1.6 Limitations and Improvements

Cloud computing and data centres are currently major trends in network systems. One of the most crucial priorities in these applications is that decreasing power consumption is good for global warming [101]. Mitigation techniques for SER, therefore, must be power conscious. Trade-offs among power, speed, cost and reliability must be considered to determine the mitigation techniques to be applied. In this sense, space domain redundancy techniques are becoming less applicable in recent generations of supercomputers and data centres [102].

In addition to the redundancy approaches explained above, there are many cases where stand-by units are operated to replace failed units. Attention must be paid to

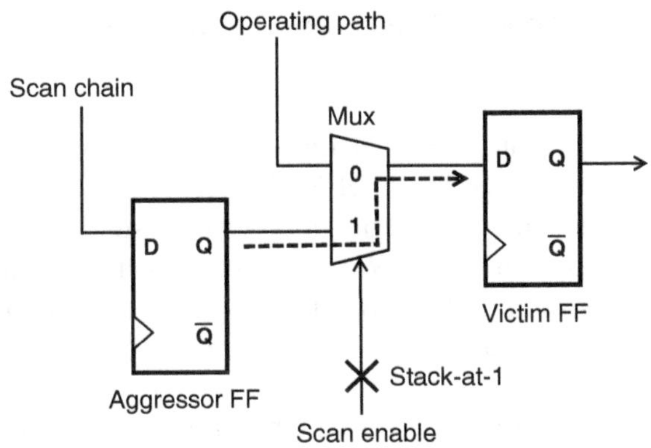

Figure 8.16 SEE caused by stack at 1 error in a scan FF. (Ballan *et al.* [93] ©2013 IEEE)

these units to see if errors are accumulated in the memory parts. MBUs can take place unless the memory content needs to be refreshed in such a system. To this end, regular monitoring of the memory parts in the stand-by unit may be necessary.

Space or time domain checkpointing-rollback methods generally have a limitation of frequency of checkpoints. If the frequency of checkpoints is high, performance of the electronic system inevitably deteriorates [103].

Moreover, in very highly integrated systems, like supercomputers and data centres, the frequency of errors may exceed the frequency of checkpoints, making comparisons among resultant outputs meaningless [104].

In FPGA systems, reconfiguration causes the performance of the system to be poor, so it may be better to employ partial or small-grain reconfiguration as is proposed in [105]. Azambuja *et al.* demonstrate a recovery time reduction of 1/50 by using a combination of large-grain TMR and selective reconfiguration [47].

A similar approach using small-grain checkpointing is possible as shown in CEC (Continuous Eventual Checkpointing) [4, 106].

It is necessary to develop failure- or reliability-aware EDA (Electric Design Automation) tools. Current EDA tools which are higher than circuit level abstraction take no account of the layout of memory cells or logic gates in circuits related to a fault/error/failure. The failure rate of a circuit may vary according to the layout and combination of logic gates [107–109].

The partial irradiation method of electronic components and systems can be a very effective way to identify the vulnerable chips and terrestrial neutron soft errors [13, 84, 110].

The radiation test cost is very high and, therefore, comprehensive estimation methods of chip-level SER are necessary for robust design. The partial irradiation test can also be used to validate the estimation methods.

8.2 Challenges for Hyper Mitigation Techniques

8.2.1 Co-operation of Hardware and Software

System level failures are categorised using three metrics: RAS (Reliability, Availability and Serviceability). System level RAS analyses are required when accurate mapping of failure modes in components, metrics of system failures and requirement for system reliability is needed [111]. Won *et al.* demonstrate the top-down RAS analysis of internet switches and point out that the modelling of system level RAS analyses requires a common hierarchical system modelling format when failure modes in components are reported.

The RIIF (Reliability Information Interchange Format) is being proposed as common format/language in EDA tools in order to realise such RAS analyses [83, 111]. Device, component and processor level fault/error/failure modes are defined and their quantitate data are given in the format together with environmental conditions. Fault injection techniques may be applied in order to obtain such figures for complicated components.

Wen *et al.* propose the evaluation metric in the two-dimensional space of failure rate f_c (in FIT) and recovery/outage time t_c (in seconds) with four classes such as no outage, low (0.15 s max.), medium (3 s max.), high (180 s max.) and unrecoverable (1000 s max.) [112]. They also propose that SEU target metrics DPM (Defects Per Million) is defined as,

$$\mathrm{DPM} = \frac{t_c \times f_c}{3.6 \times 10^6}. \tag{8.3}$$

Fixed DPM value can be plotted in a straight line in the log-log plot of the two-dimensional space.

The DPM of a system can be easily obtained using a simple sum of each component in the system.

They emphasise that tight cooperation of HW (HardWare) and SW (SoftWare) teams is necessary to realise target DPM.

Slayman *et al.* employ SECDED (Single Error Correction and Double Error Detection) to DRAM chips to map out soft and hard error in the DRAM chips. They apply dynamic memory reconfiguration with MPR (Memory Page Retire) and bit steering to eliminate the impact of hard errors as a part of a 'RAS intelligent' way to look at device reliability as DRAM technology scales below 100 nm [17]. In MPR, the page where the error resides is copied correctly by applying ECC to an intact page and removing it from the list of available memory. The DRAM pin is eliminated from the data path where hard error resides by bit steering.

8.2.2 Mitigation of Failures under Variations in SEE Responses

Even if space redundancy techniques are applied, as mentioned before, they are intrinsically vulnerable against MNUs. Effective countermeasures listed in Tables 8.1–8.3 and implemented in a single stack layer (device, chip, board ...) seem to be almost impossible as the device scaling proceeds.

Class	Concept		Merit	Demerit
Design on average (DOA)	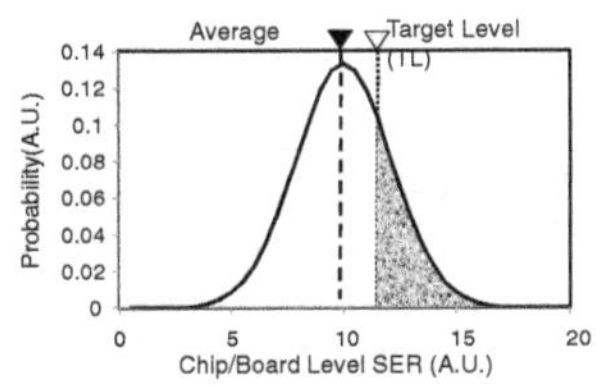	Target failure rate for mitigation design is determined upon only average rate without knowledge of variation.	• Relatively simple	• Risk of under design • Depends on circuits and applications
Design on average and variation (DOAV)	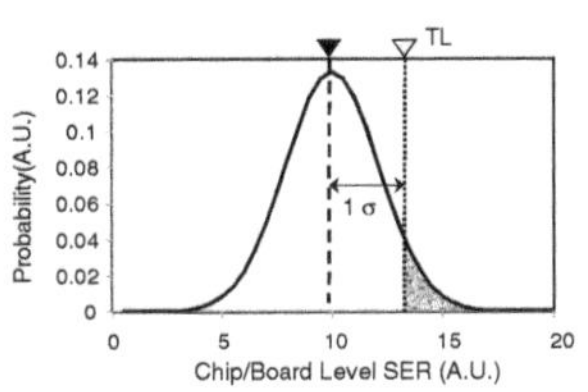	Target failure rate for mitigation design is determined upon average rate with knowledge of variation.	• Just design	• Depends on circuits and applications • High cost • Long time to market or no goal
Design on upper bound (DOUB)	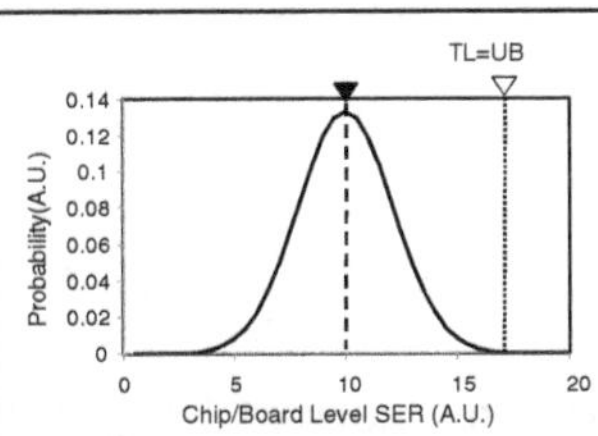	Target failure rate for mitigation design is determined upon upper bound rate irrelevant to average rate.	• Simple • Does not depend on circuits and application • low cost	• Risk of over design

Figure 8.17 Comparison of chip-level mitigation approaches: DOA (design on average), DOAV (design on average with variation) and DOUB (design on upper bound)

System-level mitigation strategies can be threefold as illustrated in Figure 8.17. First, the current major strategy is DOA (Design On the Average), which determines the target SER level without the knowledge of variation. From a reliability viewpoint, mitigation based on average chip-level SER estimation with masking factors (MFs) is somewhat dangerous if the variation due to circuits and applications (i.e. masking effects) is large. Without a certain knowledge of the variation (not necessarily Poisson distribution), one may design an electronic product with too low a SER tolerable level, which leads to *under-design*.

Second, is DOAV (Design On Average with knowledge of Variation). *Just design* (cost, power, speed and reliability) may be realised through this approach. This strategy includes an evaluation of *AVF* (Architectural Vulnerability Factor. See Section 7.2) /*DF* (Derating Factor)/*MF*. Accurate evaluation is, however, quite difficult and the time/cost consumed variable because of depends on the circuits and applications.

Ibe *et al.*, therefore, are working on a different and novel approach named DOUB (Design On Upper Bound) as illustrated in Figure 8.18, by which the upper-bound of failure rate can be estimated explicitly [113, 114].

Calculation of the upper-bound failure rate FLR_{UB} starts with FLR^{0}_{UB},

$$FLR^{0}_{\mathrm{UB}} = \sum_{\mathrm{j}} N_{\mathrm{j}} \times UBFR_{\mathrm{j}} \tag{8.4}$$

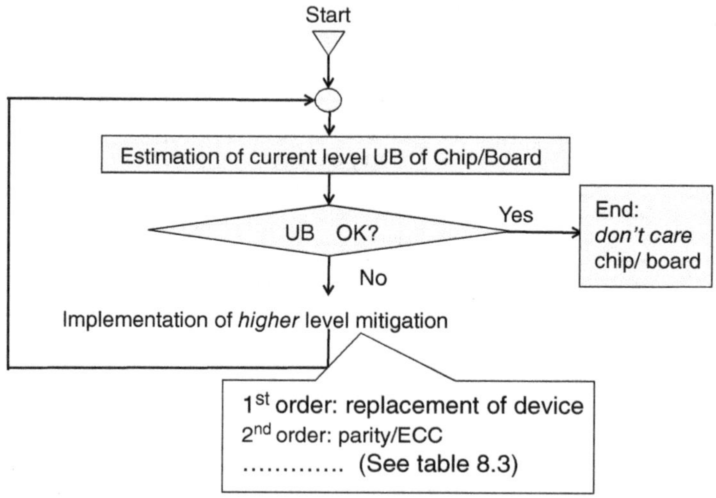

Figure 8.18 General flow of design on upper bound

where,

N_j: Number of j-th device;
$UBFR_j$: Upper-bound fault rate of j-th device (one memory cell or gate).

$UBFR_j$ can be calculated by using the simulator CORIMS as introduced in Chapter 6.

Equation (8.4) would give a safer design than other approaches but would also result in *over-design*.

Strategies to obtain a realistic and small enough FLR_{UB} must be explored in the manner exemplified below.

If the upper-bound (not average) MF of the k-th mechanism ω_k^j is known, the upper-bound failure rate FLR_{UB}^1, which is smaller and more realistic than FLR_{UB}^0, can be obtained by,

$$FLR_{UB}^1 = \sum_j N_j \times \prod_k \omega_k^j UBFR_j$$

$$0 \le \omega_k^j \le 1 \tag{8.5}$$

If there is another correction factor c_n exists due to a certain physical constraint, the next stage FLR_{UB}^{n+1} can be obtained using the following (conceptual) equation,

$$FLR_{UB}^{n+1} = c_n \times FLR_{UB}^n < FLR_{UB}^n \tag{8.6}$$

The following equation is an example of such a physical constraint.

$$UBFR_j = \phi \times \sigma_{UB}^j < \phi \times A_j, \tag{8.7}$$

where,

ϕ: flux of a certain particle (particles/cm^2/s);
σ^{j}_{UB}: upper-bound fault cross section of the j-th device to the particle (cm^2/cell);
A_j: two-dimensional area of the j-th device normal to the particle stream (cm^2).

Smaller and more realistic FLR_{UB} can be obtained in this way.

This kind of physical constraint may be obtained from other factors like operating frequency, critical charge Q_{crit}, logic depth, number of cache layers, logic depth, and so on. For example, if Q_{crit} is larger than 1 fC, it is not necessary to implement any mitigation techniques for muons, electrons and photons for any fault modes as demonstrated in Chapter 6. If the FLR_{UB} is acceptable from the viewpoints of not only reliability but also other metrics like safety, cost, power consumption and performance, the chip/board design can be completed.

If the FLR_{UB} is not acceptable, a simple countermeasure would be applied stepwise. As mentioned in Section 8.1.2, interleaving and ECC are the easiest way to mitigate memory SER if applicable.

The simplest mitigation technique is to exchange weak gates/memory devices for stronger ones based on the raw SER database as also introduced in Section 8.1.1.

FinFET or TriGate devices have reportedly higher resiliency than bulk SRAMs [115] so that replacement bulk SRAM by FinFET/TriGate devices may be effective.

8.2.3 Cross-Layer Reliability (CLR) /Inter-Layer Built-In Reliability (LABIR)

It is commonly recognised that highly integrated mission critical electronic systems (network servers/routers, supercomputers and datacentres) and safety critical real-time systems (avionics and cars) cannot be protected by a single mitigation method in a single stack layer. CLR (Cross Layer Reliability) [117, 118] and LABIR (inter-Layer Built-In Reliability) [82, 118], in which multiple stack-layers must co-operate to protect the system, are being proposed from this viewpoint.

CLR is proposed by a USA group of 82 organisations consisting of national laboratories, universities and industries led by the CCC (Computer Community Consortium). Reliability/resilience tips are communicatively developed in more than dual stack layers ranging from device to software aimed at crystallisation into comprehensive SER mitigation techniques. Five task levels are assumed: detection, diagnosis, reconfiguration, recovery and adaptation.

Herkersdorf *et al.* propose a probabilistic and bottom-up cross-layer approach RAP (Resilience Articulation Point) to the software layer in SoC (System on Chip) [119]. It is a part of the DFG (Deutsche Forschungs Gemeinschaft, Declare objective of the German Research Foundation Priority Program SRP1500 'Dependable Embedded Systems') to develop new cross-layer design methods.

Psychou *et al.* develop the probabilistic CLR suitable for BAN (Body Area Network) that delivers lifestyle and health care services, in particular, medical monitoring where absolute reliability is required [120]. Implantable devices must be able to cater for all possible error modes. Remapping of an instruction which has failed in a core to another intact core is applied to a multi-thread processor system. At middleware level, a time redundancy technique like checkpointing may be applied, but care must be paid to a BAN (Body Area Network) system because delay by re-execution may not be allowed in order to deliver the service in time. Since the BAN system has a loop operation nature, recovery from the loop beginning is effective.

Wong *et al.* demonstrate that program crashes can be drastically decreased by simple revisions such as the sanity check (elimination or correction of unreasonable values like a probability lower than zero or larger than one and unrealistic moving distance of a robot), exception handling (negligence or the skipping of an exception interrupt) and checkpointing and restart (from loop beginning) by applying probabilistic sample programs like stereo (graph program), localise (program for robot control) and decode-hmm (speech recognition engine) [121].

Lindoso *et al.* demonstrate the SHE (Software Hardening Environment) applied to a PicoBlade™ embedded microprocessor by using a high-performance emulation-based fault injection system like AMUSE (Autonomous MUltilevel emulation system for Soft Error evaluation) as the first step towards hardware/software co-design of an ASIC (Application Specific Integrated Circuit) embedded system which is most desirable to mitigate SET, indicating that a software-based approach is more cost-effective than a hardware-based approach, but cannot achieve the same performance and reliability [122]. Several versions of the system using combinations of hardened (by TMR and time redundancy) components including output port, program counter, ALU register/flags and register files are evaluated. SDCs and hangs, which are named as ACE (Architectural Correct Execution) faults, are shown to be drastically reduced by increasing hardened components. They also show the dependency of ACE faults on the SET pulse width and operation frequency.

8.2.4 Symptom-Driven System Resilient Techniques

8.2.4.1 Hardware Symptom

LABIR (inter-LAyer Built-In Reliability) proposes interactive or communicative mitigation techniques in which a recovery action such as rollback to the checkpoint is triggered when a layer finds any error symptom, not necessarily error or fault itself. BIST (Built-In Self Test) [119], BICS (Built-In Current Sensor), BIPS (Built-In Pulse Sensor) [8, 124], SAW (Surface Acoustic Wave) [124], on-chip monitor [8] can be used in this kind of technique. By using BIPS, a pulse propagated from an MCBI zone in a p-well can be detected in I_{dd} line as demonstrated in Figure 7.3 and ignite the rollback operation in the ULSI (Ultra Large Scale Integration) chip.

As illustrated in Figure 8.19, immediately after capturing such a symptom by applying a sense amp between two adjacent p-wells, for example errors or failures can be

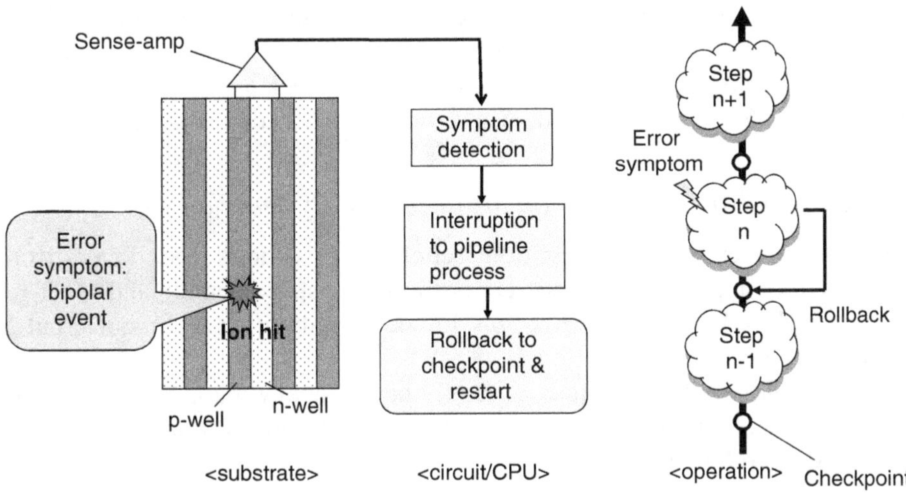

Figure 8.19 An example of LABIR (inter-LAyer Built-In Reliability)

resumed by the rollback and replication operation in CPU (Central Processing Unit) level of the ULSI chip. Other sources of noise like EMI (Electro-Magnetic Interference) and noise in the power supply line [125] propagate in a wider area than soft error over many wells which allows them to be eliminated by the differential method between adjacent wells.

8.2.4.2 Software Symptom

Li *et al.* propose a software-hardware co-designed resilient super-scalar system where the failure symptom is detected mainly by software and recovery and diagnosis (software bug/permanent error/transient) are made during rollback from the checkpoint [88]. Reconfiguration is made in the fine grain based on the diagnosis. Fault injection is applied to Solaris-9 OS with the UltraSparc-III ISA (Instruction Set Architecture). Application and OS crashes and hangs are monitored and classified by an injected component. They report that a large fraction of the fault results in OS failures and a recovery method of OS must be developed.

Hari *et al.* propose a symptom detector to detect fault including SDC in software behaviour [89].

Wang *et al.* propose SBRM (Symptom Based Redundant Multithreading) where symptoms of failure in software are monitored, like deadlock, exception handling, mis-instruction in control flow and error in the cache/translation look-aside buffer [90]. Instructions are executed simultaneously in two threads. Symptom monitoring is applied only to the redundant thread and results in two threads are compared. If the results do not match, an error signal is delivered and a restore process is initiated. Checkpointing supported by input replication, symptom detector and control logic for pipeline flows are used for the restore process. Fault injection in the RTL (Register Transfer Level) model of the Alpha microprocessor is made while SPEC2000 Integer benchmarks are operated. With choosing deadlock and exception handling as symptoms, only 2% speed loss and 75% error recovery is demonstrated.

8.2.5 Comparison of Mitigation Strategies for System Failure

Table 8.5 compares various mitigation techniques in various stack layers from the viewpoints of cost, area penalty, power consumption, speed penalty and MNT toughness. It is fairly relatively easy to apply low-level mitigation techniques, but MNT toughness is weak.

High-level mitigation techniques generally have a heavy load. Thinking of further scaling of semiconductor devices, any mitigation techniques in a single stack layer are obviously unreliable and unsafe.

8.2.6 Challenges in the Near Future

Table 8.6 shows future trends, conventional approaches towards failures, challenges and global standards relevant to fault/error/failure in various industry fields.

In safety critical systems or real time systems like avionics and cars, reducing the weight of structural material is one of the major trends because it reduces fuel and power consumption. The use of GPUs will be extended to much wider areas [76]. Millisecond recovery is one of the most challenging themes in real-time systems. In railway systems, further digitisation will take place in order to realise cost effective, communicative and resilient railway systems [126]. In an exascale supercomputer system, paradigm shifts in dependability techniques, including HW and SW, must be realised if failure due to very frequent errors including SDC and DUE (Detected Unrecoverable Error) is to be avoided.

In a network system as well as a supercomputer system, high power consumption will dictate which mitigation techniques are employed.

Table 8.5 Comparison of mitigation techniques of failures in electronic components and systems

Layer level	Example	Cost	Area penalty	Power consumption	Speed penalty	MNT toughness
Low	Replacement by tough device	Low	Low	Low	Low	Weak
Low	Parity/ECC	Low	Middle	Some	Middle	Weak
Middle	Hardened DICE	Middle	Some	Low	Middle	Strong
Middle	LABIR	Middle	Some	Low	Low	Strong
High	DMR/TMR	High	Large	Large	Some	Weak
High	Replication and rollback	High	Some	Large	Large	Strong

Table 8.6 Future trends, challenges and relevant gloval standards in various industries with regard to fault/error/failure

Industry	Application	Trend	Conventional approach against failures	Challenges	Standards relevant to fault/error/failure caused by SEE
Real time system	Plane/avionics	Lightening and X-by-wire	Hardened device and TMR	Millisecond recovery	IEC62396
	Railway	High speed	Redundancy of power units	Digital wireless train control	IEC62278, IEC62279, IEC62280 and IEC62425
	Automobiles	Lightening, X-by-wire, electronisation and GPU[a]	Lockstep micro controller	Millisecond recovery Perfect fail-safe	ISO26262, AEC-Q100-rev.G, Euro NCAP[b] rating review
Power system		Smart grid, UHV[c]			IEC60038
Supercomputer		Exaflop,100PB and low-power	TMR, RCC and DMR	100% SDC detection and recovery Cross layer reliability/LABIR Resilient software	
Network	Data centre	100PB data, low power and cloud computing	ECC, TMR, RCC and DMR	Fault aware system Prompt failure confinement	ISO27001 DMTF[d] Standards
	Server	Virtualisation, low power	ECC, TMR, RCC and DMR	Cross layer reliability/LABIR	
	Router	Performance and low power	ECC, TMR, RCC and DMR	Cross layer reliability/LABIR	
Media	Desktop PC	Cloud computing	Power cycle		
	Tablet PC	Lightening and high performance	Power cycle		
	Smart phone	High performance	Power cycle		
		Remote control of digital applications			

[a]Graphic processing unit.
[b]New car assessment programme.
[c]Ultra high voltage (1100 kV).
[d]Distributed Management Task Force, Inc.

PDAs (Portable Digital Applications) are being utilised to control digital applications remotely. Noise or intentional attack may cause serious accidents [127].

LABIR/CLR which is one of the best combinations of mitigation techniques should vary by industry and be carefully structured alongside developing key technologies. A list of global standards with their objectives and type such as IEC61508, ISO26262, JESD89A, and so on are included in Table 8.6.

8.3 Summary of Chapter 8

As semiconductor device scaling continues far below the 100 nm design rule, terrestrial neutron-induced soft error, typically in SRAMs, is predicted to be worsen in the future.

Moreover, new failure modes that may be more serious than those in memory soft error have been reported recently. Therefore, of the need to implement mitigation techniques with marginal penalties including power dissipation is rapidly growing at the design phase, together with the development of advanced detection and quantification techniques. The most advanced of these techniques are reviewed and discussed alongside proposed new mitigation strategies such as DOUB and LABIR.

A generic strategy for the low-cost and low-power consumption mitigation techniques of chip and board-level SER based upon the stepwise upper bound reduction is proposed. The partial irradiation method is one of the key techniques in this strategy. LABIR in which communicative mitigation techniques are applied to stack layers is also proposed as another key technology. LABIR can be robust against MNTs in logic parts.

Future directions and challenges in various industry fields are discussed.

References

[1] Nicolaidis, M. (2005) Design for soft error mitigation. *IEEE Transactions of on Device and Materials Reliability*, **5** (3), 405–418.

[2] Slayman, C. (2005) Cache and memory error detection, correction, and reduction techniques for terrestrial servers and workstations. *IEEE Transactions on Device and Materials Reliability*, **5** (3), 397–404.

[3] Ibe, E. (2001) Current and future trend on cosmic-ray-neutron induced single event upset at the ground down to 0.1-micron-device. The Svedberg Laboratory Workshop on Applied Physics, Uppsala, Sweden, May, 3 (1).

[4] Wang, L., Kalbarczyk, Z., Iyer, R. and Iyengar, A. (2010) Checkpointing virtual machines against transient errors. 16th IEEE International On-Line Testing Symposium, Corfu Island, Greece, 5–7 July (5.2), pp. 97–102.

[5] Neto, E.H., Kastensmidt, F.L. and Wirth, G.I. (2008) A built-in current sensor for high speed soft errors detection robust to process and temperature variations. Proceedings of the 20st Annual Symposium on Integrated Circuits and System Design, September 2007, pp. 190–195.

[6] Bota, S.A., Torrens, G., Alorda, B. *et al.* (2010) Cross-BIC architecture for single and multiple SEU detection enhancements in SRAM memories. 16th IEEE International On-Line Testing Symposium, Corfu Island, Greece, July 5–7 (7.1), 141–146.

[7] Siskos, S. (2010) A new built-in current sensor for low supply voltage analog and mixed-signal circuits testing. International On-Line Testing Symposium 2010, Corfu Island, Greece, July 5–7.

[8] Noguchi, K. and Nagata, M. (2007) An on-chip multichannel waveform monitor for diagnosis of systems-on-a-chip integration. *IEEE Transactions of on Very Large Scale Integration (VLSI) Systems* **15** (10), 1101–1110.

[9] Azais, F., Larguier, L., Bertrand, Y. and Renovell, M. (2008) On the detection of SSN-induced logic errors through on-chip monitoring. International On-Line Testing Symposium 2008, Greece, July 6–9, 2008 (10.2), 233–238.

[10] Yoshikawa, K., Hashida, T. and Nagata, M. (2011) An on-chip waveform capturer for diagnosing off-chip power delivery. International Conference on IC Design and Technology, Kaohsiung, Taiwan, May 2–4, 2011.

[11] Ibe, E., Chung, S., Wen, S. *et al.* (2006) Spreading diversity in multi-cell neutron-induced upsets with device scaling. The 2006 IEEE Custom Integrated Circuits Conference, San Jose, CA, 10–13 September, 2006, pp. 437–444.

[12] Bastos, P.R., Sicard, G., Kastensmidt, F. *et al.* (2010) Evaluating transient-fault effects on traditional C-element's implementations. 16th IEEE International On-Line Testing Symposium, Corfu Island, Greece, 5–7 July (2.2), pp. 35–40.

[13] Black, J.D., Sternberg, A.L., Alles, M.L. *et al.* (2005) HBD layout isolation techniques for multiple node charge collection mitigation. *IEEE Transactions on Nuclear Science*, **52** (6), 2536–2541.

[14] Narasimham, B., Gambles, J.W., Shuler, R.L. *et al.* (2008) Quantifying the effect of guard rings and guard drains in mitigating charge collection and charge spread. *IEEE Transactions on Nuclear Science*, **55** (6), 3456–3460.

[15] Valderas, G.M., Portela-Garcia, M., Lopez-Ongil, C. and Entrena, L. (2009) In-depth analysis of digital circuits against soft errors for selective hardening. 15th IEEE International On-Line Testing Symposium, Sesimbra-Lisbon, Portugal, 24–26 June (7.3), pp. 144–149.

[16] Shimbo, K., Toba, T., Nishii, K. *et al.* (2011) Quantification and mitigation techniques of soft-error rates in routers validated in accelerated neutron irradiation test and field test. 2011 IEEE Workshop on Silicon Errors in Logic – System Effects, Champaign, IL, 29–30 March, pp. 11–15.

[17] Slayman, C., Ma, M. and Lindley, S. (2006) Impact of error correction code and dynamic memory reconfiguration on high-reliability/low-cost server memory. IEEE International Integrated Reliability Workshop Final Report, 2006, pp. 190–193.

[18] Schindlbeck, G. and Slayman, C. (2007) Neutron-induced logic soft errors in DRAM technology and their impact on reliable server memory. IEEE Workshop on Silicon Errors in Logic – System Effects 3, Austin Texas, April 3, 4.

[19] JEDEC Standard JESD89A. (2006) Measurement and Reporting of Alpha Particle and Terrestrial COSMIC Ray Induced Soft Errors in Semiconductor Devices. JEDEC, 1–93.

[20] Allen, G., Madias, G., Miller, E. and Swift, G. (2011) Recent single event effects results in advanced reconfigurable field programmable gate arrays. 2011 IEEE Radiation Effects Data Workshop, Las Vegas, 25–29 July (W-15), pp. 1–6.

[21] Ibe, E., Kameyama, H., Yahagi, Y. *et al.* (2004) Distinctive asymmetry in neutron-induced multiple error patterns of 0.13umocess SRAM. The 6th International Workshop on Radiation Effects on Semiconductor Devices for Space Application, Tsukuba, 6–8 October, 2004, pp. 19–23.

[22] Gutsche, M., Seidl, H., Luetzen, J. *et al.* (2001) Capacitance enhancement techniques for sub-100nm trench DRAMs. International Electron Device Meeting, Washington, DC. 3–6 Dec., pp. 18.6.1–18.6.4.

[23] Hirose, K., Saito, H., Kuroda, Y. *et al.* (2002) SEU resistance in advanced SOI-SRAMs fabricated by commercial technology using a rad-hard circuit design. *IEEE Transactions on Nuclear Science*, **49**, 2965–2968.

[24] Zhou, Q., Choudhury, M.R. and Mohanram, K. (2006) Design optimization for robustness to single-event upsets. The Second Workshop on System Effects of Logic Soft Errors, Urbana-Champain, IL, 11–12 April, 2006.

[25] Almukhaizim, S., Makris, Y., Veneris, A. and Yang, Y-S. (2008) On the minimization of potential transient errors and ser in logic circuits Using SPFD. International On-Line Testing Symposium 2008, Greece, 6–9 July, 2008 (6.2), pp. 123–128.

[26] Sierawski, B.D., Bhuva, B. and Massengill, L. (2006) Reducing soft error rate in logic circuits through approximate logic functions. *IEEE Transactions on Nuclear Science*, **53** (6), 3417–3421.

[27] Makihara, A., Midorikawa, M., Yamaguchi, T., Yokose, T., Tsuchiya, Y., Matsuda, S., Arimitsu, T., Asai, H., Iide, Y., Shindou, H. and Kuboyama, S. (2005) Hardness-by-design approach for 0.15 μm fully depleted CMOS/SOI digital logic devices with enhanced SEU/SET immunity. *IEEE Transactions on Nuclear Science*, 52 (**6**), 2524–2530.

[28] Wood, A., Jardine, R. and Bartlett, W. (2006) Data integrity in HP nonstop servers. The Second Workshop on System Effects of Logic Soft Errors, Urbana-Champain, IL, 11–12 April, 2006.

[29] Subramanian, V., Avirneni, N.D. and Somani, A. (2008) Conjoined processor: a fault tolerant high performance microprocessor. IEEE Workshop on Silicon Errors in Logic – System Effects, University of Texas at Austin, 26, 27 March.

[30] Baumeister, D. and Anderson, S.G.H. (2012) Evaluation of chip – level irradiation effects in a 32-bit safety microcontroller for automotive braking applications. 2012 IEEE Workshop on Silicon Errors in Logic – System Effects, Champaign-Urbana, IL, 27–28 March (2.2).

[31] Calin, T., Nicolaidis, M. and Velazco, R. (1996) Upset hardened memory design for submicron CMOS technology. *IEEE Transactions on Nuclear Science*, **43** (6), 2874–2878.

[32] Hazucha, P., Karnik, T., Maiz, J. *et al.* (2003) Neutron soft error rate measurements in a 90-nm CMOS process and scaling trends in SRAM from 0.25-micron to 90-nm generation. 2003 IEEE International Electron Devices Meeting, Washington, DC, 7–10 December, 2003 (21.5).

[33] Mitra, S., Zhang, M., Seifert, N. *et al.* (2007) Built-in soft error resilience for robust system design. International Conference on IC Design and Technology, Austin, TX, 18–20 May, pp. 263–268.

[34] Uemura, T., Tosaka, Y. and Matsuyama, H. (2010) SEILA: soft error immune latch for mitigating multi-node-SEU and local-clock-SET. IEEE International Reliability Physics Symposium 2010, Anaheim, CA, 2–6 May, pp. 218–223.

[35] Aggarwal, N., Jouppi, N., Ranganathan, P. *et al.* (2008) Reducing overhead for soft error coverage in high availability systems. IEEE Workshop on Silicon Errors in Logic – System Effects, University of Texas at Austin, 26, 27 March.

[36] Lee, H.-H.K., Lilja, K., Relangi, P. *et al.* (2010) LEAP: layout design through error-aware placement for soft-error resilient sequential cell design. IEEE

International Reliability Physics Symposium 2010, Anaheim, CA, 2–6 May, pp. 203–212.
[37] Alexandrescu, D. (2011) A comprehensive soft error analysis methodology for SoCs/ASICs memory instances. 17th IEEE International On-Line Testing Symposium, Athens, Greece, 13–15 July (10.1), pp. 175–176.
[38] Inoue, T., Henmi, H., Yoshikawa, Y. and Ichihara, H. (2011) High-level synthesis for multi-cycle transient fault tolerant datapaths. 17th IEEE International On-Line Testing Symposium, Athens, Greece, 13–15 July (1.3), pp. 13–18.
[39] Yang, E., Huang, K., Hu, Y. *et al.* (2013) HHC: Hierarchical Hardware Checkpointing to accelerate fault recovery for SRAM-based FPGAs. 19th IEEE International On-Line Testing Symposium, Chania, Crete, 8–10 July (6.3), pp. 193–198.
[40] Furuta, J., Kobayashi, K. and Onodera, H. (2010) An area/delay efficient dual-modular flip-flop with higher SEU/SET immunity. IEICE Transactions of on Electronics, Honolulu, HI, 16–18 June, E93-C (3), pp. 340–346.
[41] Ernst, D., Kim, N.S., Das, S. *et al.* (2003) Razor: a low-power pipeline based on circuit-level timing speculation. The 36thIEEE/ACM International Symposium on Microarchitecture, pp. 7–18.
[42] Ernst, D., Das, S., Lee, S. *et al.* (2004) Razor: circuit-level correction of timing errors for low-power operation. *IEEE Micro*, **24** (6), 10–20.
[43] Blaauw, D., Kalaiselvan, S., Lai, K. *et al.* (2008) Razor II: in situ error detection and correction for PVT and SER tolerance. 2008 IEEE International Solid-State Circuit Conference, San Francisco, CA, February 4–6, pp. 400, 401, 622.
[44] Roche, P. (2010) Industrial impacts of SER on today's consumer electronic arena. 16th IEEE International On-Line Testing Symposium, Corfu Island in Greece, July 5–7.
[45] Skarin, D. and Karlsson, J. (2009) Software mechanisms for tolerating soft errors in an automotive Brake-controller. Third Workshop on Dependable and Secure Nanocomputing, Estoril, Lisbon, Portugal, 29 June, pp. D34–D38.
[46] Polian, I., Reddy, S.M. and Becker, B. (2008) Scalable calculation of logical masking effects for selective hardening against soft errors. IEEE Workshop on Silicon Errors in Logic – System Effects, University of Texas at Austin, March, 26, 27.
[47] Azambuja, J.R., Sousa, F., Rosa, L. and Kastensmidt, F. (2009) Evaluating large grain TMR and selective partial reconfiguration for soft error mitigation in SRAM-based FPGAs. 15th IEEE International On-Line Testing Symposium, Sesimbra-Lisbon, Portugal, 24–26 June (5.3), pp. 101–106.
[48] Nikolov, D., Ingelsson, U., Singh, V. and Larsson, E. (2011) Level of confidence evaluation and its usage for roll-back recovery with checkpointing optimization. 5th Workshop on Dependable and Secure Nanocomputing, Hong Kong, China, 27 July, 59–64.
[49] Grosso, M., Reorda, M.S., Portela-Garcia, M. *et al.* (2010) An on-line fault detection technique based on embedded debug features. 16th IEEE International On-Line Testing Symposium, Corfu Island, Greece, July 5–7, 167–172.
[50] Maiz, J., Hareland, S., Zhang, K. and Armstrong, P. (2003) Characterization of multi-bit soft error events in advanced SRAMs. 2003 IEEE International Electron Devices Meeting, Washington, DC, 7–10 December, 2003 (21.4), pp. 21.4.1–21.4.4.
[51] Abbas, S.M., Baeg, S. and Park, S. (2011) Multiple cell upsets tolerant content-addressable memory. 2011 IEEE International Reliability Physics Symposium, Monterey, CA, 12–14 April, pp. SE.1.1–SE.1.5.

[52] Ibe, E., Taniguchi, H., Yahagi, Y. *et al.* (2009) Scaling effects on neutron-induced soft error in SRAMs Down to 22nm process. Third Workshop on Dependable and Secure Nanocomputing, Estoril, Lisbon, Portugal, 29 June (2.1).

[53] Ibe, E., Taniguchi, H., Yahagi, Y. *et al.* (2010) Impact of scaling on neutron-induced soft error in SRAMs from a 250 nm to a 22 nm design rule. *IEEE Transactions on Electron Devices*, **57** (7), 1527–1538.

[54] Seifert, N. (2008) Soft error rates of hardened sequentials utilizing local redundancy. International On-Line Testing Symposium 2008, Greece, 6–9 July, 2008 (S1.3), pp. 49–52.

[55] Seifert, N. and Zia, V. (2007) Assessing the impact of scaling on the efficacy of spatial redundancy based mitigation schemes for terrestrial applications. IEEE Workshop on Silicon Errors in Logic – System Effects 3, Austin, TX, 3 April, 4.

[56] Furuta, J., Hamanaka, C., Kobayashi, K. and Onodera, H. (2010) A 65nm Bi-stable cross-coupled dual modular redundancy flip-flop capable of protecting soft errors on the c-element. 2010 IEEE Symposium on VLSI Circuits, Honolulu, HI, 16–18 June, pp. 123–124.

[57] Cabanas-Holmen, M., Cannon, E.H., Kleinosowski, A. *et al.* (2009) Clock and reset transients in a 90 nm RHBD single-core tilera processor. *IEEE Transactions on Nuclear Science*, **56** (6), 3505–3510.

[58] Seifert, N., Shipley, P., Pant, M.D. *et al.* (2005) Radiation-induced clock jitter and race. 2005 IEEE International Reliability Physics Symposium Proceedings, San Jose, CA, 17–21 April, 2005, pp. 215–222.

[59] Maillard, P., Loveless, T.D., Holman, W.T. and Massengill, L.W. (2011) A radiation-hardened Delay-Locked Loop Design (DLL) utilizing differential delay lines topology. The Conference on Radiation Effects on Components and Systems, Sevilla, Spain, 19–23 September, pp. 675–682.

[60] Kim, S., Tsuchiya, A. and Onodera, H. (2013) Perturbation-immune radiation-hardened PLL with a switchable DMR structure. 19th IEEE On-Line Testing Symposium, Chania, Crete, 8–10 July, pp. 128–132.

[61] Srinivasan, V., Sternberg, A.L., Robinson, W.H. *et al.* (2005) Single event mitigation in combinational logic using targeted data path hardening. *IEEE Transactions on Nuclear Science*, **52** (6), 2516–2523.

[62] Reddy, K.K., Amrutur, B. and Parekhji, R. (2008) False error study of on-line soft error detection mechanisms. International On-Line Testing Symposium 2008, Greece, 6–9 July, 2008 (3.1), pp. 53–58.

[63] Avirnen, N.D.P. (2012) Low overhead soft error mitigation techniques for high-performance and aggressive designs. *IEEE Transactions on Computers*, **61** (4), 488–501.

[64] Prasanth, V., Singh, V. and Parekhji, R. (2011) Reduced overhead soft error mitigation using error control coding techniques. 17th IEEE International On-Line Testing Symposium, Athens, Greece, 13–15 July (9.3), pp. 163–168.

[65] Pilotto, C, Azambuja, J.R. and Kastensmidt, F.L. (2008) Synchronizing triple modular redundant designs in dynamic partial reconfiguration applications. 21st Annual Symposium on Integrated Circuits and System Design, Gramado, September 1–4, pp. 199–204.

[66] Nakka, N., Pattabiraman, K., Kalbarczyk, Z.T. and Iyer, R.K. (2006) Processor-level selective replication. The Second Workshop on System Effects of Logic Soft Errors, Urbana-Champain, IL, 11–12 April, 2006.

[67] Noji, R., Fujie, S., Yoshikawa, Y. *et al.* (2010) An FPGA-based fail-soft system with adaptive reconfiguration. 16th IEEE International On-Line Testing Symposium, Corfu Island, Greece, 5–7 July (6.3), pp. 127–132.

[68] Nikolov, D., Ingelsson, U., Singh, V. and Larsson, E. (2011) Level of confidence evaluation and its usage for roll-back recovery with checkpointing optimization. 5th Workshop on Dependable and Secure Nanocomputing, Hong Kong, China, 27 July.

[69] Carmichael, C., Fuller, E., Blain, P. *et al.* (1999) SEU Mitigation Techniques for Virtex FPGAs in Space Applications, Military and Aerospace Applications of Programmable Devices and Technologies Conference, Sep. 28–30, 1999, Laurel, Maryland.

[70] Pratt, B., Carroll, J.F., Graham, P. *et al.* (2007) Fine-grain SEU mitigation for FPGAs using partial TMR. *IEEE Transactions on Nuclear Science*, **55** (4), 2274–2280.

[71] Quinn, H., Morgan, K., Graham, P. *et al.* (2007) Domain crossing events: limitations on single device triple-modular redundancy circuits in xilinx FPGAs. *IEEE Transactions on Nuclear Science*, **54** (6), 2037–2043.

[72] Rusu, C., Grecu, C. and Anghel, L. (2008) Communication aware recovery configurations for networks-on-chip. International On-Line Testing Symposium 2008, Greece, 6–9 July, 2008 (9.1), pp. 201–206.

[73] Bizot, G., Chaix, F., Zergoinoh, N.-E. and Nicolaidis, M. (2013) Variability-aware and fault-tolerant self-adaptive applications for many-core chips. 19th IEEE International On-Line Testing Symposium, Chania, Crete, 8–10 July (6.3), pp. 37–42.

[74] Wunderlich, H.J., Braun, C. and Halder, S. (2013) Efficacy and efficiency of algorithm-based fault-tolerance on GPUs. 19th IEEE International On-Line Testing Symposium, Chania, Crete, 8–10 July (S4.3), pp. 240–243.

[75] Rech, P. and Carro, L. (2013) Experimental evaluation of GPUs radiation sensitivity and algorithm-based fault tolerance efficiency. 19th IEEE International On-Line Testing Symposium, Chania, Crete, 8–10 July (S4.4), pp. 244–247.

[76] Rech, P., Aguiar, C., Ferreira, R. *et al.* (2012) Neutrons radiation test of graphic processing units. IEEE International On-Line Testing Symposium, Sitges, Spain, 27–29 June, 2012 (3.3), pp. 55–60.

[77] Kopetz, H. (2006) Mitigation of transient faults at the system level – The TTA approach. The Second Workshop on System Effects of Logic Soft Errors, Urbana-Champain, IL, 11–12 April, 2006.

[78] Gold, B., Smolens, J.C., Falsafi, B. and Hoe, J.C. (2006) The granularity of soft-error containment in shared-memory multiprocessors. The Second Workshop on System Effects of Logic Soft Errors, Urbana-Champain, IL, 11–12 April, 2006.

[79] Vaskova, A., Lopez-Ongil, C., Portela-Garcia, M. *et al.* (2011) Study on the effect of multiple errors in robust systems based on critical task distribution. The Conference on Radiation Effects on Components and Systems, Sevilla, Spain, 19–23 September (G-1), pp. 463–466.

[80] Silva, D., Bolzani, L. and Vargas, F. (2011) An intellectual property core to detect task scheduling-related faults in RTOS-based embedded systems. 17th IEEE International On-Line Testing Symposium, Athens, Greece, 13–15 July (2.1), pp. 19–24.

[81] Pellegrini, A., Smolinski, R., Chen, L. *et al.* (2011) Crash test' ing SWAT: accurate, gate – level evaluation of symptom – based resiliency solutions. 2011 IEEE Workshop on Silicon Errors in Logic – System Effects, Champaign, IL, March 29–30.
[82] Ibe, E., Shimbo, K., Toba, T. *et al.* (2011) LABIR: inter – layer built – in reliability for electronic components and systems. 2011 IEEE Workshop on Silicon Errors in Logic – System Effects, Champaign, IL, 29–30 March.
[83] Evans, A., Nicolaidis, M. Wen, S.-J. *et al.* (2012) RIIF – reliability information interchange format. IEEE International On-Line Testing Symposium, Sitges, Spain, 27–29 June, 2012 (6.2), pp. 103–108.
[84] Wen, S. (2008) Systematical method of quantifying SEU FIT. International On-Line Testing Symposium 2008, Greece, 6–9 July, 2008, pp. 109–116.
[85] Kumar, R. (2011) Stochastic computing: embracing errors in architecture and design of processors and applications. 5th Workshop on Dependable and Secure Nanocomputing, Hong Kong, China, 27 July, pp. 169–180
[86] Raghunathan, A. and Roy, K. (2013) Approximate computing: energy-efficient computing with good-enough results. 19th IEEE International On-Line Testing Symposium, Chania, Crete, 8–10 July (S2.1), p. 258.
[87] Alexandrescu, D., Lhomme-Perrot, A., Schaefer, E. and Beltrando, C. (2009) Highs and lows of radiation testing. 15th IEEE International On-Line Testing Symposium, Sesimbra-Lisbon, Portugal, 24–26 June (S3.1).
[88] Li, M.-L. and Ramachandran, P. (2007) Towards a software-hardware co-designed resilient system. IEEE Workshop on Silicon Errors in Logic – System Effects 3, Austin Texas, 3, 4April.
[89] Hari, S.K.S., Naeimi, H., Ramachandran, P. and Adve, S.V. (2011) Relyzer: application resiliency ananlyzer for transient faults. 2011 IEEE Workshop on Silicon Errors in Logic – System Effects, Champaign, IL, 29–30 March, pp. 22–26.
[90] Wang, N. and Patel, S.J. (2006) Sympton based redundant multithreading. The Second Workshop on System Effects of Logic Soft Errors, Urbana-Champain, IL, 11–12 April, 2006.
[91] Theodorou, G., Kranitis, N., Paschalis, A. and Gizopoulos, D. (2010) A software-based self-test methodology for in-system testing of processor cache tag arrays. 16th IEEE International On-Line Testing Symposium, Corfu Island, Greece, 5–7 July (7.4), pp. 159–164.
[92] Carvalho, M., Bernardi, P., Sanchez, E. *et al.* (2013) Increasing fault coverage during functional test in the operational phase. 19th IEEE International On-Line Testing Symposium, Chania, Crete, 8–10 July (3.1), pp. 43–48.
[93] Ballan, O., Bernardi, P., Yazdani, B. and Sanchez, E. (2013) A Software-based self-test strategy for on-line testing of the scan chain circuitries in embedded microprocessors. 19th IEEE International On-Line Testing Symposium, Chania, Crete, 8–10 July (5.2), pp. 79–84.
[94] Yamakita, K., Yamada, H. and Kono, K. (2011) Phase-based reboot: reusing operating system execution phases for cheap reboot-based recovery. 2011 International Conference on Dependable Systems and Networks, Hong Kong, China, 28–30 June.
[95] Bronevetsky, G. and deSupinski, B. (2007) Soft error vulnerability of iterative linear algebra methods. IEEE Workshop on Silicon Errors in Logic – System Effects 3, Austin TX, 3, 4 April.

[96] Kellington, J. and McBeth, R. (2007) IBM POWER6 processor soft error tolerance analysis using proton irradiation. IEEE Workshop on Silicon Errors in Logic – System Effects 3, Austin TX, 3, 4 April.

[97] Roy, K. (2013) Approximate computing for energy-efficient error-resilient multimedia systems. 2013 IEEE 16th International Symposium on Design and Diagnostics of Electronic Circuits and Systems, Karlovy Vary, Czech Republic, 8–10 April, pp. 5–6.

[98] Bradley, D. and Penton, J. (2008) A perspective on developing IP for embedded reliability. IEEE Workshop on Silicon Errors in Logic – System Effects, University of Texas at Austin, 26, 27 March.

[99] Mariani, R. (2012) Designing safe and availableintegrated circuits according to functional safety standards. 2012 IEEE Workshop on Silicon Errors in Logic – System Effects, Champaign-Urbana, IL, 27–28 March (1.1).

[100] Mizan, E. (2008) Reliability improvements enabled by self-imposed temporal redundancy. IEEE Workshop on Silicon Errors in Logic – System Effects, University of Texas at Austin, 26, 27 March.

[101] Falsafi, B. (2011) Reliability in the dark silicon Era. 17th IEEE International On-Line Testing Symposium, Athens, Greece, 13–15 July.

[102] Nassif, S.R. (2013) Silicon today, silicon tomorrow. The 9th Workshop on Silicon Errors in Logic – System Effects, Palo Alto, CA, 26, 27 March.

[103] Nikolov, D., Ingelsson, U., Singh, V. and Larsson, E. (2011) Level of confidence evaluation and its usage for roll-back recovery with checkpointing optimization. 5th Workshop on Dependable and Secure Nanocomputing, Hong Kong, China, 27 July, pp. 59–64.

[104] Geist, A. (2012) Exascale monster in the closet. 2012 IEEE Workshop on Silicon Errors in Logic – System Effects, Champaign-Urbana, IL, 27–28 March (5.1).

[105] Niknahad, M., Sander, O. and Becker, J. (2011) QFDR-an integration of quadded logic for modern FPGAs to tolerate high radiation effect rates. The Conference on Radiation Effects on Components and Systems, Sevilla, Spain, 19–23 September (B-1), pp. 119–122.

[106] Sebepou, Z. and Magoutis, K. (2011) CEC: Continuous Eventual Checkpointing for data stream processing operators. 2011 International Conference on Dependable Systems and Networks, Hong Kong, China, 28–30 June, pp. 145–156.

[107] Limbrick, D., Robinson, W. and Bhuva, B. (2010) Reliability-aware synthesis: XOR logic function case study. IEEE Workshop on System Efects of Logic Soft Errors, Stanford University, 23, 24 March.

[108] Alexandrescu, D. (2010) Reflections on a SER-aware design flow. International Conference on IC Design and Technology, Kaohsiung, Taiwan, 2–4 May, 2011, pp. 215–219.

[109] Rivers, J.A., Bose, P., Kudva, P. *et al.* (2008) Phaser: phased methodology for modeling the system-level effects of soft errors. *IBM Journal of Research and Development*, **52** (3), 293–306.

[110] Michalak, S., DuBois, A., Storlie, C. *et al.* (2011) Neutron beam testing of triblades. 2011 IEEE Nuclear and Space Radiation Effects Conference, Las Vegas, 25–29 July (W-28).

[111] Wong, R., Bhuva, B.L., Evans, A. and Wen, S.-J. (2012) System-level reliability using component-level failure signatures. IEEE International Reliability Physics Symposium, Anaheim, California, USA, 15–19 April, pp. 4B.3.1–4B.3.7.

[112] Wen, S., Silburt, A. and Wong, R. (2008) IC component SEU impact analysis. IEEE Workshop on Silicon Errors in Logic – System Effects, University of Texas at Austin, 26, 27 March.

[113] Ibe, E., Toba, T., Shimbo, K. and Taniguchi, H. (2012) Fault-based reliable design-on-upper-bound of electronic systems for terrestrial radiation including muons, electrons, protons and low energy neutrons. IEEE International On-Line Testing Symposium, Sitges, Spain, 27–29 June, 2012 (3.2), pp. 49–54.

[114] Ibe, E., Shimbo, K., Toba, T. *et al.* (2012) State-of-the-art study on mitigation techniques of single event effects in terrestrial applications. The 10th International Workshop on Radiation Effects on Semiconductor Devices for Space Applications, Tsukuba, Japan, 10–12 December, 2012.

[115] Seifert, N., Gil, B., Jahinuzzaman, S. *et al.* (2012) Soft error susceptibilities of 22nm Tri-gate devices. 2012 IEEE Workshop on Silicon Errors in Logic – System Effects, Champaign-Urbana, IL, 27–28 March (3.1).

[116] Carter, N. (2010) Cross-layer reliability. IEEE Workshop on System Efects of Logic Soft Errors, Stanford University, 23, 24 March.

[117] Quinn, H. (2011) Study on cross – layer reliability. 2011 IEEE Workshop on Silicon Errors in Logic – System Effects, Champaign, Illinoi, 29–30 March.

[118] Ibe, E., Shimbo, K., Taniguchi, H. *et al.* (2011) Quantification and mitigation strategies of neutron induced soft-errors in CMOS devices and components-the past and future. 2011 IEEE International Reliability Physics Symposium, Monterey, CA, 12–14 April, pp. 3C.2.1–3C.2.8.

[119] Herkersdorf, A., Kleeberger, V., Wehn, N. *et al.* (2013) Cross-layer dependability modeling and abstraction in systems on chip. The 9th Workshop on Silicon Errors in Logic – System Effects, Palo Alto, CA, 26, 27 March (7.1).

[120] Psychou, G., Schleifer, J., Huisken, J. *et al.* (2012) Cross-layer reliability exploration proposal for body area networks. 2012 IEEE Workshop on Silicon Errors in Logic – System Effects, Champaign-Urbana, IL, 27–28 March (9.4).

[121] Wong, V. and Horowitz, M. (2006) Soft error resilience of probabilistic inference applications. The Second Workshop on System Effects of Logic Soft Errors, Urbana-Champain, IL, 11–12 April, 2006.

[122] Lindoso, A., Entrena, L., Millan, E.S. *et al.* (2012) A co-design approach for SET mitigation in embedded systems. *IEEE Transactions on Nuclear Science*, **59** (4), 1034–1039.

[123] Wang, T., Zhang, Z., Chen, L. *et al.* (2010) A novel bulk built-in current sensor for single-event transient detection. IEEE Workshop on System Efects of Logic Soft Errors, Stanford University, 23, 24 March.

[124] Upasani, G., Vera, X. and Gonzalez, A. (2013) Achieving zero DUE for L1 data caches by adapting acoustic wave detectors for error detection. 19th IEEE International On-Line Testing Symposium, Chania, Crete, 8–10 July (5.2).

[125] Kanekawa, N., Ibe, E., Suga, T. and Uematsu, Y. (2010) Dependability in Electronic Systems-Mitigation of Hardware Failures, Soft Errors, and Electro-Magnetic Disturbances. Dependability in Electronic Systems.

[126] Hitachi http://www.hitachi.com/rev/archive/2001/__icsFiles/afieldfile/2004/06/08 /r2001_04_101.pdf (accessed 22 November 2013).

[127] Wacks, K. (2003) Home Electron System, http://hes-standards.org/sc25_wg1 _introduction.pdf (accessed 22 November 2013).

9

Summary

9.1 Summary of Terrestrial Radiation Effects on ULSI Devices and Electronic Systems

Here the history of soft-error research is reviewed and five or six paradigm shifts, mainly relevant to drastic change in faulty modes and device/systems in history are highlighted. Three hierarchies in faulty conditions, fault/error/failure, are proposed as a basis for evaluation of terrestrial radiation effects and their mitigation techniques. Current concerns over failures in industrial fields such as avionics, the motor industry and networks/data centres are summarised. Sources of terrestrial radiation include photons (gamma-ray), electrons (beta-ray), helium ions (alpha-ray), muons, protons and neutrons with wide energy ranges (thermal, epi-thermal to gigaelectron volts). The properties of secondary cosmic particles produced by nuclear spallation reaction of cosmic protons with atmospheric nuclei in inner space are summarised. The spectra of terrestrial radiation, such as proton, electron, muon, pion, neutron and alpha are evaluated. Radioisotopes, observable at ground level, are surveyed as radiation sources of alpha (helium nucleus), beta (electron) and gamma (photon) rays.

The fundamentals of electronic devices DRAM (Dynamic Random Access Memory), CMOS (Complementary Metal Oxide Semiconductor) inverter, SRAM (Static Random Access Memory), floating body memory, sequential and combinational logic devices) and systems FPGA (Field Programmable Gate Array) and processors with OS(Operating System) are also summarised.

Detailed simulation techniques and algorithms are introduced for nuclear spallation reactions, automatic device model construction, particle tracking techniques, charge collection models and bipolar mode fault models.

Typical simulation results are introduced with some sample codes in Visual Basic™. Upper bound fault rate simulation techniques, with a 22 nm SRAM model as an example, are proposed as a common tool for evaluating susceptibility to

Terrestrial Radiation Effects in ULSI Devices and Electronic Systems, First Edition. Eishi H. Ibe.

SEE(Single Event Effect) regardless of memory or logic devices, enabling analysis of the susceptibility of a SOC (System On Chip) to SEE.

Mechanisms of the radiation effects of – SEE and total ionisation dose effect – due to each terrestrial radiation source are summarised, together with a number of irradiation facilities, including underground and field facilities worldwide, with actual records of irradiation tests. A number of fault/error/failure modes are surveyed and prediction/estimation, detection and classification techniques are introduced to extract the underlying mechanisms for each mode. A variety of the mitigation techniques proposed are summarised and classified for each stack-layer in the silicon industry. It is pointed out that resiliency in electronic components and systems cannot be established by using mitigation techniques in only one stack layer. Importance of communication beween stack layers and a combination of mitigation techniques over multiple stack layers is addressed in some detail.

9.2 Directions and Challenges in the Future

As technologies progress at a greater rate in the future, resiliency in electronic components and systems will face unprecedented difficulties and require novel approaches. New directions are discussed and proposed to overcome the challenges and conditions generated by the paradigm shifts in mitigation techniques. Communication between stack layers – substrate, device, circuit, CPU (Central Processing Unit), OS and application/software is important. DOUB (Design On Upper Bound) and LABIR (inter LAyer Built-In Reliability) are proposed as examples of such a challenging direction. Mitigation techniques based on failure symptoms which appeared at the hardware or software level are highlighted.

Appendices

A.1 Hamming Code

Hamming code ([89] in Chapter 7) is the most commonly used in computer applications. For given integer m, code length n is given by $n = 2m - 1$, while data length is given by $k = n - m$ in Hamming code.

Typically, a check matrix H and a generator matrix G are utilized and should be defined in Hamming Code to detect and correct a data vector.

$$H = \begin{bmatrix} 1 & 0 & 1 & 1 & 1 & 0 & 0 \\ 1 & 1 & 0 & 1 & 0 & 1 & 0 \\ 0 & 1 & 1 & 1 & 0 & 0 & 1 \end{bmatrix} \tag{A.1}$$

Generator matrix G is given from H as,

$$G = \begin{bmatrix} 1 & 0 & 0 & 0 & 1 & 1 & 0 \\ 0 & 1 & 0 & 0 & 0 & 1 & 1 \\ 0 & 0 & 1 & 0 & 1 & 0 & 1 \\ 0 & 0 & 0 & 1 & 1 & 1 & 1 \end{bmatrix} \tag{A.2}$$

All column vector in matrices H and G must be independent and must not be null vector (all elements are ‘0’s).

When the data bits are $x = [1\ 1\ 0\ 1]$, the codeword can be calculated by multiplying x and G,

$$Y = x \times G = [1\ 1\ 0\ 1] \times \begin{bmatrix} 1 & 0 & 0 & 0 & 1 & 1 & 0 \\ 0 & 1 & 0 & 0 & 0 & 1 & 1 \\ 0 & 0 & 1 & 0 & 1 & 0 & 1 \\ 0 & 0 & 0 & 1 & 1 & 1 & 1 \end{bmatrix} = [1\ 1\ 0\ 1\ 0\ 1\ 0] \tag{A.3}$$

It is noteworthy that all additional calculations are exclusive or (XOR) in this matrix calculation.

Terrestrial Radiation Effects in ULSI Devices and Electronic Systems, First Edition. Eishi H. Ibe.

If Y is correctly converted,

$$Y \times H^{\mathrm{T}} = 0. \tag{A.4}$$

If not, the result of the above calculation, syndrome, gives the same vector as the one of the column vectors of the check matrix H where one bit error is introduced. When two-bit errors take place, the calculated results indicate the wrong column vector. In that case, the following check matrix in the extended Hamming code is utilised,

$$H^* = \begin{bmatrix} 1 & 0 & 1 & 1 & 1 & 0 & 0 & 0 \\ 1 & 1 & 0 & 1 & 0 & 1 & 0 & 0 \\ 0 & 1 & 1 & 1 & 0 & 0 & 1 & 0 \\ 1 & 1 & 1 & 1 & 1 & 1 & 1 & 1 \end{bmatrix} \tag{A.5}$$

H^* is made by adding an all '0' column vector in the last column to the original H matrix, and then by adding an all '1' raw in the bottom of the matrix.

The yield $Y \times H^{\mathrm{T}}$ gives the same column vector in the matrix H for one bit error or other vector that is not included in columns of the matrix H for two-bit errors. This is single error correction double error detection (SECDED)

A.2 Marching Algorithms

Stack-at faults, transition faults, a part of coupling faults and a part of address decoder faults can be detected by Marching algorithms. Typical sequences are:

1. March C-

```
1. W0 ↑↓
2. R0, W1↑
3. R1, W0↑
4. R0,W1 ↓
5. R1,W0 ↓
6. R0 ↑↓,
where,
 W0: write "0";
 R0: read "0";
 W1: write "1";
 R1: "read "1";
 ↑:upflow from address 0 to n-1;
↓:downflow from address n-1 to 0;
↑↓:any direction.
```

2. MATS+

```
1. W0 ↑↓
2. R0,W1 ↑
3. R1, W0 ↓
```

A.3 Why VB Is Used For Simulation?

Visual Basic (VB) is a very versatile simulation language that can be used for complicated matrix calculations, constructing graphic interfaces, database handling through SQL (Structured Query Language) server and direct documentations to most Windows^{TM} PC applications such as Microsoft Office^{TM} applications and Adobe Acrobat^{TM}. Programs written in VB can be easily installed in Windows^{TM} PCs. CORIMS and MUCEAC work effectively on a Windows^{TM} PC though GUI (Graphic User Interface) interfaces and results can be stored in Excel^{TM} spreadsheets. These are only part of the reason why we use VB for SEE (Single Event Effect) simulations. In this Appendix, some essence in using VB and sample codes are introduced.

A.4 Basic Knowledge of Visual Basic

VB is an object oriented language by which interactive GUI interfaces can be easily constructed.

The structure of VB is very much suitable for programming physical models.

The first preferable feature in VB is 'Type' and 'properties' in class: One can define an object and its properties freely. The following is one example:

```
Public Fault() as SoftErr
TYPE Fault()
  Global number as long
  xAddress as Long
  yAddress Long
  mcuFamily as String
END TYPE
```

Fault() is a matrix or a vector which is in the SoftErr TYPE. Fault() has grovel number, x and y addresses in a chip. mcuFamily is a string that consists of a set of the global numbers of soft errors in one MCU (Multi-Cell Upset).

mcuFamily can be expressed as follows:

```
Fault(1).mcuFamily="1_4_10".
```

This means that faults with global numbers 1, 4 and 10 make one MCU and Fault(1) is included in the family.

These kinds of comprehensible properties of variables can be clearly defined and this makes programming much more effective and free of bugs.

There is no limit to the number of characters in one line of the program code.

A.5 Database Handling by Visual Basic and SQL

Visual Basic can handle the SQL Server with an object database and the SQL Server utilises database handling software like Microsoft Access^{TM}.

With Microsoft Access™, two dimensional (rows and columns (*fields*)) tables such as *tables* in Figure 6.6 can be defined and make one *database* altogether. In the program package CORIMS, all the *tables* in Figure 6.6 are contained in one *database* 'CORIMS.mdb'.

In Visual Basic, one can make out a *recordset* that consists of data selected from a table under specific conditions. The following is one example.

```
Private Sub(rsTemp as Recordset, tabName as String, varName as String,
strIDTemp as String)
Dim tempDB As database
Dim strSQL as String
Dim strPath as String
strPath = "C:\ffff\filename"  "Difine path of the database
Set tempDB = wrkDefault.OpenDatabase(strPath)  'Define temporary
database
strSQL = "SELECT p.* FROM " & tabName & " p WHERE p." & varName &
" = '" & strIDtemp Set rsTemp = tempDB.OpenRecordset(strSQL,
dbOpenSnapshot)
End Sub
```

In the example above, strSQL gives conditions of data extraction, which send instructions to select data in the column with the field name varname and actual string in the row is strIDtemp from the table with the name tabname. All the data in the row are extracted and stored in the recordset reTemp.

There are a number of useful instructions in SQL Server™. Introduction to all the instructions is not in the scope of this book, so only one example is shown here.

```
strSQL = "SELECT DISTINCT p.Field FROM table p"
```

In this case, strSQL gives instructions to select unique strings or data in the column without overlap (distinctly).

A.6 Algorithms in Text Handling and Sample Codes

VB eases handling of codes. There are a number of functions that can be used for such handling, for example:

```
Public Sub DecompText(strText As String, keyChara As String, _
          intCount As Long, strTemp() As String, trimBefore As Long,
trimAfter As Long)
                    '
                    Dim strWord As String
                    Dim lenKey As Long
                    Dim lenAll As Long
                    Dim intStart As Long
```

```
Dim intEnd As Long
Dim intLength As Long


Dim i As Long

On Error GoTo check
lenAll = Len(strWord)
lenKey = Len(keyChara)

intCount = 0
intStart = 1
For i = 1 To lenAll
    DoEvents
    If Mid(strWord, i, lenKey) = keyChara Then
        intCount = intCount + 1
        intLength = i - intStart
        ReDim Preserve strTemp(intCount)
        strTemp(intCount) = Mid(strWord, intStart,
intLength)

        intStart = i + 1
    End If
Next
intCount = intCount + 1
intLength = lenAll - intStart + 1
ReDim Preserve strTemp(intCount)
strTemp(intCount) = Mid(strWord, intStart,
intLength)
Exit Sub
check:
Watch.curSub = "Decomptext"
ShowError 'Error managing program

End Sub
```

In this program, a sentence strText is divided by characters keyChara and character blocks are divided and stored in variables strTemp(intCount).

A.7 How to Make a Self-Consistent Calculation

Self-consistent calculation is made typically for the following type of equation,

$$x = f(x, a, b, c, \ldots), \tag{A.6}$$

in which the variable x whose value is to be obtained is contained in the right and left terms in an equation. Variables a, b, c, … can be determined from other conditions or equations.

In this case, Equation (A.6) is replaced by,

$$x_{n+1} = f(x_n, a, b, c, \dots) \tag{A.7}$$

Starting with x_0 to obtain x_1, the left term is calculated with x_n to obtain x_{n+1} in the right term.

If Abs($x_{n+1} - x_n$) converge to a predetermined small enough value, the x_{n+1} can be regarded as the solution of Equation (A.6). Stating value x_0 must be close enough to the real value depending on the largeness of variation of the function in the right term.

A.8 Sample Code for Random Selection of Hit Points in a Triangle

In the following program, random position (xShot, yShot, zShot) is selected in a trigonal prismatic that has a top and bottom triangle 'triTile' and thickness of zThick. The nodeAnchor is a starting apex of triangle with which a triangle is defined by the lengths of two sides from the nodeAnchhor and angle between the two sides, tritile.faiBetween or oAngle.

Initially, a virtual parallelogram is made from the triangle by adding a virtual apex on the other side of the nodeAnchor. Random position is selected in the parallelogram. If the position is inside the virtual triangle (flg > 1), position symmetry to the opposite side is selected.

Watch is a public variable for error handling subroutine ShowError.

```
Public Sub TriShot(triTile As TriAngle, nodeAnchor As NodePoint,
zThick As Double, _
                xShot As Double, yShot As Double, zShot As Double)

                'nodeAnchor は top 面の頂点
                Dim flg As Double
                Dim X1 As Single
                Dim Y1 As Single
                Dim z1 As Single
                Dim oAngle As Double
                Dim nodeCenter As NodePoint
                Dim nodeMove As NodePoint
                On Error GoTo check
                nodeCenter.xPosGloval = triTile.lenStart / 2#
                nodeCenter.yPosGloval = triTile.lenEnd / 2#
                nodeCenter.zPosGloval = nodeAnchor.zPosGloval
                   - zThick / 2#
                oAngle = triTile.faiBetween
                Randomize
```

```
                nodeMove.xPosGloval = Rnd * triTile.lenStart
                Randomize
                nodeMove.yPosGloval = Rnd * triTile.lenEnd
                Randomize

                nodeMove.zPosGloval = nodeAn-
chor.zPosGloval - Rnd * zThick

                flg = nodeMove.yPosGloval / triTile.lenEnd +
nodeMove.xPosGloval / triTile.lenStart

                If flg > = 1# Then
                moveSymmetry nodeMove, nodeCenter
                End If

                X1 = nodeMove.xPosGloval + nodeMove.yPosGloval
* Cos(oAngle)
                Y1 = nodeMove.yPosGloval * Sin(oAngle)
                z1 = nodeMove.zPosGloval
                AffinRotateZ X1, Y1, z1, triTile.faiBegin
                xShot = X1 + nodeAnchor.xPosGloval
                yShot = Y1 + nodeAnchor.yPosGloval
                zShot = z1


                Exit Sub
check:
                Watch.curSub = "TriShot"
                ShowError

                End Sub
```

Index

Terrestrial Radiation Effects in ULSI Devices and Electronic Systems, First Edition. Eishi H. Ibe.

www.ingramcontent.com/pod-product-compliance
Lightning Source LLC
LaVergne TN
LVHW062208200926
848484LV00006B/1347